Frontiers in Analytical Spectroscopy

Frontiers in Analytical Spectroscopy

Edited by

D. L. Andrews
University of East Anglia

A. M. C. Davies
Norwich Near Infrared Consultancy

The Proceedings of the International Conference: Spectroscopy Across the Spectrum IV: Techniques and Applications of Analytical Spectroscopy, held at the University of East Anglia, Norwich, UK on 11–14 July 1994.

Special Publication No. 163

ISBN 0-85404-730-1

A catalogue record for this book is available from the British Library

Published by The Royal Society of Chemistry,
Thomas Graham House, The Science Park,
Cambridge CB4 4WF, UK

Printed and bound by Hartnolls Ltd., Bodmin

Preface

Frontiers of Knowledge

In everyday use, *frontier* generally signifies a territorial limit, typically conjuring up images of border patrols and barriers to travel, though the word may also have the connotations of a barrier to information. In the context of this book, the term *frontier* is primarily employed to signify limits of knowledge or instrumental sophistication, but it is interesting to pursue the semantic connection a little further.

When travelling it is highly useful to know where frontiers exist, so that they can be prepared for, or circumvented. It may be helpful to know if these frontiers have proved difficult for other travellers. Are they 'easy' or 'hard'? Are there any proven strategies for passing through them or is progress going to require persistence and determination or some new lateral thinking? Most frontiers have more than one official entry point; is one more congenial than another (JFK or Dulles for entry to the USA?). The main import of all these questions and many others that arise in the context of travel, can also be applied to the current frontiers in analytical spectroscopy. The need to address such issues highlights the purpose of this book – not just to provide information on the present limits of a range of spectroscopies, but also to show how their frontiers are being approached.

It may be hoped that the observation of certain procedural parallels between protocols applying to different frontiers will afford the reader some assistance in overcoming other problematic barriers. It has, for example, been said that FTNMR was developed without any knowledge of FTIR and the possibility of FTMS was not realized by either group. Apocryphal though this comment may be, it does seem worth the effort to make knowledge of different spectroscopies readily available to all spectroscopists. This has been the *raison d'être* behind the series of conferences called **Spectroscopy Across the Spectrum**, and this volume is a progeny of the fourth such conference held at the University of East Anglia, Norwich, UK in July 1994. As on previous occasions a group of spectroscopists was invited to present a current view of analytical spectroscopy from many perspectives, but with certain common themes. We have now collected, edited, and re-arranged the written accounts under more obvious headings. In some cases the judgement on which section a particular paper should belong to was somewhat arbitrary – the reader is urged to consult the Contents and Subject and Author Index pages in order that our decisions should not create another quite artificial kind of boundary.

Perhaps one of the most enduring contemporary phrases based on the word *frontier* is 'The final frontier'. In the present context, the final frontier in analytical spectroscopy might be considered to have been reached when each spectroscopy can record a spectrum from a single molecule and when we are able to extract and comprehend all the information that each spectrum contains. For most spectroscopies we are a long way from reaching such a frontier, but we hope this volume will assist all those intent on the quest. The case study provided by Norman Sheppard in the first chapter nicely illustrates just how obstacles can eventually be overcome by perseverance, imagination, and instrument refinement over a long period of time. Subsequent chapters are concerned with current frontiers, discussing new ways of obtaining spectra and extracting salient information. We hope that this combination of example and knowledge will help all those engaged in the challenge and excitement of analytical spectroscopy.

We are grateful to all our contributors, who have helped us with the best use of current word-processing software to produce manuscripts in a good quality camera-ready format. Thanks are also due to the diligence of the editorial staff at The Royal Society of Chemistry for keeping the project on time.

D. L. Andrews
A. M. C. Davies

Dedication

This volume is dedicated to Professor Norman Sheppard, FRS our friend, colleague, and teacher who has inspired several generations of spectroscopists.

Contents

Historical Frontiers

Vibrational Spectroscopic Contributions to Chemisorption and Catalysis — 3
N. Sheppard, FRS

Instrumentation

Recent Advances in Kinetic Infrared Spectroscopy — 13
J. J. Turner, M. W. George, and M. Poliakoff

Spectroscopic Concentration Determination without Calibration: Scope and Limitations of Concentration-modulated Absorption Spectroscopy — 20
W. Jeremy Jones

Electrochemically Modulated Infrared Spectroscopy Using a Step-scanning FTIR Spectrometer: Application to the $Fe(CN)_6^{3-}/Fe(CN)_6^{4-}$ Couple — 31
C. M. Pharr, B. O. Budevska, and P. R. Griffiths

Applications of Raman Microscopy and Raman Imaging — 42
K. P. J. Williams, A. Whitley, and C. D. Dyer

An Error Model for Near Infrared Spectroscopic Instruments — 51
C. G. Eddison and A. M. C. Davies

The New Possibilities of Luminescence Spectroscopy of Microscopic Matter: Characterization of Single Hydrocarbon Fluid Inclusions by Fluorescence Excitation–Emission Microspectroscopy — 56
J. Kihle

Spectroscopic Imaging of Polyethylene — 61
S. F. Parker, C. Chai, and C. Baker

Ozone Measurements with Star-pointing Spectrometers — 66
H. K. Roscoe, W. H. Taylor, J. D. Evans, E. K. Strong, D. J. Fish, R. A. Freshwater, and R. L. Jones

Nuclear Magnetic Resonance

Recent Advances in High-resolution Chemical NMR Spectroscopy of Solids — 77
R. K. Harris

SFC/NMR On-line Coupling — 86
K. Albert and U. Braumann

Structural Determinations of Organic Compounds Found in the Environment by NMR Spectroscopyand Mass Spectrometry — 94
S. T. Belt, D. A. Cooke, S. J. Hird, E. J. Wraige, P. Donkin, and S. J. Rowland

Solid State Proton NMR and Dynamic Mechanical Analysis Studies of Polymer Latex Blends — 100
R. Ibbett, M. James, D. Hourston, and I. Aucott

Biological Applications

Speciation of Selenium in Human Serum by Size Exclusion Chromatography and Inductively Coupled Plasma Mass Spectrometry — 109
T. M. Bricker and R. S. Houk

The Investigation of Varietal Differences Among Sorghum Crop Residues Using Near Infrared Reflectance Spectroscopy 117
S. J. Lister, M. S. Dhanoa, I. Mueller-Harvey, and J. D. Reed

The Influence of Energy Migration on Fluorescence Kinetics in Photosynthetic Systems 123
D. L. Andrews and A. A. Demidov

Flow Injection Procedures with Spectrophotometric Detection for the Determination of Nitrate and Nitrite in Riverine, Estuarine, and Coastal Waters 129
T. McCormack, A. R. J. David, and P. J. Worsfold

Quantitative Determination of Chlorophyll A 136
K. Kavianpour, P. W. Araujo, and R. G. Brereton

Spectrometric Determinations

Use of Mass Spectrometry Aboard United States Spacecraft 147
T. F. Limero

Determination of Rare Earth Elements in Their Mixtures Using Inductively Coupled Plasma Atomic Fluorescence Spectroscopy 157
A. A. Galkin, G. N. Maso, and G. G. Glavin

Chemiluminescence Based Detection of Aldehydes and Carboxylic Acids in Engine Oils 164
A. N. Gachanja and P. J. Worsfold

Chemiluminescence Based Detection of Hydrogen Peroxide in Seawater 171
D. Price, P. J. Worsfold, and R. F. C. Mantoura

The 'Co-master' Concept in Developing Robust Near Infrared Calibrations 175
I. A. Cowe, C. G. Eddison, N. Hewitt, and A. M. C. Davies

The Application of Inelastic Neutron Scattering to Advanced Composites 184
S. F. Parker and J. N. Hay

A Fibre Optic Detecting System for the Simultaneous Determination of Oxygen and Carbon Dioxide Based on Absorption Spectroscopy 189
M. F. Choi and P. Hawkins

Use of GC-MS for the Detection of Metabolites in the Urine of Doped Horses: Application to Quinine Breakdown Products 196
C. Demir and R. G. Brereton

Chemometrics

Modelling Non-linear Data Using Neural Networks Regression in Connection with PLS or PCA 209
C. Borggaard

Rugged Spectroscopic Calibration Using Neural Networks 218
P. J. Gemperline

The Regression Model Comparison Plot (REMOCOP) 225
P. Geladi

Studies in Near Infrared Spectroscopic Qualitative Analysis 237
M. Coene, A. M. C. Davies, and R. Grinter

Author Index 244

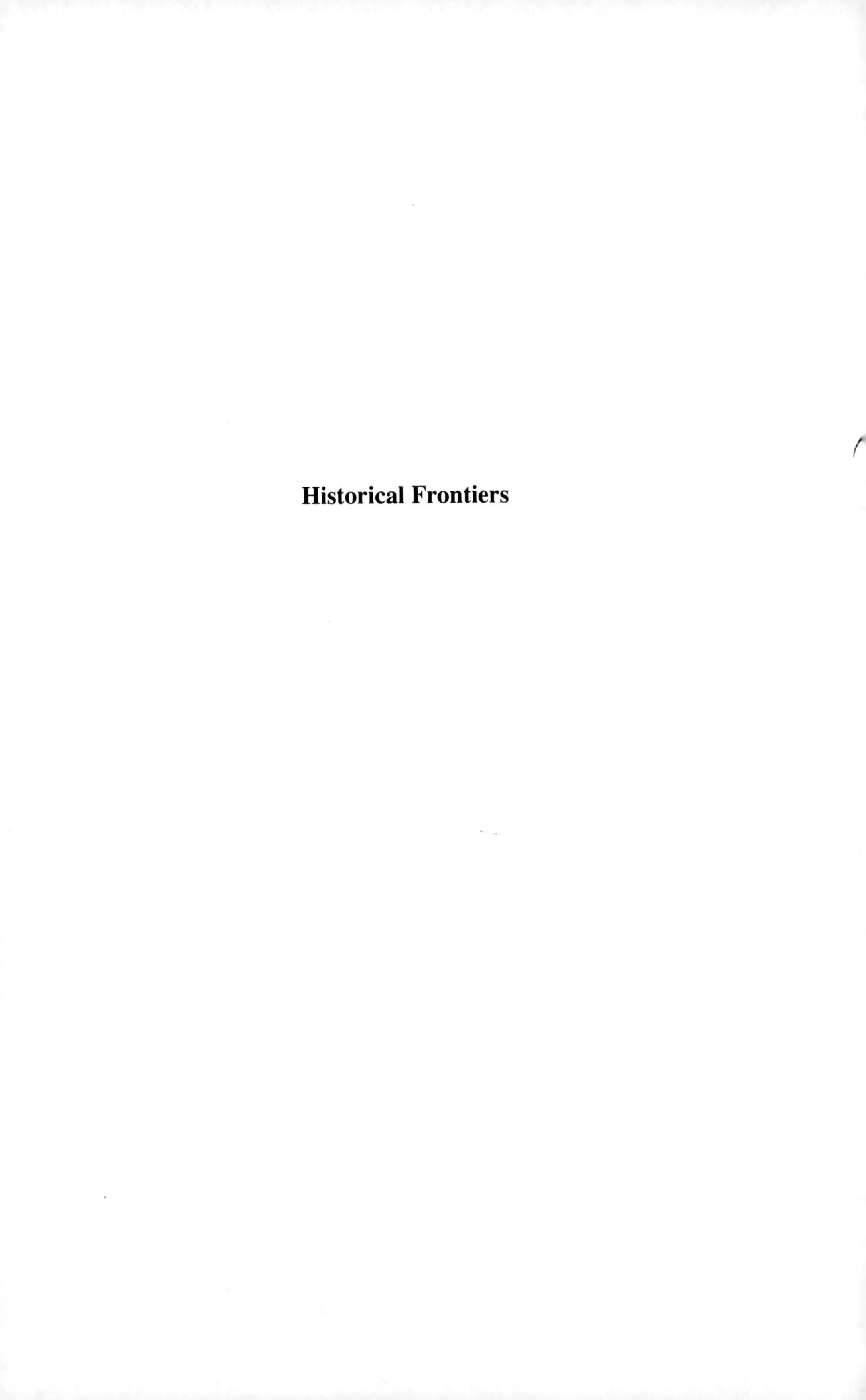

Historical Frontiers

Vibrational Spectroscopic Contributions to Chemisorption and Catalysis

Norman Sheppard, FRS

SCHOOL OF CHEMICAL SCIENCES, UNIVERSITY OF EAST ANGLIA, NORWICH NR4 7TJ, UK

1. INTRODUCTION

Since the early 1940's vibrational spectroscopy has been one of the major physical methods used for the elucidation of molecular structures. Infrared spectroscopy has long been used routinely for this purpose within organic chemistry and many structure/wavenumber (frequency) correlations have also been established for inorganic and organometallic molecules. Nuclear magnetic resonance and mass spectroscopy have also been of major importance for the chemical structure determination of liquid or solid samples. The similar investigation of adsorbed molecules, at the level of monolayers or less, obviously provides a much greater challenge. In this area it is primarily the vibrational spectroscopies, sometimes using new probes and experimental techniques to be described below, that have provided the requisite sensitivity. Only in exceptional cases has it been possible to obtain definitive NMR spectra from chemisorbed monolayers. Mass spectroscopy can only provide information about molecules or molecular fragments driven off the surface so that the deduction of the structures of the adsorbed species themselves is necessarily an indirect, if usefully-indicative, procedure.

2. ADSORPTION ON FINELY-DIVIDED SURFACES

2.1 Infrared Spectroscopy

It can readily be estimated that at single-monolayer thickness the strongest infrared spectral features of most molecules should absorb only a small fraction of one percent of the incident radiation. It was therefore to be expected that the first successful experiments would involve the use of adsorbents of very high specific area so as to maximise the number of monolayers traversed by the infrared beam. Professor Terenin of Leningrad in 1940 pioneered the spectroscopic study of physical adsorption in this way using silica-gel or (optically preferable) porous silica-glass.[1] He and his colleagues used substantial thicknesses of sample and measured the overtones of the bond-stretching vibrations of chemically-bound OH groups in the *near infrared region* where detectors were at that time particularly sensitive. The presence of the adsorbed species was monitored by the modifications to the positions and strengths of the surface νOH overtone absorptions brought about by hydrogen-bonding or other intermolecular interactions, as well as by the occurrence of new absorptions. Subsequently work was extended into the vibrational fundamental or *mid-infrared region* by Pimentel, Garland and Jura[2], by Sidorov[3], and by Yates and Sheppard [4, 5] during the earlier part of the 1950's. This work was concerned with *physical adsorption* and investigated aspects such as the presence of different adsorption sites, the distortions of molecular symmetry by surface forces, and qualitative studies of rotational motions of small adsorbed molecules.

The first successful study of chemisorption was made in 1954 by Eischens, Pliskin and Francis.[6] They studied the bond-stretching fundamental absorptions of CO chemisorbed on finely-divided surfaces of several different metals, the metal particles themselves being supported on high-area silica particles in the manner used in the preparation of metal catalysts. Success in this very favourable case (νCO bands have exceptionally high intensities) led to the gradual development of infrared spectroscopy into a major technique for research into chemisorption and heterogeneous catalysis using first prism and then diffraction-grating spectrometers.

Infrared studies of adsorption on high-area adsorbents, today extended to cover metals, oxides including zeolites, and halides etc., were greatly facilitated in the early 1970's by the development and introduction of *Fourier-transform infrared* spectrometers in the mid-infrared region (FTIR). For the purposes of structure-determination it is important to study as much of a mid-infrared spectrum as possible. Under these conditions the FT spectrometers provided greater sensitivity by considerably more than one order of magnitude using the multiplex and throughput advantages of the FT method. Further enhancements were possible on replacing thermal detectors by photoconductive ones, usually of mercury/cadmium telluride which is sensitive down to 800 to 600 cm^{-1}. The higher sensitivity could alternatively be traded for high resolution.

Additionally FTIR spectrometers require built-in mini/microcomputers for the purpose of Fourier-transformation of the original interferometric data. These can also be used to give much clearer and more complete spectra of the adsorbates themselves through the point-by-point compensations for the spectrum of the original clean adsorbate achieved either by spectral subtraction in the absorbance mode or by ratioing in the transmission mode.

After the attainment of adequate signal/noise the principal remaining experimental limitation is that in certain regions of the spectrum the adsorbent itself, usually an oxide, adsorbs virtually all the radiation to give 'black-out' conditions. For the lighter oxides, silica and alumina this in effect reduces the accessible region to above 1300 or 1100 cm^{-1} respectively; for heavier-metal oxides wider regions are available. Fortunately however many of the group-characteristic infrared absorption bands of adsorbates occur in the higher-wavenumber ranges. Today between 400 and 500 papers a year report vibrational spectra of adsorbed species.

Given good spectral experimental data the remaining challenge is to find correct interpretations of the spectra in structural terms. This is often a demanding task for a number of reasons. It became clear early that a given reagent adsorbed on finely-divided adsorbents frequently gave rise to several surface species derived from different adsorption sites. The interpretation of these complex spectra with overlapping components proved to be greatly helped by a capability that arose in the mid-1970's to obtain spectra on flat single-crystal faces with known atomic arrangements. In these simplified systems it often proved to be possible to identify the spectra of particular adsorbed species one at a time. Furthermore by combining spectroscopic and diffraction measurements, some quite unexpected surface species were identified. Ethylidyne CH_3CM_3 (M = metal), derived from the adsorption of ethene on metal surfaces at room temperature, was a prime example and this turned out to give rise to some of the strongest absorptions in the spectra on finely-divided catalysts! As in the ethylidyne case,[7] the interpretation of the spectra on the single-crystal surfaces was in turn helped by comparison with the infrared spectra of metal cluster-compounds with known hydrocarbon ligands that had previously been identified by X-ray crystallography.

Another form of interpretational assistance that developed from the single-crystal work on metal surfaces was the realisation that a metal-surface selection rule (Section 3.1) was also very effective in simplifying spectra of adsorbates on large metal particles.[8]

2.2 Raman Spectroscopy

Although infrared spectroscopy remains the most widely-used method for work with finely-divided surfaces others have made significant, if often more specific, contributions. *Raman Spectroscopy* is usually the first to come to mind and it has been particularly effective in studying adsorption on oxide surfaces.[9] There is a well-known general complementarity in the intensities of features in the infrared and Raman spectra from a given vibration of an adsorbed species. Strength in one spectrum often goes with weakness in the other. Hence putting together the data from both techniques gives more complete information on the vibration fundamentals of the adsorbed molecules or adsorbents concerned. For example the highly polar metal oxides, of surface interest either as primary adsorbents or as supports, lead to very strong infrared absorptions from their lattice modes but very weak corresponding Raman features. Hence a wider wavenumber range for adsorbate bands is usually available from the Raman spectra.

Less Raman success has been achieved with oxide-supported metal catalysts. When these have been obtained the spectra have often been unexpectedly complex. This may have arisen from decomposition of adsorbates such as through heating effects arising from absorption of the laser beam used to excite the Raman spectrum from these black or highly-coloured samples. However, some successes have been achieved.[10]

2.3 Surface Enhanced Raman Spectroscopy (SERS)

Very great surface-enhancement of Raman spectra, by up to a factor of a million, can occur when such spectra are obtained from roughened metal surfaces such as electrodes, cold-deposited metal vapours, or colloids. Whereas the potential analytical applications of this technique are obvious unfortunately, for excitation of Raman spectra in the normal visible region, the effect is largely limited to the coinage metals Cu, Ag and Au. It is to be hoped that extensions of the technique to the transition metals of catalytic interest such as Ni, Pd or Pt, will become possible by excitation in the ultraviolet region and/or by the preparation of samples of these metals with particles of the appropriate shapes or sizes. In the first instance these factors determine whether a plasmon resonance of the metal particle is excited. Other contributions to enhancements can come from resonance Raman contributions of the adsorbate/metal complex.[11]

2.4 Inelastic Neutron Scattering (INS)[12]

This technique has given useful vibrational spectroscopic information, particularly in the lower wavenumber ranges, from adsorbed molecules on working catalysts such as Raney nickel. The method is not confined to metals but does require the availability of large samples. Although INS has no symmetry-based selection rules, in practice much the strongest signals arise from vibrations involving motions of hydrogen atoms. These vibrations can occur within a group such as CH_3, or through the hydrogen atoms being 'carried' during vibrational motions of heavier atoms. INS has the advantage that the relative intensities associated with different vibrational modes can be calculated from the atomic amplitudes involved.

2.5 Inelastic Electron Tunnelling Spectroscopy (IETS)[13]

This technique involves the tunnelling of electrons between two different metal electrodes such as aluminium and lead separated by a thin layer of finely-divided oxide. The Fermi levels of the two metals differ but the difference can be bridged by appropriate vibrational quanta from within the oxide layer. As the voltage-difference between the two electrodes is scanned different vibration frequencies facilitate the electron tunnelling and cause increased currents to flow. Vibrations of molecules adsorbed on the finely-divided alumina (or even on metal particles supported on the oxide) can give rise to IETS spectra. Although such spectra can be obtained with resolution comparable to those in infrared spectra, this is only so when the overall probe is held at liquid-helium temperatures. This limits the general usefulness of the method, but we shall return to the topic again below in the conclusion section.

3. ADSORPTION ON FLAT, INCLUDING SINGLE-CRYSTAL, SURFACES

3.1 Reflection-Absorption Infrared Spectroscopy (RAIRS)

In this technique the infrared beam is reflected off the flat surface of a metal so that it twice traverses a thin film of any adsorbate that is present. In 1959 Francis and Ellison[14] succeeded in measuring the very weak spectrum anticipated from a monolayer of an adsorbed long alkyl-chain carboxylate. They also demonstrated that the spectrum obtained is simplified by the operation of a *metal-surface selection rule* (MSSR) caused by the response of the free conduction electrons of the metal to electrical dipole moment changes associated with vibrations of the adsorbate. The effect of the selection rule is only to allow measurable intensity from these modes of vibration of the adsorbed species which cause vibrational changes in the component of the dipole moment that is perpendicular to the metal surface. Such dipole changes are best excited by a beam of radiation at near-grazing incidence polarised in the plane of incidence so that the electric vector is close to perpendicular to the surface.

In 1970 Chesters, Pritchard and Sims[15] successfully coupled RAIRS to the ultra-high vacuum techniques of surface science so as to enable such measurements to be made using the surfaces of metal single-crystals. They used CO as an adsorbate. This, like the long-chain carboxylate studied by Francis and Ellison, was a very favourable case. It led to some very nice studies of CO on other metal surfaces but sensitivity remained too limited for general use. In due course this difficulty was overcome as described below in Section 3.3.

3.2 Vibrational Electron Energy-Loss Spectroscopy (VEELS)[16]

In the mid-1970's this technique was introduced and provided the first capability of obtaining vibrational spectra from single monolayers from a wide range of adsorbed species. It is alternatively described in the literature with the adjective 'high-resolution' (HREELS). However, as such spectra are normally of very low resolution (~30 cm^{-1}) compared with infrared spectra (~1 cm^{-1}) the designation VEELS is more suitable in this context. The term 'high resolution' is relative to the resolution achieved in electronic EELS.

A beam of monoenergetic electrons is caused to be reflected off the surface of a metal which has an adsorbed monolayer. A small proportion of the electrons transfer a portion of their energy to excite vibrations of the adsorbed species. Their energies are reduced by the magnitude of the vibrational quantum. The method is analogous to Raman spectroscopy except that energetic electrons rather than photons are used. The method has high sensitivity. It was shown that electrons reflected near the specular direction (angle of incidence = angle of reflection) excite perpendicular vibrations in accordance with the MSSR described above for RAIRS in Section 3.1. However, a second so-called 'impact' mechanism can cause wider deflections of electrons into 'off-specular' directions which provide information, but with less sensitivity, about vibrational modes forbidden under the MSSR. In this respect VEELS gives information not available from RAIRS. However, the method has the disadvantage, shared by other electron-based spectroscopies, that a high vacuum is required over the sample.

Ibach in particular[17] applied VEELS to a wide range of chemically-interesting systems to yield valuable information about atomic adsorbates (H, O, N etc.), and diatomic (CO, NO, N_2 etc.) and polyatomic ones (hydrocarbons, alcohols etc.).

3.3 RAIRS rejuvenated

The success of VEELS spurred the infrared spectroscopists to greater efforts and in the mid-1980's Chesters and his colleagues[17, 18] combined RAIRS with Fourier-transform techniques so as to greatly enhance both spectral sensitivity and resolution. Because many of the absorption bands of adsorbed species turned out to be intrinsically very narrow, of half width ~1 cm^{-1} (they looked very wide in VEELS!), substantially improved peak signal/noise was achieved compared with what could have been anticipated on sensitivity grounds alone. RAIRS has also the advantage compared with VEELS that photon experiments can be carried out with substantial pressures of gases over the metal surface. This is of considerable importance for experiments simulating catalysis. However, the MSSR is strict in RAIRS so that, unlike with VEELS, 'parallel' vibrations cannot be studied. Also for routine work the mercury/cadmium telluride photoconductive detectors restrict the available wavenumber region to above 800-600 cm^{-1}. However, many adsorption systems have, and many more would, benefit from RAIRS experiments, particularly from the point of view of resolution when investigating polyatomic adsorbates. Papers using this technique are rapidly growing in numbers and in importance. The method has also been successfully applied to research on single-crystal electrodes in solution as well as to gas/solid interfaces.

3.4 Sum-Frequency Spectroscopy (SFS)

This laser-spectroscopy technique, pioneered recently by Shen,[19] has a capacity for fruitful future development. It has applicability to gas/liquid and liquid/liquid interfaces as well as to those of the gas/solid nature. It is a higher-order technique with properties similar to those of second-harmonic generation in that it normally only generates signals from the interface itself and not from the bulk phases on either side. As the name implies, the sum-frequency technique involves the interaction of two laser beams at the interface. One of these is a monochromatic laser generating visible radiation and the other is a variable-frequency infrared laser. The 'sum-frequency' radiation generated by their interaction occurs in the visible region where detectors are of high sensitivity. The signal is enhanced when the wavenumber of the infrared laser matches that of a molecular vibration which has both infrared and Raman activity. The intensity depends on the product of the infrared and Raman intensities. This method has so far mainly been applied to the νCH bond-stretching region of long-chain alkyl groups within adsorbates. These have the favourable advantages of considerable intensity in both the 'parent' spectra so that again we have a 'favourable case' situation. However, the method is so 'surface' sensitive that only the νCH modes from the ends of the alkyl chains, i.e. from the CH_3- or -CH_2X groups, give good signals. Most flat-surface spectroscopies provide opportunities, using polarised radiation, to investigate the orientation of the axes of adsorbed species relative to the surface. Sum-frequency spectroscopy has the unique capability of telling the direction, up or down relative to the surface, in which a molecule or group is pointing. However, this leads to complete loss of signal from finely-divided adsorbents such as colloidal particles, micelles etc. because of mutual cancellation of the signals from the adsorbed species on opposite sides of the particle.

SFS appears to be a promising technique for studying the spectra of molecules at the flat liquid/liquid interface provided one or other of the bulk phases is laser-transparent. At present it is limited by the accessible wavenumber ranges of infrared lasers which so far rarely go below 2500 cm^{-1}. However, developments to lower wavenumbers can be anticipated.

3.5 Other Techniques Applicable to Flat Surfaces

The Raman Effect is intrinsically a low-intensity technique and at this time such spectra can only be obtained with difficulty from monolayers on flat metal,[20] non-metal or liquid surfaces.[21] These 'favourable-case' successes have mostly involved the use of the recently introduced array-type charge - coupled detectors (CCD). Using integrated optics some further improvements in sensitivity can be achieved by 'waveguide' optical excitation methods.[22] Matters are much more favourable if a *Resonance Raman spectrum* can be excited. Monolayers of dyestuffs have been very successfully studied on liquid surfaces in such circumstances.[23]

Infrared transparent materials such as silicon and germanium which can be fabricated in the forms of prisms for multiple attenuated total internal reflection (ATR) experiments can give high-quality monolayer spectra.[24] Also partially-metallised germanium electrodes have promise in relation to this experimental method.

4. CONCLUSIONS

The semi-historical approach adopted in this article serves to illustrate the very great progress that can be made during a period of some 50 years in a field of research within a mature science. In the 1940's it was a struggle to obtain spectra of adsorbed species at monolayer level even on very high area adsorbents. Today many such spectra have been obtained with good resolution and reasonable signal/noise even for weakly absorbing surface species on a few cm^2 of flat surfaces of single crystals with known atomic arrangements. Again and again a 'favourable case' allows a 'breakthrough'. Once signals have been obtained experimental attention can be directed with efficiency to improving the situation. In this particular field it has principally been improvements in experimental techniques that have generated progress although new theoretical understandings, such as the significance of the MSSR, have also contributed.

FTIR will continue to be the most widely-used technique for the study of adsorption on (or the surfaces of) finely-divided materials. Raman spectroscopy will probably increasingly contribute to work on clear or weakly-coloured polar materials such as oxides. The more recently introduced near-infrared FT Raman method may enable better results to be obtained with highly coloured adsorbents such as oxide-supported metals.

For flat metal surfaces VEELS and RAIRS will continue to play the major roles and both have much scope for wide applications. Large crystals of oxides are more difficult to obtain and variable-temperature work with these is difficult. However, the epitaxial growth of thin oxide layers on metal surfaces may provide the simplified model adsorption systems analogous to those that have been so valuable on metals for spectral interpretation purposes.

The study of the structures of adsorbed species by vibrational spectroscopic methods has, during the past few decades, revolutionised our understanding of chemisorption in molecular terms. As chemisorption is involved in most heterogeneous catalysis the latter field has also been greatly advanced. Of course not all the adsorbed species detected by spectroscopic methods are catalytically active; some can cause site-blocking catalytic poisons, or just be unreactive. But where more than one reagent is involved it is not difficult to find out by qualitative experiments which are the catalytically labile ones. However the study of catalytic mechanisms in more details requires analyses of reaction kinetics. Much remains to be done in this area. A very good example is the study of the kinetics of the ethene/hydrogen reaction over a finely-divided Pt surface by Soma[25] using infrared methods. On metal surfaces the RAIRS method, which can be applied in the presence of considerable pressures of gas-phase reagents, should in the future enable kinetic analysis of the elementary reactions on particular sites on well-defined flat surfaces.

I repeat a challenge that I offered recently.[26] Perhaps with the help of low-temperature scanning tunnelling microscopy it might be possible to obtain the inelastic electron tunnelling spectrum (IETS) of a single particular adsorbed molecule by varying the voltage bias between the tip and the molecule adsorbed on the surface!

REFERENCES

1. A.N. Terenin, *Zh. Fiz. Khim.*, 1940, **14**, 1362; A.N. Terenin and N.G. Yaroslavsky, *Izv. Akad. Nauk SSSR*, 1945, **2**, 203.
2. G.C. Pimentel, C.W. Garland and G. Jura, *J. Amer. Chem. Soc.*, 1953, **75**, 803.
3. A.N. Sidorov, *Dokl. Akad. Nauk SSSR*, 1954, **95**, 1235.
4. D.J.C. Yates, N. Sheppard and C.L. Angell, *J. Chem. Phys.*, 1955, **23**, 1980.
5. N. Sheppard and D.J.C. Yates, *Proc. Roy. Soc. A.*, 1956, **238**, 69.

6. R.P. Eischens, W.A. Pliskin and S.A. Francis, *J. Chem. Phys.*, 1954, **22**, 1786.
7. P. Skinner, M.W. Howard, I.A. Oxton, S.F.A. Kettle, D.B. Powell and N. Sheppard, J. *Chem. Soc., Faraday Trans. 2*, 1981, **77**, 1203.
8. H.A. Pearce and N. Sheppard, *Surface Sci.*, 1976, **59**, 205.
9. T.A. Egerton, A.H. Hardin, Y. Kozirovski and N. Sheppard, *J. Catal.*, 1974, **32**, 343.
10. W. Krasser and A.J. Renouprez, *Solid State Comm.*, 1982, **41**, 231.
11. J.A. Creighton, in 'Spectroscopy of Surfaces' (Advances in Spectroscopy, volume 16) (R.J.H. Clark and R.E. Hester, eds), Wiley, Chichester, 1988, p.37.
12. 'Chemical Applications of Thermal Neutron Scattering', (B.T.M. Willis, ed.) Oxford U.P., Oxford, 1973.
13. 'Tunneling Spectroscopy', (P.K. Hansma, ed.), Plenum Press, New York (1982).
14. S.A. Francis and A.H. Ellison, *J. Opt. Soc. Amer.*, 1959, **49**, 131.
15. M.A. Chesters, J. Pritchard and M.L. Sims, *Chem. Comm.*, 1970, 1454.
16. H. Ibach and D.L. Mills, 'Electron Energy Loss Spectroscopy and Surface Vibrations', Academic Press, New York, 1982.
17. M.A. Chesters, *J. Electron Spect. & Rel. Phenom.*, 1986, **38**, 123.
18. M.A. Chesters, S.F. Parker and R. Raval, *J. Electron Spect. & Rel. Phenom.*, 1986, **39**, 155.
19. P. Guyot-Sionnest, J.H. Hunt and Y.R. Shen, *Phys. Rev. Lett.*, 1987, **59**, 1597.
20. V.M. Hallmark and A. Campion, *J. Chem. Phys.*, 1986, **84**, 2933.
21. T. Kawai, J. Umemura and T. Takenaka, *Chem. Phys. Lett.*, 1989, **162**, 243.
22. J.F. Rabolt and J.D. Swalen in 'Spectroscopy of Surfaces' (see reference 11), 1988, p.1.
23. T. Takenaka and H. Fukuzaki, *J. Raman Spec.*, 1979, **8**, 151.
24. Y.J. Chabal, G.S. Higashi and S.B. Christman, *Phys. Rev. B*, 1983, **28**, 4472.
25. Y. Soma. *J. Catal.*, 1982, **75**, 267.
26. N. Sheppard, Special 50th anniversary issue of *Spectrochimica Acta*, 1990, p.149.

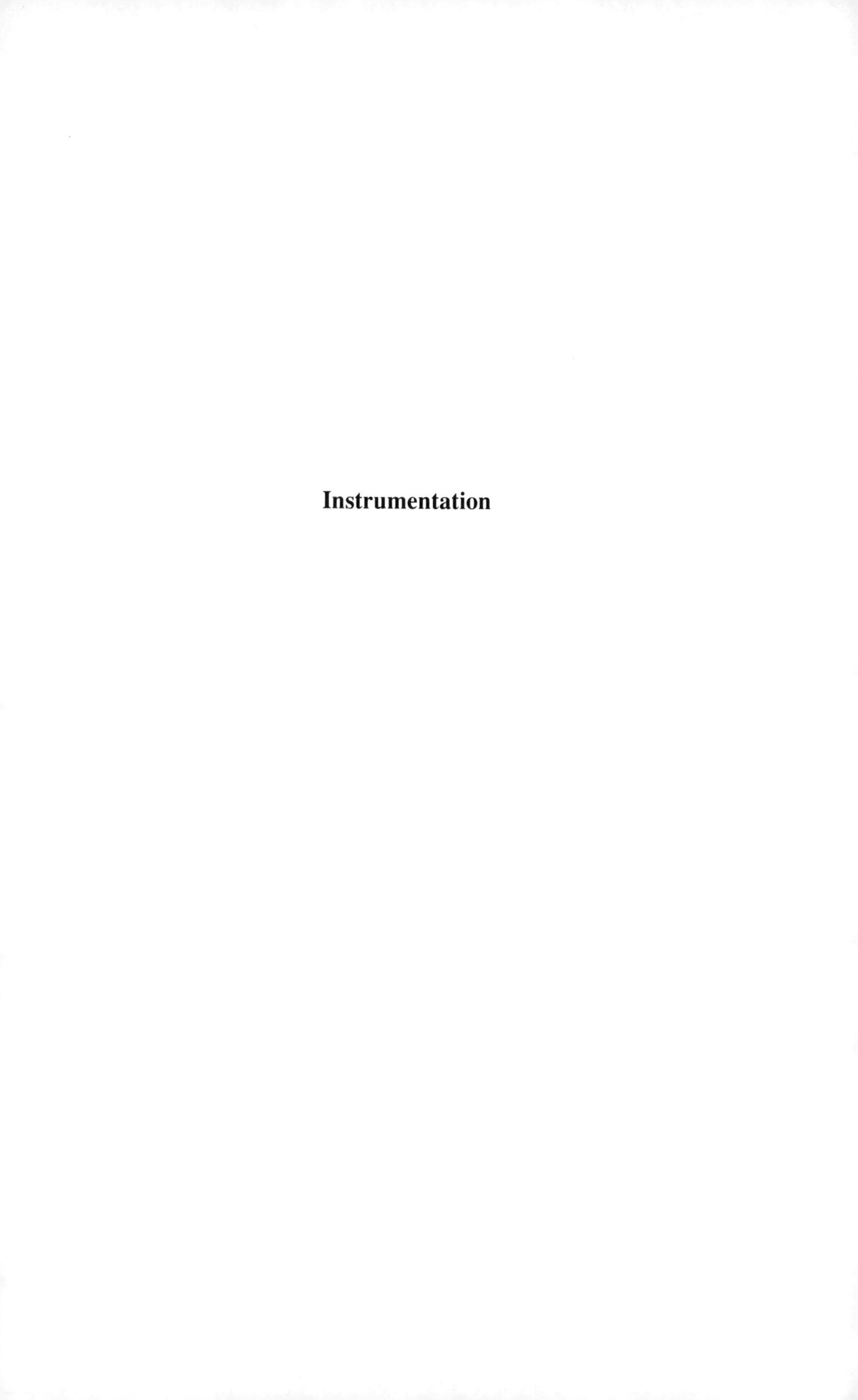

Instrumentation

Recent Advances in Kinetic Infrared Spectroscopy[+]

J. J. Turner, M. W. George, and M. Poliakoff

DEPARTMENT OF CHEMISTRY, UNIVERSITY OF NOTTINGHAM, UNIVERSITY PARK, NOTTINGHAM NG7 2RD, UK

1 INTRODUCTION

"Slow" chemical reactions can easily be monitored by infrared (IR) spectroscopy; if the whole spectrum is required as a function of time, then the rate of reaction must be slow compared with either the scan time of a dispersive instrument or the mirror scan time of a Fourier Transform (FT) instrument. With modern rapid scan FTIR instruments, the timescale can be of the order 10 ms with moderate resolution. As the rate of reaction increases, then several possibilities for time-resolved IR (TRIR) present themselves:

(1) Monitor a single IR frequency as a function of time and then repeat at different *frequencies;*

(2) Take the whole spectrum in a single shot, and repeat at different *delay times* after the "event";

(3) Modify the method of obtaining the FTIR spectrum.

Each of these modifications presents novel experimental features, depending on the time scale of the event to be monitored. This article will consider techniques on time scales varying from ms to fs with examples drawn from the chemistry of transition metal carbonyl and related species[1]. The ν(CO) IR bands observed in metal carbonyls are extremely intense and, usually, well removed from other vibrations. Thus they have provided very important structural information on such species. It is not surprising therefore that most of the major developments in instrumentation have employed these carbonyls to test their effectiveness. This includes biological molecules such as carboxymyoglobin, which on a simple view is just a complicated iron carbonyl.

2 SINGLE FREQUENCY IR DETECTION

2.1 Timescales down to ~50 ns

The simplest method is to replace the usual detector in a conventional dispersive instrument with a fast detector - such as a cooled MCT - and to set the grating to a particular frequency. The first successful experiments of this kind were performed by Siebert et al[2] and by Schaffner and colleagues[3]. It proved possible to photolyse, with a flash lamp, solutions of $M(CO)_6$ (M=Cr, Mo,W) and generate the species $M(CO)_5$...S, where the $M(CO)_5$ fragment is weakly coordinated to the hydrocarbon solvent. The fragments were identified, and their structures determined, by repeating the flash experiment over a range of IR frequencies around 2000 cm^{-1}, and comparing the results[3] with the IR spectra obtained for $M(CO)_5$ in low temperature matrices[4].

[+]Dedicated to Professor Norman Sheppard, FRS.

This method, of using matrix data to underpin the TRIR results, has proved to a particularly powerful one[5]. The limitation of a dispersive instrument is largely connected with the low intensity of the IR source, although elegant TRIR has been done in Mülheim[6] and Japan[7,8]. Most of the recent work has used IR lasers. These have usually been of two types; the CO laser is particularly valuable for the study of transition metal carbonyls because the output of the laser can scan the range ~2000 cm^{-1} to 1700 cm^{-1}. Incorporation of a grating permits the tuning of the output over a series of very narrow lines ~4 cm^{-1} apart with power in each line of 50-500 mW. This resolution is adequate for much solution chemistry, where the width of the IR bands may be say 10-20 cm^{-1}, but is less useful for low pressure gases. The other common IR laser is based on semiconductor diodes operating at cryogenic temperatures. These lasers can give highly monochromatic, continuously tunable, IR output over a range of ~150 cm^{-1} with power ~1mW. The advantage of the diode laser is clearly the continuous tunability, but this is somewhat offset by the low power, the occurrence of "multimoding" and the general difficulty of obtaining reproducible output.

Although the shortest timescales currently obtainable with IR detectors is ~5 ns, this requires careful matching of preamplifiers and amplifiers and in practice the time resolution is ~50 ns. This imposes a limitation on the photochemical source used to generate transients. Early experiments used flash lamps with lifetimes of several μs; more recent experiments have all used UV or visible lasers. The "workhorses" have been of two kinds: excimer lasers, usually operating with XeCl, with energies up to several hundred mJ at 308 nm, and a lifetime of ~20 ns; or Nd/YAG lasers with output at 532, 355 and 266 nm, with a wide energy range and with a lifetime of ~10 ns.

A great deal of work on organometallic intermediates, both in solution and in the gas phase, has been carried out with such instrumentation[9-13]. We give just two illustrations from our own recent work on excited states[14,15] and intermediates[16].

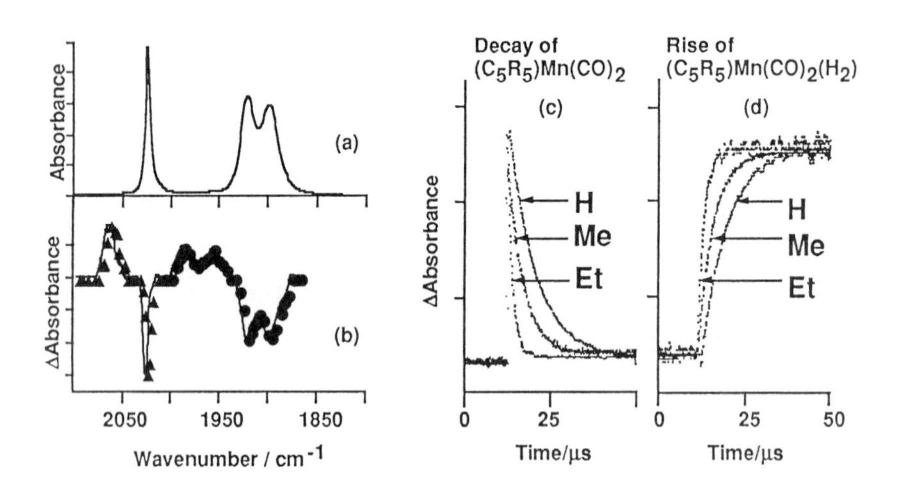

Figure 1 *(a) FTIR and (b) TRIR spectra (obtained 50 ns after UV excitation) of* $ClRe(CO)_3(2,2'\text{-}bpy)$*. The positive absorptions produced upon excitation are due to the* $\nu(CO)$ *absorptions of the MLCT excited state of* $ClRe(CO)_3(2,2'\text{-}bpy)$ *which are shifted up in frequency relative to the ground state. (c) and (d) IR kinetic decay traces of* $(\eta^5\text{-}C_5R_5)Mn(CO)_2$*, in n-heptane, obtained from photolysis of* $(\eta^5\text{-}C_5R_5)Mn(CO)_3$*. The traces show, from both the decay of* $(\eta^5\text{-}C_5R_5)Mn(CO)_2$ *(c) and the formation of* $(\eta^5\text{-}C_5R_5)Mn(CO)_2(H_2)$ *(d), that the rate of reaction between* $(\eta^5\text{-}C_5R_5)Mn(CO)_2$ *and* H_2 *increases in the order R = Et > Me > H. In these complexes the rate appears to be governed by steric rather than electronic factors.*

As explained above, the timescale is limited by the IR detectors to ~ 50 ns, although, with the new detectors appearing on the market, this is probably extendable to ~10 ns. Many interesting events occur on timescales much shorter than this and hence there has been a substantial effort to extend this to the ps and even fs regime by special techniques. We consider these next.

2.2 Timescales down to femtoseconds

2.2.1 Pulsed IR Lasers. In the techniques described above, the IR sources are continuous wave; the IR detectors monitor the changing intensity during and following the "event", usually UV/visible flash photolysis. If an IR source with a very short pulse (say 1 ps) is used then, in principle, this can monitor the change in the sample cell at a fixed time (say 10 ps) after the flash, provided the IR detector, with its slow risetime, can integrate the whole of the intensity of the 1 ps pulse. To obtain *spectral* information this must be repeated at other IR wavelengths; to obtain ***kinetic*** information it must be repeated at different time delays after the flash. There is no difficulty with the IR detectors but it is not trivial to generate a picosecond IR pulse. For example Spears[17] obtained a 10 ps IR pulse created by difference frequency techniques and in this way probed[18] the early stages in the solution photochemistry of $Cr(CO)_6$. One of the interesting results of these, and other ps experiments on metal carbonyls[19,20], is that at early times following the laser flash, the population of high vibrational levels is so large that it is difficult to interpret the data until the species are vibrationally cold; this time varies considerably but can be as long as 200 ps. Woodruff and colleagues obtained ~2 ps IR pulses by difference frequency mixing of a tunable visible dye laser output and a 532 nm pulse from a Nd/YAG laser. He has used this method to probe very rapid electron transfer processes in bridged metal species[15,21]. Kaiser[22] and Laubereau[23] and their colleagues use parametric generators employing $LiNbO_3$ crystals. Heilweil and colleagues[24] generated IR pulses by down conversion in $LiIO_3$ between the output of two tunable red dye lasers and the Nd/YAG second harmonic.

2.2.2 Upconversion to the Visible The laser mixing techniques take advantage of the fact that detectors in the UV/visible region can be made very much faster than IR detectors. Thus a pulse from a visible laser (frequency ν_{vis}) can be mixed in a crystal with the diode laser signal at a particular IR frequency (ν_1) which has passed through the sample. The output is a pulse at the visible frequency $\nu_{vis} + \nu_1$ with an intensity which depends on the intensity of the IR signal. Thus in principle, the changing IR signal can be monitored continuously in the visible; in practice, it is more common to use the pump/delay/probe method. The experiment is repeated at a different IR frequency (ν_3) to give a visible signal at $\nu_{vis}+\nu_2$. With this technique, or modifications of it, it has proved possible to probe events down to subpicoseconds[19].

The disadvantage of the methods considered so far is that data about only one IR frequency is obtained at a time, and in the ps/fs regime this requires considerable signal averaging. The experiment, often photochemically very sophisticated, must be repeated for each IR frequency. Of course it is possible to automate the scanning of the IR frequencies but the principle remains the same. It would clearly be an immense advantage to obtain the whole of the relevant IR spectrum at "one shot".

3 "ONE SHOT" TECHNIQUES

3.1 Detecting in the IR

There are two problems that make it difficult to record a complete IR spectrum on a short time scale. The first is that it requires the generation of a broad band IR source of sufficient intensity; the second is that it requires a detector which can record the complete IR spectrum at the same time. As we shall see modern laser and detector technology can overcome these problems but it is worth mentioning Pimentel's Herculean efforts to obtain the same result by a different route. Well before the advent of lasers, Pimentel constructed a rapid scan IR spectrometer with a design that was, in principle, very simple[25].

A conventional single beam IR spectrometer was modified in three particular ways: the source was replaced by a powerful carbon arc; the scan rate was enormously increased by using a rapidly rotating Littrow mirror; the detector was replaced by the fastest available. With this apparatus it was possible to scan at ~1 $cm^{-1}/\mu s$ and it was used to obtain spectral and kinetic data for several species in the gas phase including CF_2.

More recently, Pimentel[26] introduced a novel nanosecond IR spectrometer in which the broad band IR source was obtained, following the lead by Sorokin and co-workers[27], via stimulated electronic Raman scattering (SERS) in a metal vapour. After passing through the sample - itself subjected to laser flash photolysis - the IR radiation was dispersed in a spectrometer and then fell on to a multi element array IR detector which recorded the intensity of 120 different spectral slits simultaneously. This method is technically very demanding but it has been possible to study such species as the CF_3 radical[28] and the C_2F_3 radical[29]. In principle it should be possible for each detector element to follow continuously the changing IR intensity at that frequency, thus - again in principle - providing complete spectroscopic and kinetic information in a single shot. In practice the broad band source was pulsed (10 ns) and the response time of each detector element was long (~500 ns) thus forbidding monitoring in real time. Time-resolved data were obtained by varying the time between flash photolysis and the pulsed IR laser.

3.2 Detecting in the visible

As described before it is possible to convert the IR signal into the visible. Sorokin and colleagues[27] upconverted their broad band IR signal back into the visible by another SERS step. This permitted detection, following dispersion, based on a photographic film or other methods. Downconversion by SERS followed by upconversion by SERS is technically very demanding and in addition time-resolution is limited. The most recent, and elegant, experiments which permit "one shot" detection in the ps regime have been carried out by Heilweil and colleagues[20,30]. The broad band IR pulses (150 cm^{-1} and 300 fs FWHM) are produced by difference mixing of two visible lasers in a $LiIO_3$ crystal. After passing through the sample, these IR pulses are upconverted by mixing with the residual energy from one of the visible lasers; this "mixed" radiation in the visible is dispersed by a spectrograph on to a CCD array detector. This apparatus produces time resolution <1 ps and spectral resolution of ~4 cm^{-1}. It has proved particularly valuable for probing the early stages of metal carbonyl solution photochemistry. In hydrocarbon solution, the complex $(\eta^5\text{-}C_5H_5)Co(CO)_2$ shows two $\nu(CO)$ IR bands (2031 and 1971 cm^{-1}); on laser photolysis and on a 150 ns timescale, a species with one $\nu(CO)$ band (1993 cm^{-1}) is produced; this is assigned to the CO-loss species $(\eta^5\text{-}C_5H_5)Co(CO)$. With the ps equipment, Dougherty and Heilweil[30] were able to show that, at very early times after the flash, the band associated with the new species is very broad due to coupling of the C-O mode with other modes which are vibrationally hot. As time proceeds the band narrows as the hot modes cool so that by ~100 ps the spectrum is identical with that observed at 150 ns. Very recently a combination of ns TRIR in Nottingham and ps TRIR in Washington has unravelled[31] much of the subtlety in the photochemistry of $[(\eta^5\text{-}C_5H_5)Fe(CO)_2]_2$.

4 FOURIER TRANSFORM METHODS

All the methods described so far rely on direct measurement of intensity versus time and versus IR frequency. The advantages of Fourier transform IR compared with dispersive instrumentation are well known: all wavelengths at once (Fellgett); absence of slits (Jacquinot); internal calibration (Connes). It is not surprising therefore that attempts have been made to couple FTIR with rapid chemical change. We shall not be concerned with those experiments in which the lifetime of the species being examined is much longer than the time of the mirror travel of the FTIR instrument; as mentioned at the beginning of this article, this time can be as short as ~ 10 ms. We shall consider much faster timescales. The first successful attempts were carried out by Murphy and colleagues[32] on the emission from mixtures of N_2 and O_2 subjected to pulsed electron bombardment. More recent, and more sophisticated, emission experiments have been carried out by Leone[33] and by Hancock and colleagues[34]. However emission measurements have considerable S/N advantages and there has been much interest in performing the more difficult absorption measurements. These can be classified in two ways: stroboscopic and step-scan methods.

4.1 Stroboscopic FTIR

This phrase is used to describe a technique which combines normal scanning FTIR - with a mirror travel time of say 1 s - with a rapidly repeating event. Suppose that we wish to determine the IR spectrum of a transient 50 μs after a 10 ns UV laser flash. If the photochemistry is 100% reversible, then it may be possible to repeat the flash at say 100 Hz, so that during a single mirror scan there are 100 flashes. What is required is to extract from all the data points, after many scans, those data points which generate a complete interferogram corresponding to 50 μs after the flash. The collected data will also include interferograms corresponding to other times after the flash. This is not a trivial task but instrument manufacturers have introduced software packages to do this. Since this is a point-by-point method, it really loses the multiplex advantage of FTIR. A number of experiments have been described, the most successful ones involving repetitive stretching of polymers[35,36]. However there are two disadvantages in this method. The first is that the event must be absolutely reproducible with every condition constant. In the absence of such reproducibility, spurious artefacts can appear in the spectrum after Fourier transformation. Such artefacts can be very misleading as has been described elsewhere[37,38]. Moreover, in an experiment specifically designed to compare stroboscopic FTIR with the IR laser technique, it was clear that the laser had enormous advantages[38]. The second disadvantage is that FTIR information is collected "on-the-fly"; better S/N could be obtained if the mirror scanned in steps, with information being collected while the mirror is stopped. Nonetheless, on biological systems, elegant [39,40] stroboscopic FTIR measurements, on a few μs timescale, have been reported.

4.2 Step-scan FTIR

In the very early days of FTIR, when technical problems limited the range to the far IR, the instruments were step-scanned. With He/Ne laser technology able to pinpoint with great accuracy the position of the moving mirror, the IR range was extended and continuous scanning became the norm. It is interesting that manufacturers are now making available highly reliable and sophisticated step-scan instruments. The use of step-scan FTIR for time-dependent phenomena is becoming more common and has recently been reviewed[42]. With all the advantages of step-scan over stroboscopic, it is still necessary to ensure that the "event" is exactly reproducible. Thus step-scan FTIR has found applications in photobiology, particularly on systems such as bacteriorhodopsin[43], carboxymyoglobin[44] and photosynthetic bacterial reaction centre[45], with time-resolution as low as 500 ns.

5 CONCLUSIONS

We have presented a very brief summary of most of the current methods of obtaining fast time-resolved IR spectra. It is clear that with the range of techniques available, kinetic IR spectroscopy promises to continue to unravel a host of novel chemical problems.

Acknowledgments

We have described a limited amount of our own work in this account, but work at Nottingham has benefited from help from many colleagues, collaborators, students and technicians. We are also particularly grateful for the advice and guidance of Professor N Sheppard.We have been generously supported by the SERC, EEC, the Paul Instrument Fund of the Royal Society, and NATO Grant 920570.

References

1. Nanosecond TRIR equipment in Nottingham has recently been described in detail: M. W. George, M. Poliakoff and J. J. Turner, *Analyst*, 1994, **119**, 551.
2. F. Siebert, W. Mäntele and W. Kreutz, *Biophys. Struct. Mech.*, 1980, **6**, 139.
3. H. Hermann, F-W. Grevels, A. Henne and K. Schaffner, *J. Phys. Chem.*, 1982, **86**, 5151.
4. R. N. Perutz and J. J. Turner, *Inorg. Chem.*, 1975, **14**, 262; *J. Am. Chem. Soc.*, 1975, **97**, 4791.
5. For a review see J. J. Turner in "Photoprocesses in Transition Metal Complexes, Biosystems and Other Molecules", NATO ASI series, ed. E. Kochanski, Kluwer, Dordrecht, 1992, 125.
6. S. T. Belt, F-W. Grevels, W. E. Klotzbücher, A. McCamley and R. N. Perutz, *J. Am. Chem. Soc.*, 1989, **111**, 8373.
7. K. Iwata, C. Kato, and H. Hamaguchi, *Appl. Spectrosc.*, 1989, **43**, 16; K. Iwata, and H. Hamaguchi, *Appl. Spectrosc.*, 1990, **44**, 1431; C. Kato, M. W. George and H. Hamaguchi, in "Time-Resolved Vibrational Spectroscopy VI" eds. A. Lau, F. Siebert and W. Werncke, Springer-Verlag, Berlin, 1994, 78.
8. S. Oishi, M. Watanabe and T. Muraishi, *Chem. Letters*, 1993, 713.
9. For reviews see: M. Poliakoff and E. Weitz, *Adv. Organomet. Chem.*, 1986, **25**, 277; E. Weitz, *J. Phys. Chem.*, 1987, **91**, 3945; F-W. Grevels, W. E. Klotzbücher and K.Schaffner, *Chem. Rev.*, in the press. Representative articles are listed below.
10. P. H. Wermer and G. R. Dobson, *Inorg. Chim. Acta*, 1988, **142**, 91.
11. S. T. Belt, D. W. Ryba and P. C. Ford, *Inorg. Chem.*, 1990, **29**, 3633.
12. Y. Ishikawa, C. E. Brown, P. A. Hackett and D. M. Rayner, *J. Phys. Chem.*, 1990, **94**, 2404.
13. B. H. Weiller, E. P. Wasserman, C. B. Moore and R. G. Bergman, *J. Am. Chem. Soc.*, 1993, **115**, 4326.
14. M. W. George, F. P. A Johnson, J. R. Westwell, P. M. Hodges and J. J. Turner, *J. Chem. Soc. Dalton Trans.*, 1993, 2977.
15. For a review on TRIR of excited states of coordination compounds see: J. J. Turner, M. W. George, F. P. A. Johnson and J. R. Westwell, *Coord. Chem. Rev.*, 1993, **125**, 101.
16. F. P. A. Johnson, V. K. Popov, M. W. George, V. N. Bagrarashvili, M. Poliakoff and J. J. Turner, *Mendeleev Commun.*, 1991, 145.
17. L. Wang, X. Zhu, and K. G. Spears, *J. Am. Chem. Soc.*, 1988, **110**, 8695; K. G. Spears, X. Zhu, X. Yang and L. Wang, *Opt. Commun.*, 1988, **66**, 167.
18. J. R. Sprague, S. M. Arrivo and K. G. Spears, *J.Phys. Chem.*, 1991, **95**, 10528; and references therein.
19. R. M. Hochstrasser, B. R. Cowen, P. L. Dutton, C. Galli, S. LeCours, S. Maiti, C. C. Moser, D. Raftery, M. Therien, G. Walker and K. Wynne, in "Time-Resolved Vibrational Spectroscopy VI", eds. A. Lau, F. Siebert and W. Werncke, Springer-Verlag, Berlin, 1994, 191.
20. T. P. Dougherty and E. J. Heilweil, *J. Chem. Phys.*, 1994, **100**, 4006.
21. S. K. Doorn, P. O. Stoutland, R. B. Dyer and W. H. Woodruff, *J. Am. Chem. Soc.*, 1992, **114**, 3133; 1993, **115**, 6398.
22. T. Elsaesser, W. Kaiser and W. Lüttke, *J. Phys. Chem.*,1986, **90**, 2901.

23. H. Graener, R. Dohlus and A. Laubereau, *Chem. Phys. Letters*, 1987, **140**, 306.
24. S. A. Angel, J. C. Stephenson and E. J. Heilweil, in "Time-Resolved Vibrational Spectroscopy V", ed. H. Takahashi, Springer-Verlag, Berlin, 1992, 239; T. P. Dougherty and E. J. Heilweil, *Opt. Letters*, 1994, **19**, 129.
25. K. C. Herr and G. C. Pimentel, *Appl. Opt.*, 1965, **4**, 25.
26. M. A. Young and G. C. Pimentel, *Appl. Opt.*, 1989, **28**, 4270.
27. J. H. Glownia, J. Misewich and P. P. Sorokin, *Opt. Letters*, 1987, **12**, 19; and earlier references therein.
28. M. A. Young and G. C. Pimentel, *J. Phys. Chem.*, 1990, **94**, 4884.
29. B. E. Wurfel, N. Pugliano, S. E. Bradforth, R. J. Saykally and G. C. Pimentel, *J. Phys. Chem.*, 1991, **95**, 2932.
30. T. P. Dougherty and E. J. Heilweil, ref. 19, p. 136.
31. T. P. Dougherty, M. W. George and E. J. Heilweil, in preparation.
32. R. E. Murphy, F. H. Cook and H. Sakai, *J. Opt. Soc. Am.*, 1975, **65**, 600.
33. S. R. Leone, *Acc. Chem. Res.*, 1989, **22**, 139; S. A. Rogers and S. R. Leone, *Appl. Spectrosc.*, 1993, **47**, 1430.
34. D. E. Heard, R. A. Brownsword, D. G. Weston and G. Hancock, *Appl. Spectrosc.*, 1993, **47**, 1438.
35. S. E. Molis, W. J. MacKnight and S. L. Hsu, *Appl. Spectrosc.*, 1984, **38**, 529.
36. W.M. Grim, J.A. Graham, R.M. Hammaker and W.G. Fateley, Amer. Lab., 1984 **16**(3) 22; and references therein.
37. A. A. Garrison, R. A. Crocombe, G. Mamantov and J. A. De Haseth, *Appl. Spectrosc.*, 1980, **34**, 399.
38. B. D. Moore, M. Poliakoff, M. B. Simpson and J. J. Turner, *J. Phys. Chem.*, 1985, **89**, 850.
39. K. Gerwert, ref. 24, p. 61.
40. D. L. Thibodeau, E. Nabedryk, R. Hienerwadel, F. Lenz, W. Mäntele and J. Breton, ref. 24, p. 79.
41. R. Hienerwadel, S. Grzybek, C. Fogel, W. Mäntele, J. Breton, E. Nabedryk and M. Y. Okamura, ref. 19, p. 231.
42. R. A. Palmer, J. L. Chao, R. M. Dittmar, V. A. Gregoriou and S. E. Plunkett, *Appl. Spectrosc.*, 1993, **47**, 1297.
43. K. Noelker, O. Weidlich and F. Siebert, ref. 24, p. 57; O. Weidlich and F. Siebert, *Appl. Spectrosc.*, 1993, **47**, 1394.
44. C. Rödig, O. Weidlich, and F. Siebert, ref. 19, p. 227.
45. J-R. Burie, W. Leibl, E. Nabedryk and J. Breton, *Appl. Spectrosc.*, 1993, **47**, 1401.

Spectroscopic Concentration Determination without Calibration: Scope and Limitations of Concentration-modulated Absorption Spectroscopy

W. Jeremy Jones

DEPARTMENT OF CHEMISTRY, UNIVERSITY COLLEGE OF SWANSEA, SINGLETON PARK, SWANSEA SA2 8PP, UK

1 INTRODUCTION

Among the most powerful methods for the determination of the concentrations of atomic and molecular species are the techniques of optical spectroscopy which provide information on chemical entities in all states of aggregation, solids, liquids and gases. Such methods rely on the absorption of radiation by chromophores, the sample absorbance being related to the species concentration by the Beer-Lambert relationship

$$\text{Absorbance} = ln\,(I_0 / I) = \sigma\ N\ L, \tag{1}$$

where the absorbance and the species concentration, N, are inter-related via the absorption path length, L, and the cross-section for the absorption process, σ. The limitations of this relationship are immediately apparent since, even if the path length of the absorbing species is known, it is not possible to determine N without knowing σ. Because of this limitation, spectroscopic concentration determination relies on the availability of a sample of the species at a known concentration, from which σ may be determined, so that concentrations of that species in an unknown sample may be determined from its measured absorbance under conditions similar to that employed for the determination of σ. Even in favourable circumstances, however, it is frequently not possible to provide a sample of the pure species for calibration purposes, *e.g.* for the provision of a sample of known concentration in an atomic absorption flame or an ICP source, or of a gaseous sample in a mixture of gases, or of a known concentration of a free radical, or an excited state atomic or molecular species.

In view of this limitation there is urgent need of a general method which will permit the measurement of absolute concentrations of absorbing species without the need for the provision of a sample of known concentration to calibrate the spectroscopic method employed.

One general method which successfully overcomes the limitations of the Beer-Lambert relationship, indeed which even takes advantage of this relationship to devise a method of absolute concentration measurement, is COncentration-Modulated Absorption Spectroscopy, COMAS.[1-3] The principle of this method takes advantage of the ability of laser sources to deliver significant levels of radiation power

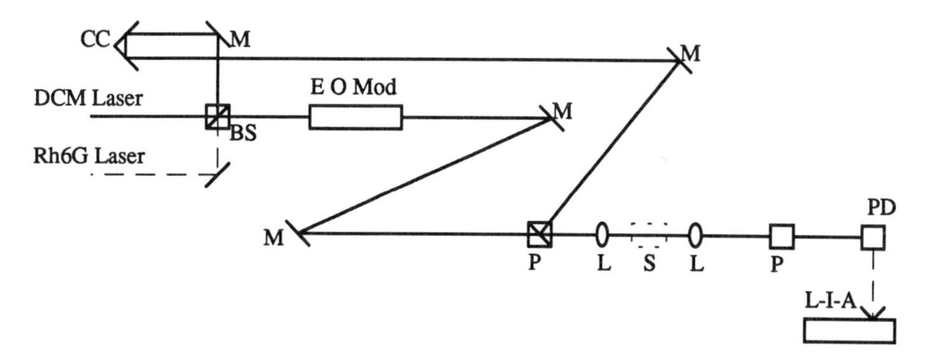

Figure 1 *Plan Layout for the COMAS Experiments. L - Lens; CC - Corner Cube Retroreflector; BS - Beam Splitter; S - Sample; E O Mod - Electro-Optic Modulator; P - Polariser; PD - Photodiode; L-I-A - Lock-In-Amplifier*

of (a) narrow spectral line width, (b) wide tuneability and (c) controllable pulse or c-w characteristics, into identifiable regions of space within a sample. Because the Bohr frequency condition $\varepsilon = h\nu$ defines the energy per photon (J photon^{-1}) at each frequency, a measurement of the power (J s^{-1}) of the radiation immediately gives a direct measure of the flux of photons (photon s^{-1}) incident on that region of space of the sample. Since there is a direct 1 : 1 correspondence between each absorbed photon and each atom/molecule removed from the lower energy state of the absorbing chromophore, measurement of the absorbed power of one beam (the pump beam) in a known region of space immediately provides a direct register of the reduction of the concentration of the lower state involved in the absorption process and the increase of concentration of the upper state. Since the absorption process is determined by the population difference between lower and upper levels coupled by the radiation, this population perturbation by the first beam, the pump beam, can be monitored by introducing a second, probe, beam which can be tuned to an optical transition within the sample that registers the reduction of the population of the lower level, the increase of population of the upper level, or both together (pump and probe are then of the same frequency and couple the same pair of energy states). In considering the characteristics of this method it is convenient initially to consider pump and probe lasers of the same frequency coupling the same pair of energy states, and therefore being controlled by the same cross section for the absorption process.

2 CONCENTRATION PERTURBATION

If a flux of pump photons i_p (considered initially as a pulse of radiation of short duration) is incident on a volume element dV and di_p of these photons are absorbed, the concentration perturbation in that volume element is di_p/dV, the ground state concentration N_1 being decreased with a corresponding increase in the excited state concentration N_2. [As an example of this situation it is convenient to consider the absorption of photons at the focus of a laser beam, simplistically considered as a cylinder of radius 10 μm and length 0.1 cm (commensurate with the confocal parameter of a typical laser beam focused by a lens of 5 cm focal length). In this volume element dV

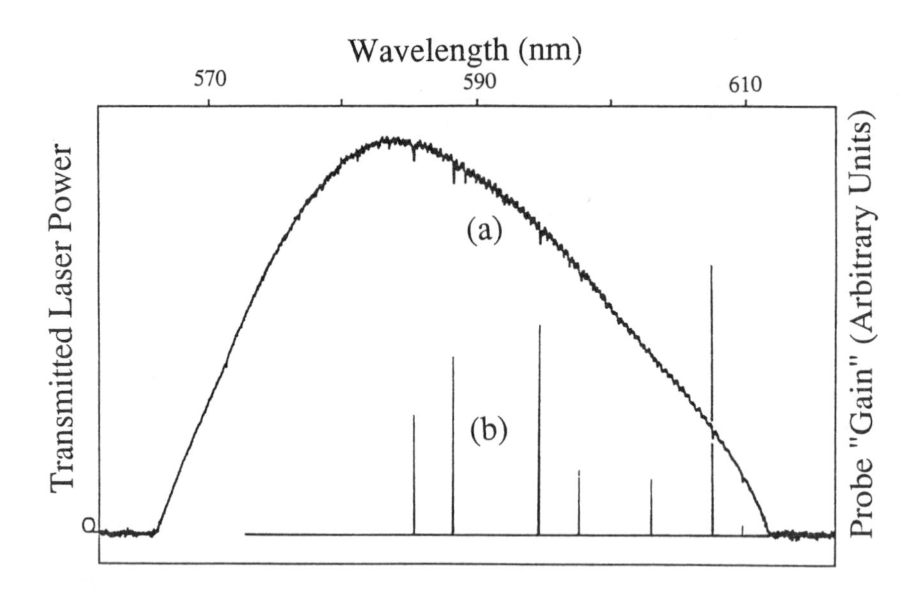

Figure 2 *Comparison of the absorption and COMAS spectra of atomic neon excited by a current of 10 mA in a see-through hollow cathode optogalvanic lamp.*[4,5] *For the absorption spectrum (a) the laser is passed 4 times through the 2 cm cathode: for the COMAS spectrum (b) pump and probe lasers are focused into the lamp. In each case the picosecond Rhodamine 6G dye laser is scanned over its tuning range. The left hand scale refers to the probe power transmitted through the cell.*

of ~ 3.1 x 10^{-7} cm^3 there may be ~ 10^4 atoms of an absorbing chromophore such as lithium (typical of the flame concentration of ~ 1 ppm lithium solution injected into a hydrocarbon flame). Assuming an absorption cross section of ~ 10^{-14} cm^2 for the 2s $\rightarrow$ 2p 671 nm Li transition there will be an absorption of 320 photons from an initial pulse of 10^7 photons incident on this volume element, providing the laser line width is less than the absorption line width. Thus the population difference between ground and excited states will be decreased by some 6.4% as compared with the original concentration following each pulse of 10^7 photons, typical of the photon flux from a mode locked picosecond dye laser of ~ 0.3 mW power at a repetition rate of 82 MHz]. This perturbation of the state populations subsequently relaxes back to the equilibrium distribution at a rate which depends on the lifetime of the excited state in this two level system. By comparing the transmission of a second probing laser of lower power (to avoid further perturbation of the state populations) with and without a preceding pump pulse it is possible to monitor the magnitude of the population perturbation and the rate of its recovery to equilibrium, thereby yielding information for the direct determination of the species concentration as well as the excited state lifetime. The experimental methods whereby this is done lie at the heart of the COMAS method.

3 CONCENTRATION PROBING

Because the absorption from the second, probing, beam is reduced following concentration perturbation by the pump pulse, the transmitted intensity of the probe beam **increases** leading to an apparent fractional gain, defined as $G = (i^*_{pr} - i_{pr}) / i^*_{pr}$, where i^*_{pr} is the transmitted probe flux following pump pulse excitation and i_{pr} is the transmitted probe flux in the absence of the perturbation. By integrating the power distributions of coincident and co-propagating pump and probe beams over all regions of space within the sample it has been shown[1,2] that the "gain" is given by the expression (for $\lambda_p = \lambda_{pr} = \lambda$)

$$G = \frac{i^*_{pr}(L) - i_{pr}(L)}{i^*_{pr}(L)} = \frac{2\pi \sigma^2 N^e_1 i_p}{\lambda}, \quad (2)$$

where N^e_1 is the equilibrium concentration of the chromophore and L the absorption path length. Because of the differing dependencies of the absorbance and the "gain" on the absorption cross section and the species concentration it is possible by combining equations (1) and (2) to obtain a direct measure of the concentration by means of the equation

$$N^e_1 = \left(\frac{2\pi i_p}{\lambda L^2}\right) \frac{\{ln(i^o_p / i_p)\}^2}{G} \quad (3)$$

4 EXPERIMENTAL CONCENTRATION-MODULATION

Because the fractional population change created by the pump laser is likely to be small (occasionally up to a few per cent, but generally many orders of magnitude smaller) it is essential to have a method capable of measuring very small fractional differences between the probe power transmitted through the sample with and without the preceding pump pulse. Low repetition lasers of high pulse power are of little or no value for this purpose because of the intrinsic noise of such lasers (with a pulse to pulse reproducibility rarely better than a few per cent) and the total saturation created by their high photon flux. By contrast mode-locked picosecond lasers (and even for some purposes cw lasers [6-7]) are ideal for this purpose. For such lasers with repetition rates of ~ 82 MHz and pulse durations of a few picoseconds the flux of photons is typically 10^6 - 10^8 photons per pulse (0.03 - 3 mW), ideal power levels to avoid population saturation and of sufficiently high pulse repetition rate to impose a further high frequency (~ 10 MHz) amplitude modulation, 4 pulses on and 4 pulses off during each pump modulation cycle. When mixed with the un-modulated probe pulses the "pump on" pulses create an increased probe transmission as compared with the "pump off" pulses: *i.e.* a 10 MHz modulation is transmitted to the probe laser as a result of the optical mixing process within the focal zone in the sample. This amplitude modulation transmitted to the probe laser is readily detected with a fast lock-in-amplifier referenced to the modulation frequency imposed on the pump laser.

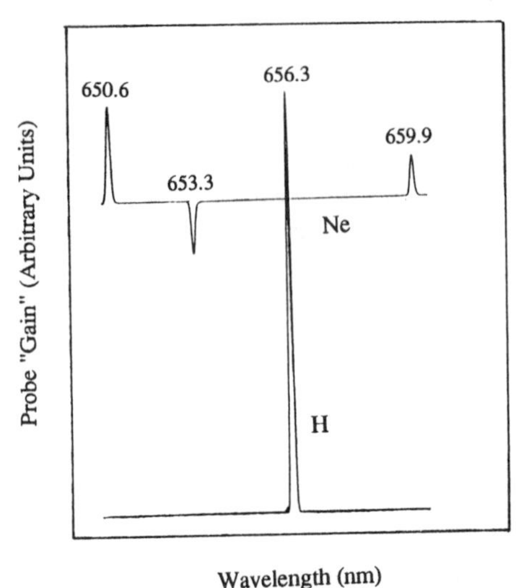

Figure 3 *COMAS spectrum of the Balmer line (n=2 → n=3) of atomic hydrogen excited by a microwave discharge in water vapour.*[8] *The upper trace is an optogalvanic spectrum of Ne recorded for wavelength calibration.*

The optical modulation methods displayed in Figure 1 were developed originally for the study of non-linear Raman spectroscopy[9,10] and because of the introduction of exceptionally fast modulation methods the fractional change of probe power which may be detected is close to the shot-noise limit (~ 1 part in 10^8 for a few milliwatts probe power on the detector). At this exceptionally low noise level the sensitivity limit for the detection of atomic species with oscillator strengths close to unity *e.g.* (Li (2s - 2p), Na (3s - 3p) *etc.* is typically ~ 10^5 atom cm^{-3} in a flame (~ 10^{-5} ppm solution-injected into a flame) or ICP source, or some one or two orders of magnitude greater for the strong transitions of a molecular species.

Examples of this exceptional sensitivity are apparent from typical spectra of atomic neon and hydrogen displayed in Figures 2 and 3. Both of these spectra were obtained by employing the COMAS technique with picosecond lasers (line width 1 cm^{-1}, one laser serving as both pump and probe, the Rhodamine 6G laser for the neon spectrum and the DCM laser for the H-atom spectrum. The spectra of atomic neon and hydrogen shown in these diagrams are absorption spectra of transient species created in an electrical discharge, the four excited neon atomic states near 134,000 cm^{-1} leading to the absorptions of Figure 2 being excited within the cathode of a see-through optogalvanic lamp, and the H-atoms in n = 2 at 82,260 cm^{-1} from a microwave discharge in water vapour. The spectra displayed correspond to absorptions from these metastable states (lifetimes less than 1 μs) to higher energy states within the systems, the 656.3 nm absorption line in Figure 3 being the first transition of the Balmer series of hydrogen, more usually being seen in emission from an electrical discharge. The atom concentrations in these samples are typically in the range 10^9 - 10^{13} atom cm^{-3}, the very high signal to noise ratios stressing the sensitivity of this approach to pump-probe spectroscopy. The limitations of conventional absorption spectroscopy for such studies are apparent from Figure 2 which, in addition to the "gain" spectrum, also displays a direct absorption spectrum obtained by monitoring the power of the laser beam quadruply passed through the 2 cm hollow cathode[4]: by contrast the COMAS spectrum is obtained from the ~ 2 mm confocal zone at the focus of the laser beams.

5 ABSOLUTE CONCENTRATION MEASUREMENT

The series of spectra of the $2s^2S_{1/2}$ - $2p^2P_{1/2,3/2}$ transitions of various concentrations of atomic lithium in a hydrocarbon flame shown in Figure 4 form part of a study of the COMAS and direct absorption spectra of this species carried out with a view to

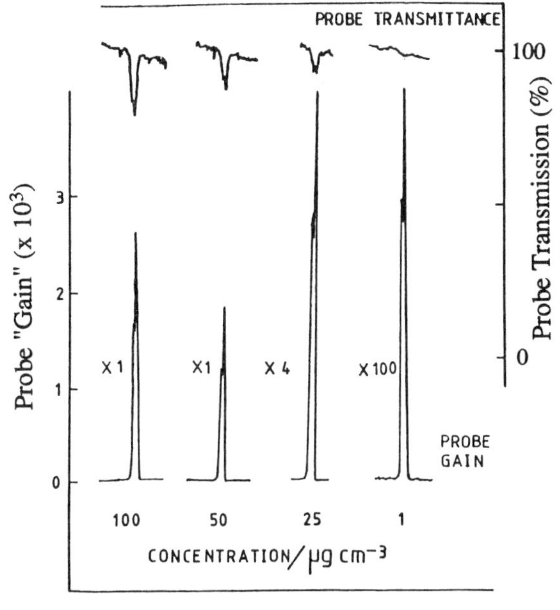

Figure 4 *Absorption and COMAS "gain" spectra of Li injected into a hydrocarbon flame*[1,2] *obtained with a picosecond DCM dye laser at 671 nm. The right hand scale refers to the absorption spectrum, the left hand scale to the "gain" spectrum.*

confirming the value of this approach to absolute concentration measurement. Series of solutions of lithium at various concentrations in the range ~ 1-500 ppm gave rise to the spectra displayed in this diagram, the absorbance and "gain" values varying over the range to 0.2 and to 6×10^{-3} for the two methods, respectively. Plots of the absorbance and "gain" values as ordinates in Figure 5A against the solution concentrations injected into the flame show the typically curved Beer-Lambert plots obtained when the radiation source employed for such studies has a significantly greater spectral width than the width of the absorption feature recorded. It is of particular interest that when displayed in this fashion the "gain" plot shows a much greater curvature than the plot of absorbance versus concentration. By contrast, the plot of $\sqrt{G}$ against $ln\,(1/T)$ (where $ln\,(1/T) = 1/L\ ln\,(i_0/i)$) shown in Figure 5B displays an ideal linear relationship between these parameters, as required by equation (3). From the slope of this plot the absolute concentration of lithium in the flame may be ascertained for each solution aspirated into the flame; *e.g.* for the 1 ppm solution the flame concentration at the beam focus is ~ 1.8×10^{10} atom cm^{-3} for this particular flame.

It is of particular interest that the plot of $\sqrt{G}$ against $ln\,(1/T)$ displayed in Figure 5B is accurately linear, even though the individual plots of G and $ln\,(1/T)$ versus the solution concentration display such a marked curvature. The basis for the self-correcting nature of such plots was demonstrated in a theoretical treatment of the problem[11] taking into account the influence of spectral hole-burning created by absorption line features very much narrower than the linewidth of the radiation source. The crucial conclusion from this latter study is that the slope and curvature of the plot of $\sqrt{G}$ versus $ln\,(1/T)$ do not depend on the relative spectral widths of the absorption feature and the radiation source, providing the sample absorption is not too great. Because of this, when measured in this fashion it is not essential to employ laser sources which are intrinsically very much narrower than an absorption feature in order to obtain an accurate measure of the species concentration. This self-correcting nature of the

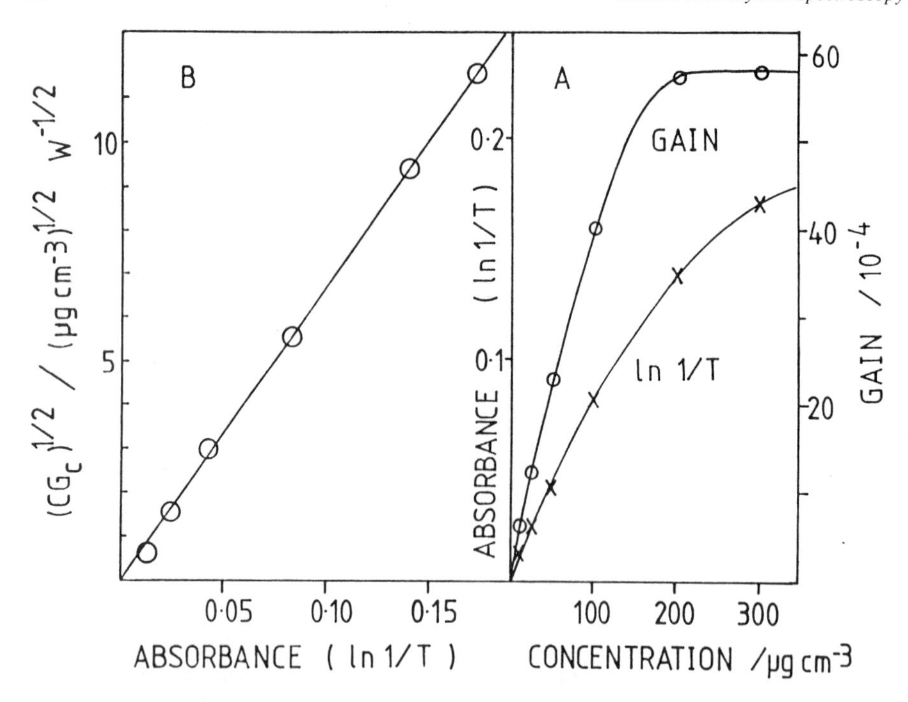

Figure 5 A - *Plots of the absorbance and "gain" data of Figure 4 against the solution concentration injected into the flame.*[1,2] *The left hand scale refers to the absorption, the right hand scale to the "gain". B - Plot of* $\sqrt{G}$ *versus* $ln(1/T)$ *for the data shown in Figure 5A.*

"gain" versus absorbance relationship for concentration measurement does not hold if these two parameters are used for determining the absorption cross section, σ, by combining equations (1) and (2). For this latter purpose, even following the measurement of N_1^e, absorbance (or "gain") measurements with a very narrow line radiation source are essential for an accurate determination of σ.

6 COMAS - ADVANTAGES & LIMITATIONS

Quite apart from the value of the COMAS method for absolute concentration measurements in atomic systems referred to above, there are other aspects of this spectroscopic method which are capable of providing information not readily yielded by other methods. Typical of some of the other unique characteristics of this spectroscopic approach are the ability:

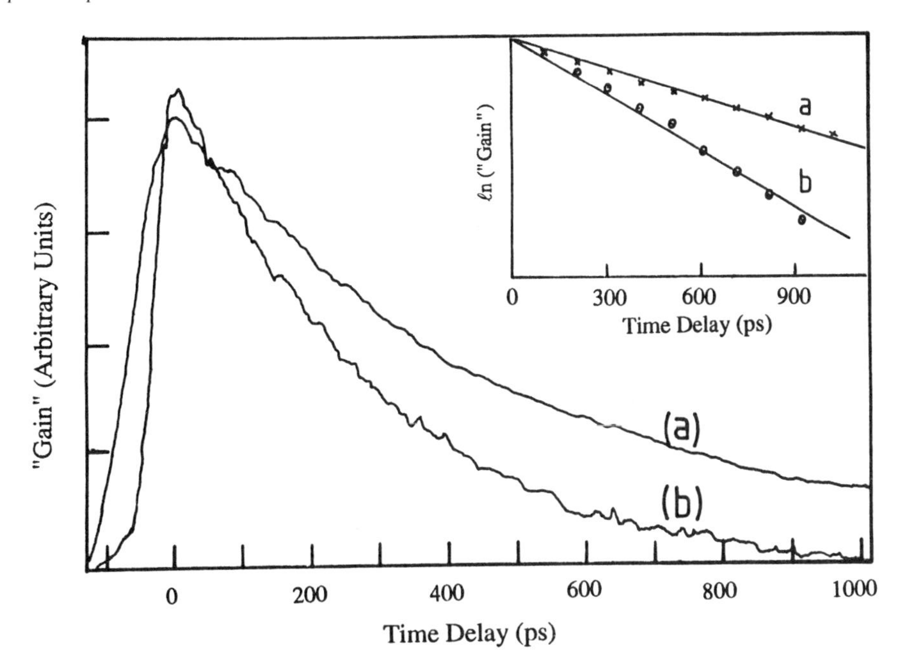

Figure 6 *Temporal decay curves of (a) the 671 nm and (b) the 611 nm transitions of Li in a hydrocarbon flame.*[3] *(a) depends only on the lifetime in the 2p state and gives a simple exponential decay, (b) depends on the lifetimes of the 2p and 3d states which gives rise to more complex decay kinetics.*

(i) to determine excited state lifetimes and ground state repopulation kinetics on a time scale from 10^{-12}s to 10^{-2}s. [A typical example for short lived excited states taken from the study of lithium in a flame is shown in Figure 6, curve (a) showing the COMAS temporal decay curve of the 671 nm 2s → 2p lithium absorption (yielding information on the lifetime of the 2p state), curve (b) showing the very much weaker transition 2p → 3d in absorption at 611 nm and yielding information on the lifetime of the 3d as well as the 2p excited states. (Interestingly, the population of the 2p state from which this latter absorption takes place is more than 10^4 times weaker than the absorption from the 2s state at the temperature of the flame.)];

(ii) to measure spectra of atomic and molecular species in all states of aggregation (solids/liquids/gases) and thereby to obtain information on their concentrations and lifetimes in a localised environment. [An example of a molecular spectrum is shown in Figure 7 which compares a portion of the conventional absorption spectrum of molecular iodine at 0° C (p = 0.03 torr) in a 10 cm cell (upper trace) with the COMAS spectrum (lower trace). These spectra were taken with a narrow line c.w. laser to resolve features in the rotational spectrum and with a

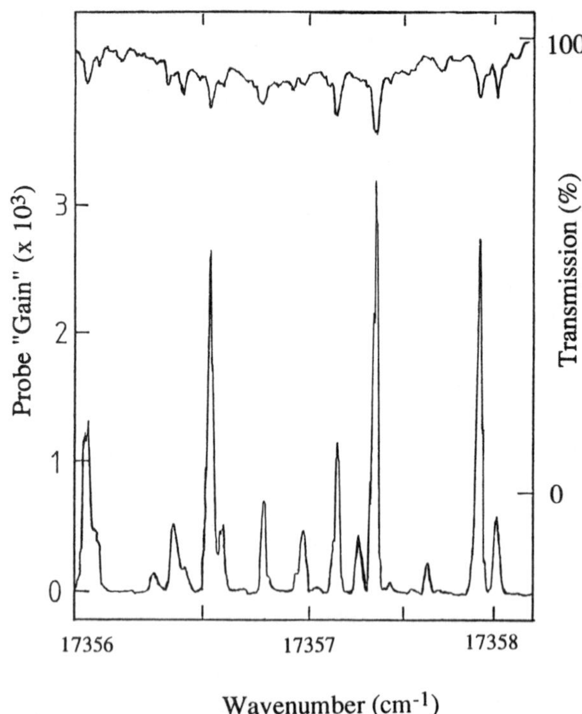

2 MHz modulation frequency, lower than usually employed because of the comparatively long lived excited states involved in these molecular transitions. The advantages of the COMAS method for the study of such absorption spectra are immediately apparent, particularly when one considers that by combination of the "gain" and absorbance values it becomes possible to determine the populations and the excited state lifetimes for each of the rotational levels involved in this very rich ro-vibronic spectrum of molecular iodine, only a very small part of which is shown in Figure 7;

Figure 7 *Portions of the absorption and "gain" spectra of molecular iodine in a 10 cm absorption cell recorded with a c.w. laser of narrow spectral line-width.*[12] *Upper trace - the absorption spectrum; Lower trace - the "gain" spectrum.*

(iii) by the use of single frequency lasers possessing very narrow spectral line-widths it becomes possible to obtain accurate profiles of individual spectral lines and thereby to extract information on the factors in the environment of the absorbing chromophore that influence the widths and the shapes of individual spectral features (see for example Reference 13 reporting the results of a study of the 3s → 3p transitions of sodium in a hydrocarbon flame which shows the dominance of the Lorentzian broadening as a result of collisions in the flame and the dramatic collapse of the Doppler contribution as a result of the constant interruptions in the movement of the species in the harsh flame environment);

(iv) to determine all the above information in localised regions of space. Since the greater part of the concentration perturbation occurs near the focus of the laser beams, the majority of the "gain" signal is created in the confocal zone at the combined focus of the beams. This region of space is determined by the laser spot size, which may be as small as ~ 1 μm, yielding a confocal parameter of ~ 10 μm and permitting 3-D species distribution profiling[5] with a lateral resolution of ~ 1 μm and a longitudinal resolution of not very much greater than an order of magnitude more than this.

The unique characteristics of the COMAS method, many of which are identified above, represent a very powerful addition to the armoury of the analyst seeking to take advantage of the spectroscopic properties of atomic and molecular species for characterisation of specific samples or to obtain information on particular environments. The methods are completely general and, unlike fluorescence methods, do not depend on the presence of fluorescence to overcome the noise limitations of conventional radiation sources. Pump and probe lasers can both interrogate one pair of energy states, or the pump laser can be used to perturb concentrations of one pair of states and the probe to monitor the perturbation to either of these states by using a completely different optical transition to or from one or other of these perturbed states. They can be used throughout the regions of the electromagnetic spectrum to obtain information on atomic and molecular systems, the infrared, visible and ultraviolet regions, although to the present they have only been employed in the yellow/red regions of the visible region where the picosecond and c.w. laser sources employed currently operate. Suitable laser and other radiation sources in different regions of the spectrum are currently available commercially and future developments will seek to obtain more suitable sources with a view to making the method more user-friendly and capable of being employed over a much broader range of the electromagnetic spectrum, thereby dramatically enhancing its value to the analytical community. In so far as the limitations of the method are concerned, apart from the availability of a radiation source with a suitably high photon flux capable of perturbing state populations in identifiable regions of space, there appears to be only one major obstacle for absolute concentration measurement, the difficulty of obtaining reliable measurements of the species absorbance to combine with the measurements of the "gain" in equation (3). Notwithstanding this limitation for absolute concentration determination, however, even without the direct absorbance value for a sample the method is still as valuable as conventional absorption spectroscopy with the added benefit of dramatically enhanced sensitivity (at least comparable with laser excited fluorescence) and the ability of obtaining lifetime information and relative species distributions in highly localised regions of space.

Acknowledgements

I am indebted to the many students and co-workers who by their efforts made much of this research possible, and to the EPSRC for the financial support which enabled the equipment essential for this work to be developed. It is also a pleasure to acknowledge my indebtedness to Professor Norman Sheppard who, so many years ago, helped to guide me onto this path.

References

1. A. J. Langley, R. A. Beaman, J. Baran, A. N. Davies and W. Jeremy Jones, *Opt. Lett.*, 1985, **10**, 327.
2. A. J. Langley, R. A. Beaman, A. N. Davies, W. Jeremy Jones and J. Baran, *Chem. Phys.*, 1986, **101**, 117.
3. R. A. Beaman, A. N. Davies, A. J. Langley, W. Jeremy Jones and J. Baran, *Chem. Phys.*, 1986, **101**, 127.
4. T. R. Griffiths, Ph. D. Thesis, University of Wales, 1990.
5. T. R. Griffiths, W. J. Jones and G. Smith, *"Optogalvanic Spectroscopy", Institute of Physics Conference Series No. 113*, IOP Publishing, Bristol, 1990.
6. W. Mallawaarachchi, A. N. Davies, R. A. Beaman, A. J. Langley and W. Jeremy Jones, *J. Chem. Soc., Faraday Trans. 2*, 1987, **83**, 707.
7. B. Csarnik-Matusewicz, R. Griffiths, P. F. Jones, W. J. Jones, W. Mallawaarachchi and G. Smith, *J. Chem. Soc., Faraday Trans. 2*, 1988, **84**, 1867.
8. W. Jeremy Jones, *European Spectroscopy News*, 1986, **No. 63**, 6.
9. J. Baran, D. Elliott, A. Grofcsik, W. Jeremy Jones, M. Kubinyi, A. J. Langley and V. U. Nayar, *J. Chem. Soc., Faraday Trans. 2*, 1983, **79**, 865.
10. W. Jeremy Jones, *"Advances in Non-Linear Spectroscopy", Eds. R. J. H. Clark and R. E. Hester*, Wiley, London, 1988.
11. W. Jeremy Jones, *J. Chem. Soc., Faraday Trans. 2*, 1987, **83**, 693.
12. R. A. Beaman, Ph. D. Thesis, University of Wales, 1985.
13. R. A. Beaman and W. Jeremy Jones, *J. Phys. B: At. Mol. Opt. Phys.*, 1994, **27**, 2139.

Electrochemically Modulated Infrared Spectroscopy Using a Step-scanning FTIR Spectrometer: Application to the $Fe(CN)_6^{3-}/Fe(CN)_6^{4-}$ Couple

Christine M. Pharr, Boiana O. Budevska, and Peter R. Griffiths

DEPARTMENT OF CHEMISTRY, UNIVERSITY OF IDAHO, MOSCOW, ID 83844-2343, USA

1. INTRODUCTION

Electrochemical detection techniques have long been renowned for their excellent sensitivity and widespread applicability. They are, however, limited by their inherent lack of selectivity and mechanistic information on interfacial electrochemical reaction products is difficult or impossible to derive. For this reason, considerable effort has been devoted to the search for techniques that provide mechanistic information complementary to the electroanalytical measurement. External reflection measurements utilizing Fourier transform infrared (FT-IR) spectrometry offer an avenue for examination of electrochemical interfacial events *in situ*. In addition to providing structural information about solution species present at the electrode/electrolyte interface, information about adsorbed species and the effects of surface modification schemes can be deduced.

There are two major limitations that have inhibited the development of infrared spectroelectrochemical measurements: strong infrared absorption by electrolytic solvents and the lack of sensitivity inherent in traditional infrared methods. Thin-layer electrochemical cells used in an external reflection configuration are used to overcome the large infrared absorption of aqueous electrolytes. The use of a variety of potential waveforms and detection schemes in conjunction with the sensitivity of rapid-scanning FT-IR spectrometers has greatly enhanced the sensitivity of infrared spectroelectrochemical measurements. Nevertheless, techniques based on conventional rapid-scanning interferometers still have certain limitations (*vide infra*). In this paper, we will describe a novel technique based on the use of a step-scanning interferometer that enables new mechanistic and kinetic information about electrochemical systems on bulk metal electrodes to be derived.

Surface enhanced Raman spectrometry (SERS) has also been used for *in-situ* vibrational spectroelectrochemistry[1] but the enhancement is confined to a few roughened electrode materials, such as silver, gold and copper. Provided that the sensitivity is adequate, infrared spectrometry overcomes this limitation, and should provide useful structural information on a wide variety of molecular systems pertinent to electrochemical analysis. The earliest infrared spectroelectrochemical measurements were made using optically transparent

---

[1] Present address: Experimental Station, E. I. du Pont de Nemours and Co., Inc., Wilmington, DE 19898

electrodes.[2,3] These experiments had limited sensitivity and suffered the disadvantage of a limited selection of electrode materials. Most contemporary infrared spectroelectrochemical techniques involve measurement of the external reflection spectra of molecules proximal to the surface of a bulk metal electrode while applying a specified potential across the electrode and the adjacent electrolyte.

The first infrared external reflection spectroelectrochemical measurements were made by Bewick *et al.* on a dispersive spectrometer using a technique they called electrochemically modulated infrared spectroscopy (EMIRS).[4,5] In this technique the electrode potential is modulated between two specified limits at a frequency of 5 to 100 Hz, while the optical signal is detected through a lock-in amplifier and recorded as a function of wavelength. Although EMIRS has been successfully used to detect species on electrode surfaces, the signal-to-noise ratio (SNR) of the spectra measured in this way is often quite low. Improved sensitivity can be achieved by subtracting two spectra measured using a conventional rapid-scanning FT-IR spectrometer at different electrode potentials. This technique, which was also first reported by Bewick's group, is known as subtractively normalized interfacial FT-IR spectroscopy (SNIFTIRS).[6] In an analogous external reflection technique involving the use of a rapid-scanning FT-IR spectrometer, spectra are obtained during a single potential excursion, rather than by modulating the potential as in SNIFTIRS. The use of kinetics software allows acquisition of spectra as the potential waveform is changed. At time intervals determined by the operator, a specified number of scans is co-added. Each scan set is ratioed against either the first or last scan set to produce an absorbance spectrum as a function of electrode potential. The potential resolution of each scan set is determined by the potential scan rate, the number of infrared scans per scan set and the scan speed of the interferometer. This approach has been dubbed single potential alteration infrared spectroscopy (SPAIRS).[7]

In principle, EMIRS is preferable to SNIFTIRS or SPAIRS because for any type of difference measurement it is more accurate to measure the amplitude of a small ac signal (potential modulation) than to measure two large dc signals (one at each electrode potential) and subsequently subtract them.[8] Conversely, SNIFTIRS and SPAIRS have the SNR advantages of Fourier transform spectroscopy, but lose the ac advantage. The optimal technique would include both advantages. The combination of conventional FT-IR spectroscopy with a true potential modulation technique is, however, very difficult because of the interaction between the optical and potential modulation frequencies. In rapid-scanning FT-IR spectrometry, the infrared beam is modulated at a frequency that is equal to the product of the wavenumber of the radiation and the optical velocity of the scanning mirror (which is typically between 1 mm and 1 cm per second). Thus the modulation frequencies generated by the interferometer (*Fourier frequencies*) are in the range of a few hundred to a few thousand Hertz for the mid infrared. The electrode potential modulation frequency must be at least an order of magnitude faster than the highest Fourier frequency,[9] if these frequencies are to be separable. Thus, the potential modulation frequency necessitated by this arrangement would be on the order of several kiloHertz, a speed at which electrochemical cells cannot respond due to the limitations of electrode kinetics, diffusion and the response time of the cell.[10,8] Conversely, to modulate the potential at lower frequencies would produce cross talk between the electrochemical modulation and the Fourier frequencies.

A technique which overcomes the cross-talk problem has been implemented in our laboratory with the use of step-scanning interferometry. With this method, the Fourier frequencies are reduced to zero by stepping the moving mirror from position to position rather than scanning it, as is done in conventional FT-IR. In this way the optimal electrode modulation frequencies (less than 20 Hz) can be used.[11] An additional advantage of the step-scan EMIRS technique is that it discriminates against processes which are constant during the course of the measurement. Thus absorptions due to atmospheric H_2O, CO_2 or processes which do not vary at the same frequency as the electrochemical modulation are effectively eliminated. The reversible nature of the step-scan EMIRS technique makes it particularly applicable to time-dependent processes. A final advantage is that the use of a modulation techniques allows acquisition of phase information which can be used to obtain dynamic information on electrochemical systems.

2. INFRARED SPECTROELECTROCHEMISTRY OF THE $Fe(CN)_6^{3-}/Fe(CN)_6^{4-}$ COUPLE

The heterogeneous charge transfer reaction of the hexacyanoferrate redox couple has been investigated extensively in the past by electrochemical and spectroelectrochemical methods in an effort to explain its non-ideal kinetic behavior. Still, the mechanism of this electrochemical system remains uncertain. Both electrochemical[12,13] and spectroelectrochemical methods,[14,15] including infrared[16,17,18,19] and Raman[14,15,20,21] spectroscopies, have been applied to the investigation of what was once thought to be a simple outer-sphere one-electron transfer.

SNIFTIRS was utilized by several authors[17,18,19] to investigate the ferri/ferrocyanide system. Solution-phase ferro- and ferricyanide show absorption bands at 2040 cm^{-1} and 2114 cm^{-1}, respectively. Pons *et al.*[17] also observed a band at 2092 cm^{-1} which was assigned to an adsorbed intermediate. A second band which is sometimes observed at 2061 cm^{-1} was assigned to CN^- adsorbed on Pt formed by decomposition of hexacyanoferrate at high positive potential. Subsequent experiments reported by Christensen *et al.*[18] showed only solution-phase bands. This difference in the spectroelectrochemistry was attributed to the different surface pretreatment procedures.

An EMIRS study of the $Fe(CN)_6^{3-/4-}$ system using a dispersive infrared spectrometer was reported by Kunimatsu *et al.*[16] These authors suggest that the intermediate of ferricyanide reduction and ferrocyanide oxidation is different. Although a potential modulation technique was used in this study, neither the modulation frequency nor the quadrature spectra were reported.

The goal of the present work is to study the ferro/ferricyanide couple with the recently introduced step-scan FT-IR spectrometry, applying potential modulation and phase-sensitive detection. In the present paper we report spectra of ferrocyanide and ferricyanide within limits that produce reversible reduction and oxidation of the redox couple.

3. EXPERIMENTAL

The instrumental configuration used for these experiments is shown in Figure 1; detailed experimental parameters are described elsewhere[22]. All spectroelectrochemical measurements were performed on an optical bench external to an Bio-Rad/Digilab FTS-60A spectrometer. The beam exits the spectrometer and is reflected from the working electrode surface by passing through a thin-layer electrochemical cell. The optical signal is detected by a narrow-band MCT detector (Graseby Infrared). A high frequency phase modulation (400 Hz)

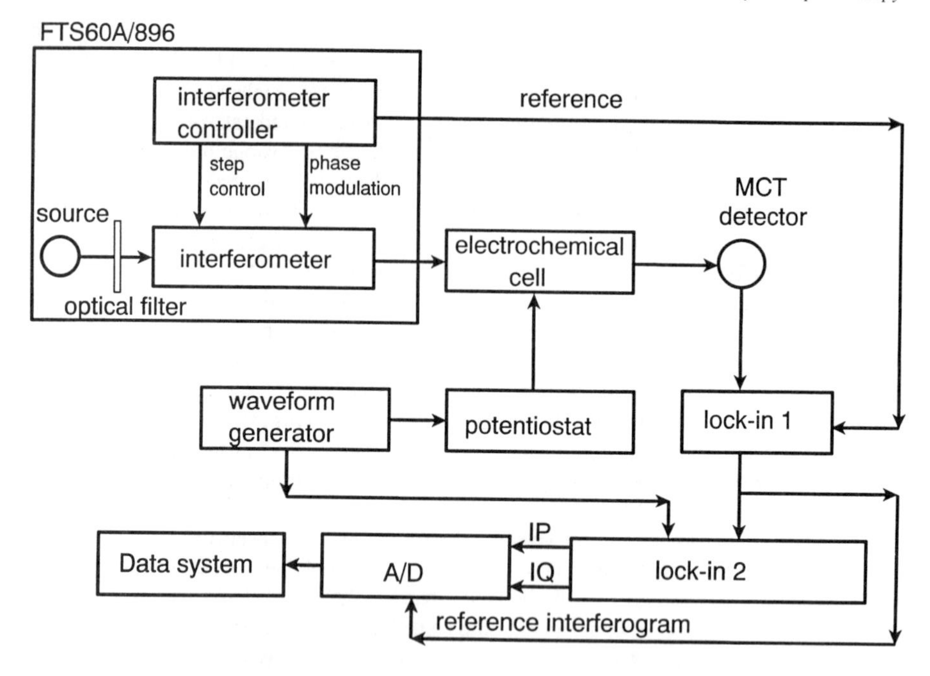

Figure 1: *Block diagram for EMIRS measurements using a phase-modulated step-scanning FT-IR spectrometer*

is used as a carrier for the lower frequency potential modulation. Two lock-in amplifiers (LIAs) are used to demodulate the double-modulated optical signal. The first LIA demodulates the spectrometer phase modulation at each mirror step and the second LIA demodulates the electrochemical potential modulation. Using the two outputs of a two-phase LIA, three interferograms and the corresponding spectra can be generated. The in-phase (IP) and quadrature (IQ) spectra, which are generated from the outputs of the second LIA, represent the optical changes which occur synchronously and asynchronously (90° out-of-phase) with the applied sinusoidal potential modulation. The in-phase and quadrature spectra are sensitive only to the spectroscopic changes which occur at the same frequency as the potential modulation. The reference spectrum, which is generated from the output of the first LIA, shows the sum of the total absorption of the system during potential modulation.

The electrode used was a 8-mm platinum disk which was polished using successively smaller diameter diamond paste followed by soaking in concentrated HNO_3/H_2SO_4 (1:1) for at least 1 hour. The spectroelectrochemical cell was filled with potassium ferrocyanide at concentrations between 0.1 M and 0.1 mM in 1-M KCl. Solutions were prepared fresh before each series of experiments. EMIRS spectra were measured using modulation frequencies between 0.5 and 5 Hz. Potential modulation limits and solutions concentrations will be described with each experiment in this paper.

4. RESULTS

The $Fe(CN)_6^{3-/4-}$ redox couple is a quasi-reversible electrochemical system with its formal electrode potential at approximately 0.18 V versus a saturated calomel electrode (SCE) in 1-M KCl. From a typical cyclic voltammogram on the 8-mm platinum disk electrode, the rate of ferrocyanide production becomes diffusion limited near 0.28 V vs SCE and the diffusion-limited production of ferricyanide occurs at 0.13 V. In the experiments presented here, the potential modulation was either centered around the formal electrode potential or near the diffusion-limited production of ferrocyanide.

Figures 2 - 5 present normalized in-phase (top) and quadrature (bottom) spectra of the ferri/ferrocyanide system using the FT-IR spectrometer in its phase-modulated, step-scan mode. The reference spectrum (not shown) represents the total absorption of the electrochemical system during the potential modulation, and indicates the equilibrium concentration of the redox system during the modulation experiment. In order to obtain an accurate representation of spectral features, both the in-phase and quadrature spectra were normalized with their corresponding reference spectrum to account for any changes in the equilibrium concentration of the redox couple. The concentration of ferrocyanide was 0.1 M in these experiments and the potential of the working electrode in the spectro-electrochemical cell was modulated between -100 mV and +450 mV vs SCE at modulation frequencies of 0.5, 1, 2 and 5 Hz. This corresponds to potential scan rates of 0.55, 1.10, 2.20 and 5.50 V/s. The spectra shown represent the difference in absorbance of the electroactive species at the two electrode potentials. Although the sign and the magnitude of features in all these spectra are

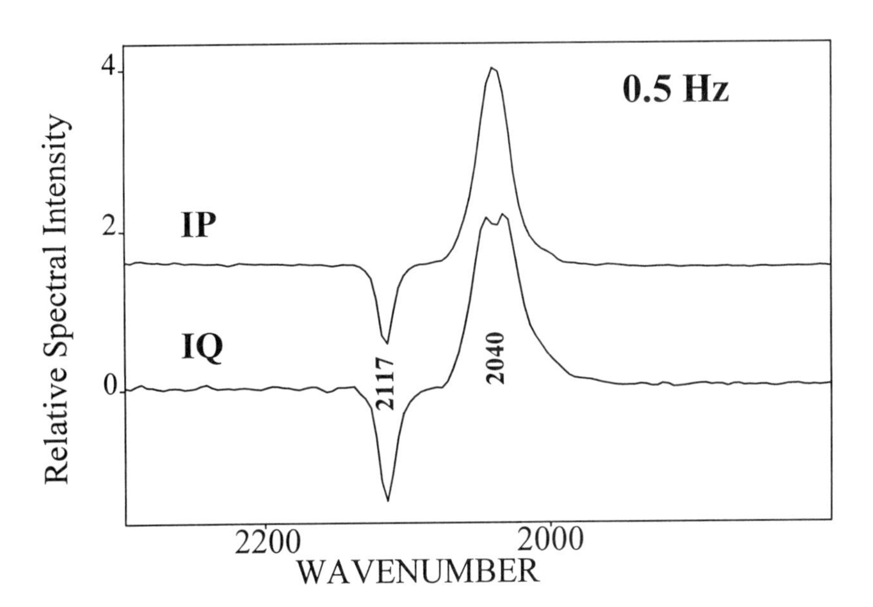

Figure 2: *FT-EMIRS spectrum of 0.1-M $K_4Fe(CN)_6$* in 1-M KCl; modulation frequency 0.5 Hz; potential limits from -100 to +450 mV; (above) in-phase spectrum, (below) quadrature spectrum.

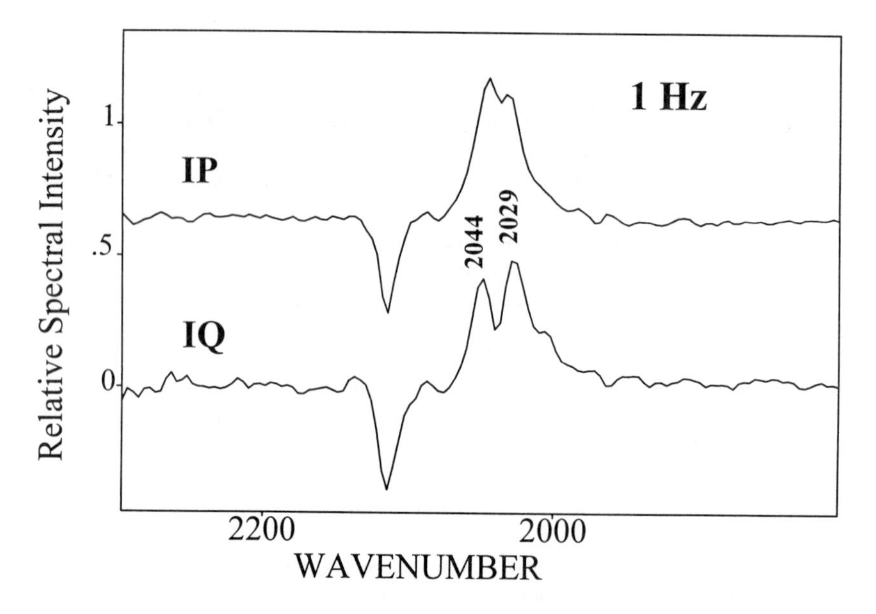

Figure 3: *FT-EMIRS spectrum of 0.1-M $K_4Fe(CN)_6$* in 1-M KCl; modulation frequency 1.0 Hz; all the other parameters the same as for Figure 2.

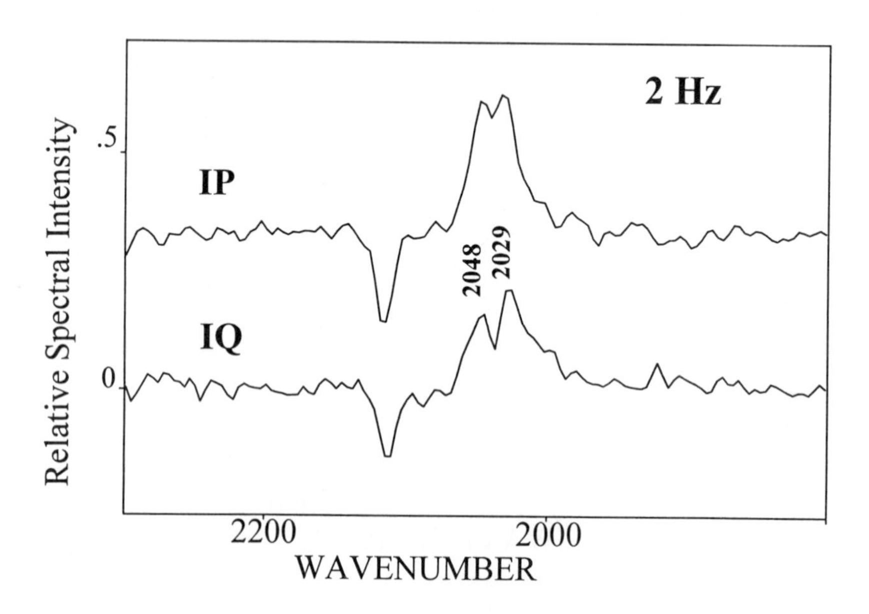

Figure 4: *FT-EMIRS spectrum of 0.1-M $K_4Fe(CN)_6$* in 1-M KCl; modulation frequency 2.0 Hz; all the other parameters the same as for Figure 2.

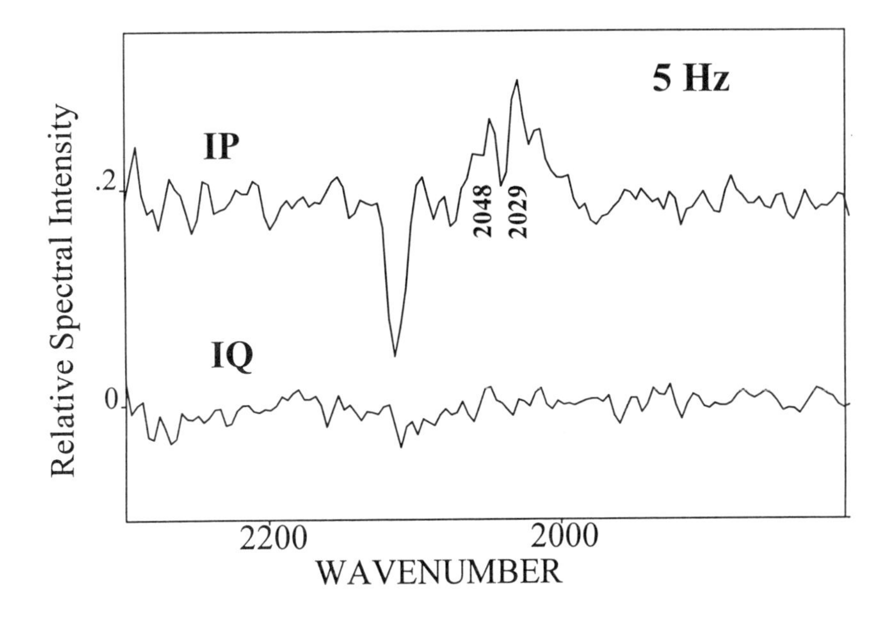

Figure 5: *FT-EMIRS spectrum of 0.1-M $K_4Fe(CN)_6$* in 1-M KCl; modulation frequency 5.0 Hz; all the other parameters the same as for Figure 2.

arbitrary, it is straightforward to assign the lower-wavenumber, positive-going bands to ferrocyanide species produced at lower potential and the negative-going bands at higher wavenumber to ferricyanide species. It must be emphasized that the dynamic spectra (both in-phase and quadrature) represent only those optical changes which occur at the modulation frequency. If the response is delayed but still at the modulation frequency, it will be detected in the quadrature channel. Any follow-up reactions that lag behind the modulation frequency will not be observed in these dynamic spectra.

The in-phase spectrum measured with a low modulation frequency (0.5 Hz) shows only two bands (see Figure 2). These bands are readily assigned to solution-phase $Fe(CN)_6^{4-}$ (2040 cm^{-1}) and $Fe(CN)_6^{3-}$ (2117 cm^{-1}). At this modulation frequency and concentration, very careful examination of the spectra is required to see any suggestion of the formation of intermediates or byproducts during the electron transfer, indicating complete conversion of the redox couple in the thin-layer cell. More complex spectral features, which were only marginally evident in the quadrature spectrum measured at 0.5 Hz, become obvious at higher modulation frequencies. The ferrocyanide band centered at 2040 cm^{-1} in solution phase is split into 2 or more bands centered around 2040 cm^{-1} at the higher modulation frequencies. The solution phase band at 2117 cm^{-1} assigned to the ferricyanide is broader and shifted to 2113 cm^{-1}. These changes are likely attributable to one of two possibilities: a decomposition product of the electron transfer which adsorbs onto the surface such as cyanide[17] or prussian blue[23] or an adsorbed intermediate in the electron transfer, possibly the redox species itself.

To further investigate the cause of the spectral changes, similar experiments were performed with a decrease in the concentration of the redox couple. Figure 6 shows the dynamic power spectrum of the ferri-/ferrocyanide couple at various concentrations. The power spectrum represents the total optical response of the redox couple to the potential modulation frequency, and is calculated as $(IP^2+IQ^2)^{1/2}$, where IP and IQ are the experimentally measurable in-phase and quadrature spectra. All bands in the power spectra are positive, but their relation to the potential can be determined by comparison with the normalized in-phase and quadrature spectra in Figures 2 - 5. These spectra were obtained at 0.5-Hz modulation frequency, with the concentration of ferrocyanide decreasing one order of magnitude in each scan. A new peak is evident at 2092 cm^{-1} in spectra measured with the analyte present at low concentration. Both prussian blue and adsorbed cyanide have absorptions in this region on silver electrodes. As the concentration is reduced from 10 mM to 1 mM both of the solution peaks at 2117 cm^{-1} and 2040 cm^{-1} decrease significantly; however, the peak at 2092 cm^{-1} remains essentially constant in intensity, indicating that a new species is adsorbed and desorbed at the modulation frequency.

Decreasing the concentration by another order of magnitude (to 0.1 mM) causes little reduction in the ferrocyanide peak, which suggests that ferrocyanide itself may adsorb during the electron transfer. This is in agreement with *ex-situ* work done using Auger spectroscopy which indicates that ferrocyanide adsorbs onto the electrode retaining its stoichiometry and ionic charge.[24] If so, the peak at 2040 cm^{-1} in the 0.1-mM ferrocyanide solution probably represents mostly adsorbed ferrocyanide while the 1-mM solution indicates mainly adsorbed ferrocyanide together with some solution-phase ferrocyanide.

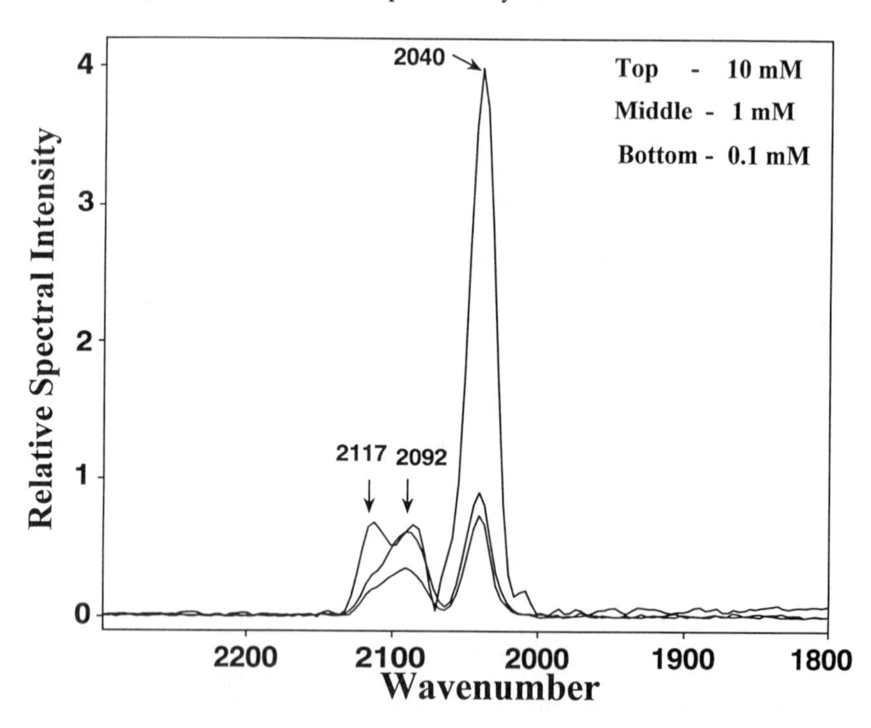

Figure 6: *Dynamic power spectrum of the $Fe(CN)_6^{3-}/Fe(CN)_6^{4-}$ couple at concentrations of 10 mM (most intense spectrum), 1 mM and 0.1 mM (least intense spectrum).*

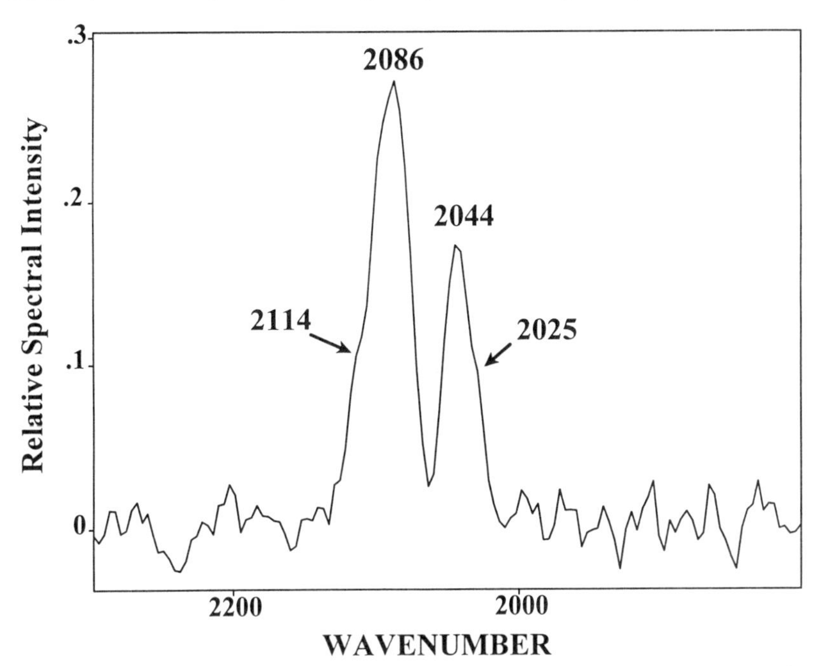

Figure 7: *Difference spectrum obtained by subtraction of 0.1-mM spectrum from 1-mM spectrum of ferri/ferrocyanide. (Original spectra shown in Figure 6.)*

In an effort to explain the divided ferrocyanide peak in Figures 3 - 5, the bottom spectrum (0.1 mM) from Figure 6 was subtracted from the middle spectrum (1 mM). The resultant spectrum is shown in Figure 7. The subtraction of the 2092 cm^{-1} peaks produces a peak at 2086 cm^{-1} with a shoulder at 2114 cm^{-1} from the solution ferricyanide. The ferrocyanide peak appears shifted to 2044 cm^{-1} with a shoulder at 2025 cm^{-1}. While these are difficult to resolve, their wavenumbers are not incompatible with the wavenumbers of the divided peaks seen at higher modulation frequencies. This suggests that the divided peak could be due to the difference between the solution ferrocyanide absorption at one potential and an adsorbed form of a redox species at the other potential which may have only a small shift in its absorption band center.

5. DISCUSSION

From the spectra shown above, surface sensitivity appears to be enhanced at high potential modulation frequencies or low concentrations of the redox couple. The surface sensitivity at lower concentrations of the redox couple (shown in Figure 6) can be explained by the fact that the bands due to solution-phase species are much smaller and do not swamp out the signal(s) from the surface species. The ability to detect surface species at high modulation frequencies has often been attributed to a lack of diffusion across the thin-layer cell within the time frame of the potential modulation, so that only those species adsorbed to the surface or within the double layer are able to follow the potential modulation. While this explanation seems logical, it can be shown that at 5-Hz modulation frequency the root-mean-square diffusional distance is approximately 10 μm, at least twice the thickness of the thin-layer electrochemical cell. Therefore, the above explanation of surface sensitivity is not applicable at modulation frequencies at or below 5 Hz. The surface sensitivity in this case can probably be attributed to a reduction in the electron transfer rate caused by the high impedance characteristics of the electrochemical cell.

This can be understood by the following argument. In steady-state cyclic voltammetry, which is comparable to EMIRS measurements made at low modulation frequencies, a diffusion-limited reaction rate is evidenced at potentials sufficiently far from E^o by a decreasing current with increasing potential. In regions near E^o the current increases exponentially with potential and the redox process is kinetically controlled. As the potential modulation frequency increases in these experiments, the charging of the double layer is inhibited by the high impedance characteristics of the thin-layer spectroelectrochemical cell. The potential limits measured at the working electrode and sensed by the ferri/ferrocyanide couple do not reach the diffusion-limited potential values. Thus the portion of the thin layer that is reversibly electrolyzed at the modulation frequency is significantly less than at lower modulation frequencies, enhancing sensitivity near the electrode surface. This hypothesis explains the decreasing intensity of the power spectra with increasing modulation frequency, as fewer of the redox molecules are actually electrolyzed within the smaller potential limits. By analogy with the result found on decreasing the concentration of the electroactive species, increasing the modulation frequency produces fewer solution-phase species that follow the potential modulation and hence swamp out the signal from the surface species. Thus both surface species seen in the low-concentration spectrum are also present in the spectrum measured at high modulation frequency. In this case, however, the 2040-cm^{-1} band is manifested as a divided peak because of the difference in width between the solution and adsorbed redox species. As the modulation frequency increases, the spectra measured at modulation frequencies of 1 and 2 Hz show a peak at 2092 cm^{-1}, but the signal-to-noise ratio at 5 Hz has degraded such that it is impossible to see that peak.

Step-scan FT-EMIRS is only capable of detecting peaks for species that move on and off the electrode or shift wavenumber at the two different electrode potentials. A wavenumber shift results in a bipolar band, which is not evident here. The shape and size of the 2092-cm^{-1} peak indicates that while its identity is uncertain, it is clearly an adsorbed species that is capable of adsorbing and desorbing within the time frame of the modulation. Further investigations will be necessary to determine the identity of this complex and to determine whether it plays an integral role in facilitating the electron transfer or is simply a decomposition product that adsorbs onto the electrode at less positive potentials.

In summary, EMIRS spectra measured with a phase-modulated step-scanning FT-IR spectrometer have been shown to shed new light on the electrochemical mechanism of the $Fe(CN)_6^{3-}/Fe(CN)_6^{4-}$ couple on a platinum electrode. It is not unreasonable to predict that this technique will prove to be of great importance in many future investigations of electrochemical processes.

References

1. M. J. Weaver, J. T. Hupp, F. Barz, J. G. Gordon, and M. R. Philpott, *J. Electroanal. Chem.*, 1984, **160**, 321.
2. H. B. Mark and B. S. Pons, *Anal. Chem.*, 1966, **38**, 119.
3. D. R. Tallant and D. H. Evans, *Anal. Chem.*, 1969, **41**, 835.
4. A. Bewick, K. Kunimatsu and S. Pons, *Electrochim. Acta.*, 1980, **25**, 465.
5. A. Bewick and K. Kunimatsu, *Surface Sci.*, 1980, **101**, 131.

6. T. Davidson, S. Pons, A. Bewick and P. P. Schmidt, *J. Electroanal. Chem.*,1981, **125**, 237.
7. M. J. Weaver, *Anal. Chem.*, 1987, **59**, 2252.
8. M. A. Habib and J. O'M. Bockris, *J. Electroanal. Chem.*,1984, **180**, 287.
9. L. A. Nafie and D. W. Vidrine, *Fourier Transform Spectroscopy: Applications to Chemical Systems*, Academic Press, New York, 1982.
10. S. Pons, *J. Electroanal. Chem.*,1983, **150**, 495.
11. A. Bewick, K. Kunimatsu, S. Pons and J. W. Russel, *Electroanal. Chem.*,1984, **160**, 47.
12. L. M. Peter, W. Durr, P. Bindra, and H. Gerischer, *J. Electroanal. Chem.*, 1976, **71**, 31.
13. J. Kawiak, T. Jedral and Z. Galus, *J. Electroanal. Chem.*, 1983, **145**, 163.
14. M. Fleischmann, P. R. Graves and J. Robinson, *J. Electroanal. Chem.*, 1985, **182** 87.
15. M. J. Weaver, P. Gao, D. Gosztola, M. L. Patterson and M. A. Tadayyoni, 'Excited States and Reactive Intermediates', A. P. B. Lever (Ed.), ACS Symposium Series 307, American Chemical Society, Washington, DC, 1986, p. 135-149.
16. K. Kunimatsu, T. Shigematsu, K. Uosaki and H. Kita, *J. Electroanal. Chem.*, 1989, **262**, 195.
17. S. Pons, M. Datta, J. McAleer and A. S. Hinman, *J. Electroanal. Chem.*, 1984, **160**, 369.
18. P. A. Christensen, A. Hamnett and P.R. Trevellick, *J. Electroanal. Chem.*, 1988, **242**, 23.
19. J. O'M. Bockris and B. Yang, *J. Electroanal. Chem.*, 1988, **252** 209.
20. R. B. Lowry, *J. Raman Spectrosc.*,1991, **22**, 805.
21. R. B. Lowry, *Spectrochim. Acta*, 1993, **49A**, 831.
22. B. O. Budevska and P.R. Griffiths, *Anal. Chem.*, 1993, **65**, 2963.
23. K. Niwa and K. Doblhofer, *Electrochim. Acta.*, 1986, **31**, 439.
24. M. Baltruschat, F. Lu, D. Song, S. K. Lewis, D. C. Azpien D. G. Frank, G. N. Salaita and A. T. Hubbard, *J. Electroanal. Chem.*, 1987, **234**, 229.

Applications of Raman Microscopy and Raman Imaging

K. P. J. Williams, A. Whitley, and C. D. Dyer

RENISHAW PLC., OLD TOWN, WOOTON UNDER EDGE, GLOUCESTERSHIRE GL12 7DH, UK

The use of Raman spectroscopy is steadily growing in importance. The nature of the equipment, its ease of use and its cost make the method more acceptable to non-experts. This paper illustrates the capabilities of a Raman microscope system which fulfils the criteria of ease of use, speed and low cost. The system uses a single spectrograph and CCD detector. It provides confocal microscopy, high throughput and optimum sensitivity. In addition direct Raman imaging is available, which provides spatial information over large surface areas quickly and without the need for excessive amounts of data processing. The performance application is illustrated by a number of examples.

1. INTRODUCTION

Over the past two decades, Raman spectroscopy has been little used in routine analytical laboratories. The technique has largely been used by experts, working in large companies, using sophisticated and expensive lasers and spectrometers. The method has found many and varied applications. Often the natural advantages of the method such as the high spatial resolution (*ca.* 1 μm) offered by Raman microscopy have been used to good advantage, e.g. in the study of inclusions in minerals,[1] identification of contaminants in polymers[2] and the investigation of stress in single fibres.[3] Other applications have made use of the ease of cell design for *in situ* Raman experiments with glass providing an excellent medium. These have included *in situ* electrochemical experiments[4] and the study of dynamic systems at elevated temperatures and pressures.[5] Other investigations have included corrosion studies,[6] the study of polymer morphology[7] and polymer degradation.[8]

In 1986, the practical application of Raman spectroscopy using a research-grade near-infrared Fourier transform (FT) spectrometer was published.[9] This was quickly followed by the demonstration of FT-Raman spectroscopy on an inexpensive bench-top interferometer.[10] With further instrument development, this heralded a resurgence of interest in Raman spectroscopy. The method was brought into the price range and level of complexity that were acceptable in many industrial laboratories. The FT-Raman method offers largely fluorescence-free spectra (by virtue of the near-infrared excitation laser wavelength used) and an ease of use. However, it is detector noise-limited and fails to provide high-level sensitivity that the more traditional dispersive Raman (shot noise limited) provides. In addition, problems have been experienced in interfacing microscope accessories to FT-Raman spectrometers[11] and, whilst improvements in performance are still being made,[12] to date no high-quality confocal data providing good spatial resolution (1 μm) have been obtained. This means that whereas reasonable Raman scatterers can be run with relative ease, weaker scatterers and extremely small samples offer considerable challenges.

Two other major advances have had a large impact on Raman spectroscopy. Firstly, the charge-coupled device detector (CCD), developed and used extensively by astronomers, became acknowledged as a major advance for the Raman spectroscopist. This two-dimensional silicon detector has the properties of high quantum efficiencies and extremely low dark current levels which is far improved over the conventional technology.[13]

The second major advance has been the advent of the Raman holographic filter. These filters are wavelength specific and efficiently block the undesirable Rayleigh scatter, with an optical density of a *ca.*4 and a cut-off that routinely permits approach to within 100 cm^{-1} of the laser line. This means that the Raman experiment that has used high-dispersion monochromators to filter out the elastically scattered laser radiation, can be conveniently performed with a single spectrograph and a Raman holographic filter.[14] This approach provides considerable benefits. The expense is relatively low and its operation is routine. In addition the photon throughput efficiency is high.

When a single spectrograph is coupled to a CCD detector, it provides much higher sensitivity which in turn can lead to the use of lower powered lasers and shortened data acquisition times. The efficient coupling of a microscope presents no difficulties. The spot size emerging from the microscope is fully compatible with the slit widths used in dispersive Raman systems. Recent publications have detailed many of these advantages and discussed the merits of this approach over the FT-Raman method.[14] The increased levels of sensitivity coupled with similar discrimination over fluorescence when using near-infrared laser diode wavelengths, between 750 and 830 nm,[15] make the more traditional approach to Raman spectroscopy very attractive.

The purpose of this paper is to review the use of a recently commercialised Raman microscope stigmatic spectrometer and provide examples of its use. Both examples of spectroscopy and Raman imaging will be given.

2. EXPERIMENTAL

The instrument used in these studies was the Renishaw Raman imaging microscope consisting of; in brief, a low-powered (25 mW) air-cooled laser source (Ar^+, HeNe or diode laser), an Olympus microscope, a single spectrograph (250 mm focal length) for spectroscopy mode, a set of angle-tunable band pass filters for imaging mode and a Peltier-cooled CCD detector (576 x 384 pixels). The laser power delivered at the sample point in all of the experiments detailed in the paper was no more than 6 mW.

A schematic diagram of the system layout is shown in Figure 1. It can be seen that the instrument concept is relatively simple. The system has been designed to be rugged, easy to use and compact (with a total footprint size of 47 x 133 cm). Full details of the system are given elsewhere.[16]

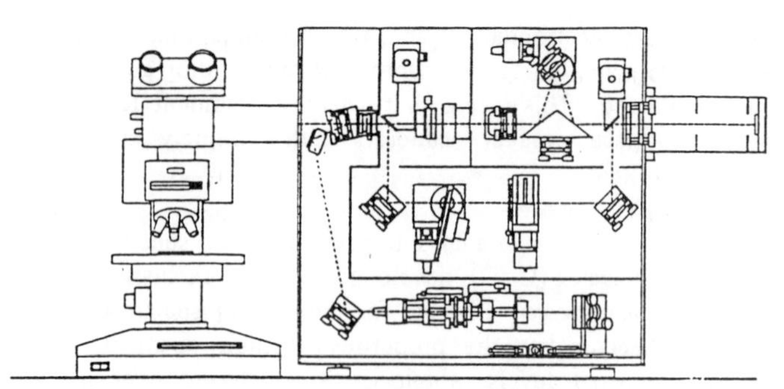

Figure1. Schematic diagram of the Renishaw imaging Raman spectrometer.

3. EXAMPLES OF RAMAN SPECTROSCOPY AND RAMAN IMAGING

The sensitivity, speed and throughput of the instrument can be demonstrated by reference to the Raman spectrum recorded from silicon shown in Figure 2. The data illustrate the fundamental vibration at 519 cm^{-1} together with the second order structure at *ca.* 950 cm^{-1} and the third and fourth order bands at *ca.* 1400 and 1900 cm^{-1} respectively. The data was acquired using a x50 objective, 5 mW laser power at the sample. The third-order spectrum required an exposure time of 30 minutes. The fourth order peak was acquired after *ca.* 3 hours and represents a signal strength of one photon every 2 seconds. Similar data of the third order peak by other workers,[17] using a conventional Raman triple spectrometer and CCD detector, required a laser power of 400 mW and 3 h acquisition time. No previous references have been found for the fourth order spectrum. The superior throughput performance of a single spectrograph is clear.

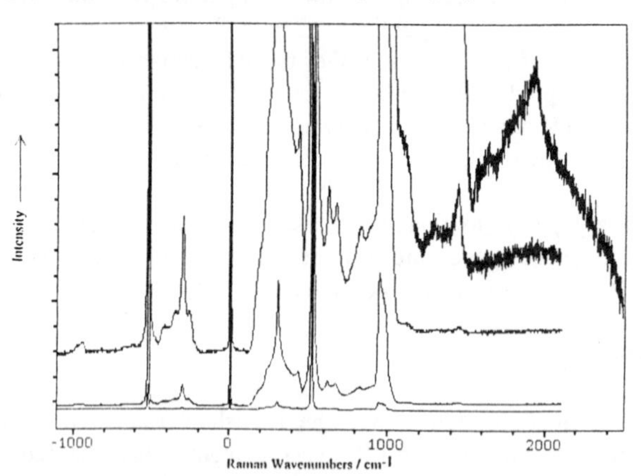

Figure 2. Raman spectrum recorded from silicon using a x50 microscope objective, 5 mW, 632.8 nm laser excitation.

The confocal arrangement for a microscope in Raman spectroscopy provides the optimum depth resolution from the sample. A spatial resolution in the lateral plane of 1 μm is achievable using most systems. However, when there is a need to minimise the depth resolution of the sample volume, a pinhole has often been installed at the back focal plane of the microscope. This limits unwanted Raman scatter from outside the laser focus in the sample[18] but requires a precise optical alignment which is often difficult and time consuming to achieve. The design of a stigmatic spectrograph coupled to a CCD detector provides a means of obtaining a confocal arrangement by simply restricting the active area of the CCD in the image height dimension.[19]

By simply restricting the detection area in the spatial direction it is possible to discriminate effectively between the Raman signal coming from the focused laser spot and that from the out-of-focus radiation. This operation of the system is largely software controlled and as such is relatively trivial to implement.

An example of the confocal Raman data that can be obtained is shown in Figure 3, taken from a polymer multi-layer. The sample comprises of a thin, 2 μm, polyethylene (PE) film over a polypropylene (PP) sub-layer. The data shows (Figure 3(a)) that by using a x100 microscope objective and only 4 pixels height on the CCD detector complete discrimination between the PE and PP can be obtained, and a depth resolution of better than 2 μm is achieved. However, when a x50 objective is used (under the same conditions) PP features become apparent. This is because the natural depth of field of this lower magnification objective is 4-6 μm which means that the Raman scatter from the PP is within the sample volume of the objective.

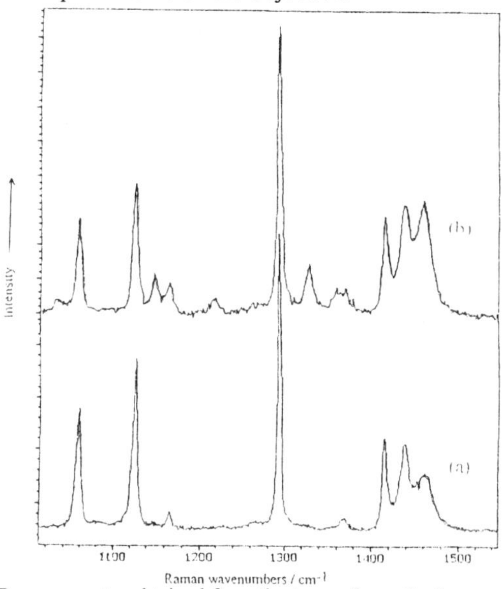

Figure 3. Raman spectra obtained from the top surface of a 2 μm polyethylene film placed over a polypropylene sub-layer, using: (a) x100 objective, 5 mW laser power, 632.8 nm and 4 pixel height on the CCD, (b) x50 objective.

Figure 4 illustrates data acquired from a raw materials and processing problem. The spectrum shown in Figure 4(b) was taken from a polyethylene terephthalate (PET) preform used in bottle production. The spectrum shown in Figure 4(a) was obtained from a small white inclusion embedded in the side wall. This data was obtained using the confocal Raman microscope arrangement (ie. limiting the depth resolution of the sample volume) to discriminate clearly between, and isolate, the inclusion from the bulk sample. An analysis of the data reveals that the inclusion is a small amount of the crystalline form of PET. Whilst the two data sets are largely similar the band widths of the 1735, 1096 and 859 cm^{-1} peaks are considerably reduced, indicative of a crystalline polymeric form.

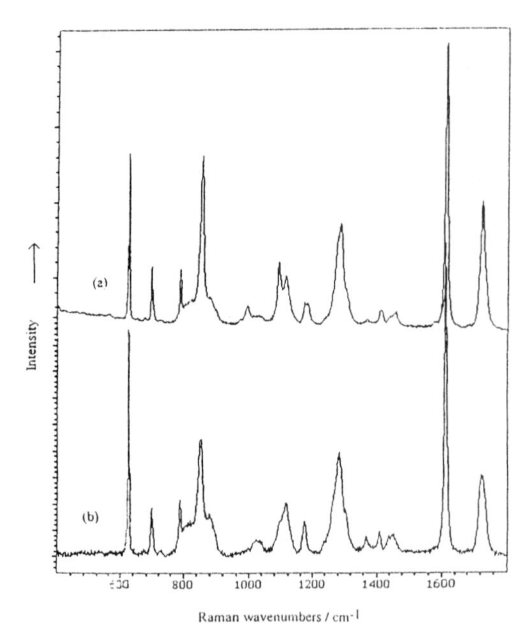

Figure 4. Raman data obtained from an inclusion in a PET preform, 632.8 nm laser excitation, 5 mW and an accumulation time of 40 seconds (a) inclusion, (b) bulk.

A second example of the use of a confocal arrangement is for discrimination between the analyte and a fluorescent matrix. Previous work[20] has shown the viability of this approach using a conventional aperture-based confocal system. The Raman spectrum shown in Figure 5 illustrates data obtained from a single carbon fibre embedded in an epoxy resin used as a composite material. Using the high spatial resolution of the system, the discrimination against the fluorescence of the matrix is excellent.

Coatings such as synthetic diamond or diamond-like carbon formed by chemical vapour deposition (CVD) are of interest for improving the durability of metal contact surfaces. Not only do these films find application in an industrial environment for hard disks but also in other fields as potentially robust window materials.[21] Raman spectroscopy has

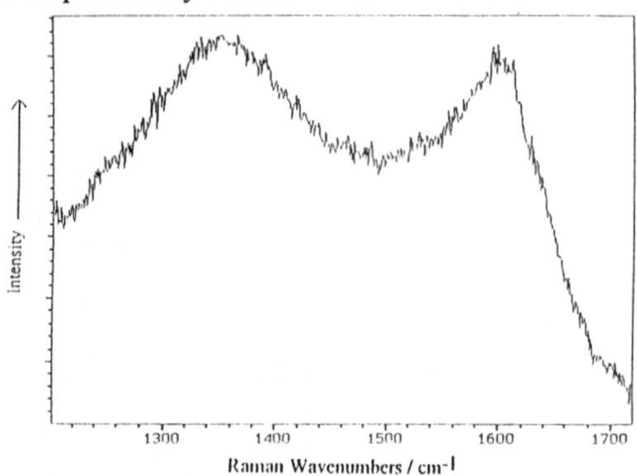

Figure 5. Raman spectrum acquired from carbon fibres in a carbon composite matrix using a x100 objective and 4 pixels height of the CCD. λ = 632.8 nm, 2.5 mW, accumulation time 4 min.

become a standard technique to characterize these diamond and DLC films.[22] The Raman method provides unequivocal evidence for the diamond structure with an intense band of 1332 cm^{-1}, together with amorphous and graphitic structures with bands at *ca.* 1370 and 1600 cm^{-1}, respectively.[22, 23] Figure 6 illustrates data recorded in only 2 s from a diamond film using 514 nm and 632.8 nm laser excitation. In keeping with the work of Wagner *et al.*,[24] we have noted large spectral changes using lower energy laser lines. These are attributed to amorphous carbon.[25]

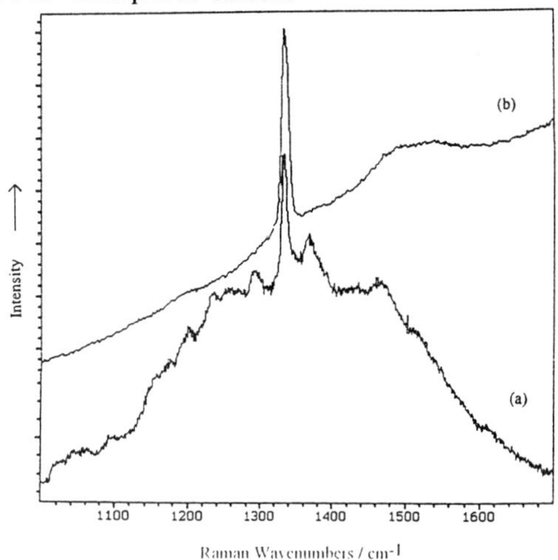

Figure 6. Raman spectra recorded from a CVD diamond film: (a) using 5 mW 632.8 nm laser excitation. (b) using 5 mW 514 nm laser excitation, both 2 minute accumulation.

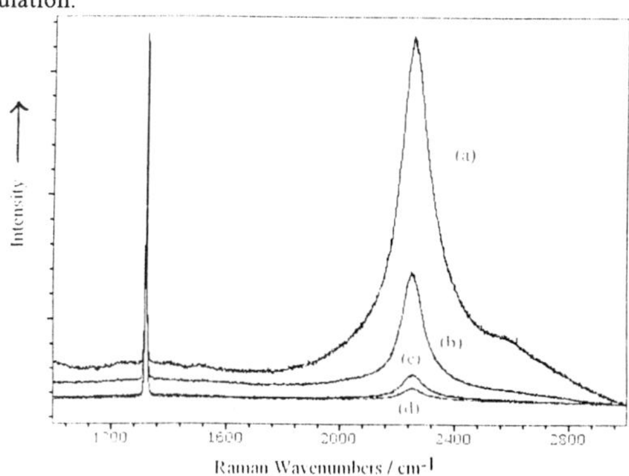

Figure 7. Confocal Raman data obtained from a diamond window: (a) top surface. (b) 6 µm. (c) 10 µm. (d) 16 µm into bulk. λ_e 633 nm acquisition time of 10 seconds.

Figure 7 illustrates the confocal arrangement for the analysis of a diamond window. The data depict an optical slicing of the window material. Figure 7(a) represents data taken from the top surface with 7(b)(c)(d) taken from 6, 10 and 16 µm, respectively, into the material. The intense feature at 2260 cm^{-1} is thought to arise from a vacancy site caused during the diamond growth from the contact sites on the silicon substrate. Clearly, these are very localised on the surface and diminish in intensity as the film thickness increases. A Raman image taken from the 2260 cm^{-1} band is shown in Figure 8. This data illustrates two dimensional information from the diamond surface and shows that not only is the photoluminescence in the diamond localised on the upper few microns of the

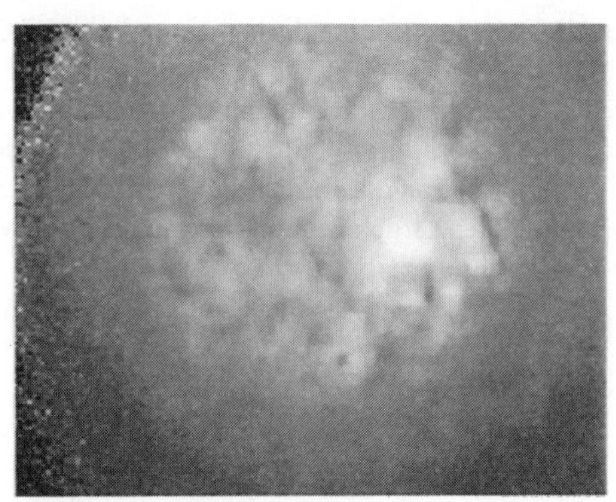

Figure 8. Imaging of the photoluminescence band at 2260 cm^{-1} from a diamond window.

window surface but also is not uniformally distributed across the surface. The Raman image taken from the 1332 cm^{-1} Raman peak of diamond in the same sample area showed no heterogeneity.

In addition to diamond film characterisation the study of DLC is increasing in significance for the hard disk coating companies. Figure 9 shows the Raman spectra acquired from DLC coated hard disks obtained from 4 different manufacturers. Using curve fitting procedures it is possible to ratio the D and G bands, at *ca.* 1345 and 1600 cm^{-1} respectively, assigned to the disordered and graphitic carbon structures, to estimate the quality of the coatings. By imaging or acquiring data from a few sample regions it is also possible to assess the coating homogeneity. Clearly, the data shown illustrate distinct differences between different sources of hard disks.

Not only is the DLC coating from hard disks amenable to analysis by Raman spectroscopy but also the hard disk magnetic heads can be studied. The heads on hard disk magnetic storage systems float on an air bearing of sub-micron thickness. The heads consist of an aerodynamically shaped ceramic body, with embedded magnetic coil. Contaminants, especially on the lower parts of the head, can drastically reduce the lifetime of the hard disk assembly, and must be avoided.[27] A Raman spectrum of the ceramic body shows that all the peaks occur below 1000 cm^{-1}. Images taken using light with greater shifts reveal contaminants. Often the contaminants are luminescent, giving a relatively wide spectral range of intense light, showing up as bright regions on the

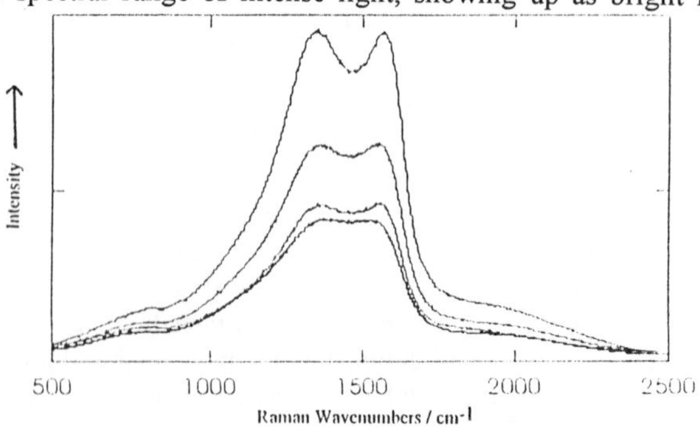

Figure 9. Raman data acquired from a diamond like carbon coating on hard disks from different manufacturers. λ_e 633 nm, 5 mW, 30 seconds accumulation time.

images (Figure 10). Once the contaminant particle is detected from the imaging experiment a spectrum may be taken to identify its composition (Figure 11). In this instance the contamination was found to be polystyrene. Other material has been detected at the edge of the glass region on the head. Raman spectra have revealed that these were traces of the ceramic body that have presumably collected during polishing.

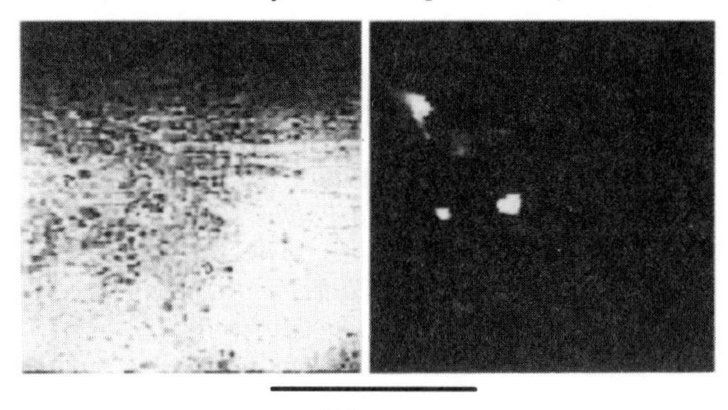

Figure 10. Raman image obtained from the contaminants present on a hard disk. (left - white light image, right - Raman image)

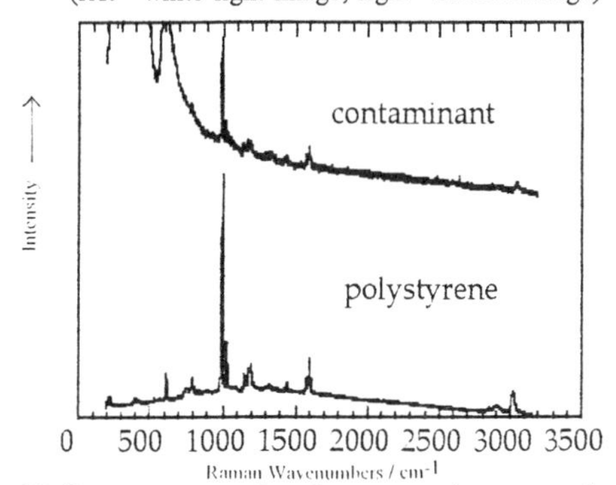

Figure 11. Raman spectra taken from contaminants on a hard disk.

Contamination can of course also be a problem on the hard disks themselves. Raman spectra of contamination seen on a hard disk are often very characteristic. A variety of species have been detected by us including long chain alkanes or wax, polymers and ceramics.

4. CONCLUSION

The examples provided in this paper illustrate the application of Raman microscopy and Raman imaging to a variety of sample types. Many other examples of applications to superconductors, semiconductors, gemology, analysis of artefacts and *in-situ* reaction monitoring exist, but are beyond the scope of this paper. The 1 μm spatial resolution offered by the microscope represents a powerful analytical capability. The efficient coupling of this to a high throughput spectrograph and CCD detector provides a highly sensitive Raman system. The direct imaging capability of the system is also a powerful analytical tool particularly for assessing sample inhomogeneities over a large surface area.

The Raman method is ever expanding its areas of influence. It will do this for some years to come as instrumentation becomes even simpler, more robust and easier to operate.

REFERENCES

1. B Wopenka and J D Pasteris, *Appl. Spectrosc.* **40**, 144 (1986).
2. H Boyer, *Microbeam Anal.* 265 (1983).
3. B J Klp, M C P Van Eljk and R J Meler, *J Polym. Sci., Polym. Phys. Ed.* **29**, 99 (1991).
4. D Masheder and K P J Williams, *J Raman Spectrosc.* **18** 391 (1987).
5. M Mehicic, M A Hazle, R L Barbour and J G Grasselli, *Microbeam Anal.* **68** (1985).
6. G Johnston, *Vib. Spectrosc.* **1**, 87 (1990).
7. M Glotin and L Mandlekern, *Colloid Polym. Sci* **260**, 182 (1982).
8. B J Kip, S M van Aaken, R J Meier, K P J Williams and D L Gerrard, *Macromolecules* **25**, 4290 (1992).
9. T Hirschfeld and B D Chase, *Appl. Spectrosc.* **40**, 133 (1986).
10. K P J Williams, S F Parker, P J Hendra and A J Turner, *Appl. Spectrosc.* **42**, 762 (1988).
11. R G Messerschmidt and D B Chase, *Appl. Specrosc.* **43**, 11 (1989).
12. A J Sommer and J E Katon, *Spetrochim. Acta, Part A* **49**, 611 (1993).
13. C A Murray and S B Dierker, *J Opt. Soc. Am.* **A3**, 2151 (1986).
14. S M Mason, N Conroy, M N Dixon and K P J Williams, *Spectrochimica Acta, Part A* **49**, 633 (1993).
15. N S Ferris and R B Bilhorn, *Spectrochimica Acta, Part A* **47**, 1149 (1991).
16. K P J Williams, G D Pitt, B J Smith, A Whitley, I P Hayward and D N Batchelder, *J Raman Spectrosc.* **25**, 131 (1994).
17. W P Acker, B Yip, D H Leach and R K Chang, *J Appl. Phys.* 64, 2283 (1988).
18. R Tabaksblat, R J Meier and B J Kip, *Appl. Spectrosc.* **46**, 60 (1992).
19. K P J Williams, G D Pitt, D N Batchelder and B J Kip, *Appl. Spectrosc.* **48**, 232 (1994)..
20. C Filou, C Galiotis and D N Batchelder, *Composites* **23**, 28 (1992).
21. M Yoder, in *Diamond and Diamond Like Films and Coatings*, edited by R Clausing, L Horton and P Koidl, Vol 266, Plenum Press, New York (1991).
22. D S Knight, R Weimer, L Pilonc and W B White, *Appl. Phys. Lett.* **56**, 1320 (1990).
23. R E Schroder, R J Nemanich and J T Glass, *Phys, Rev. B*, **41**, 3738 (1990).
24. J Wagner, C Wild and P Koidl, *Appl. Phys. Lett,* **58**, 773 (1991).
25. J Wagner, M Ramsteiner, C Wild and Koidl, *Phys. Rev. B* **40**, 1817 (1989).
26. A J Melveger, *J Polym. Sci. Part A2* **10**, 317 (1972).
27. S Chandrasekar and B Bhushan, ASME Journal of Tribology, Vol. 112, 1-16.

An Error Model for Near Infrared Spectroscopic Instruments

C. G. Eddison and A. M. C. Davies[1]

RESEARCH AND DEVELOPMENT DEPARTMENT, MULTISPEC LIMITED, WHELDRAKE, YORK YO4 6NA, UK

[1] NORWICH NEAR INFRARED CONSULTANCY, 75 INTWOOD ROAD, CRINGLEFORD, NORWICH NR4 6AA, UK

1 INTRODUCTION

It is part of the 'folklore' of analytical spectroscopy that the minimum error in concentration measurements will be obtained when the absorption is 0.434[1]. This is because of the effect of taking logs of the ratio I_t/I_0 where I_0 is the incident radiation and I_t is the radiation transmitted through the sample. When I_t is small or nearly equal to I_0 then the relative error is large. The minimum occurs at 0.434. Figure 1. illustrates this effect for a series of noise levels:- 0.1, 0.05, 0.02, 0.01, 0.005 % of I_0.

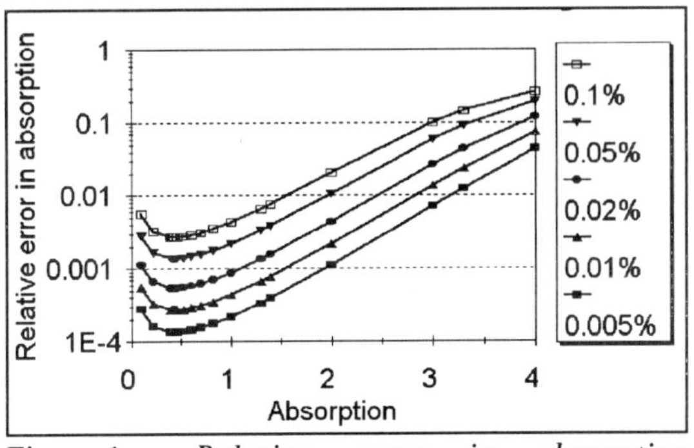

Figure 1. *Relative error in absorption measurements for a series of different levels of detector noise.*

Near infrared (NIR) instruments for whole grain analysis have become very popular in the last five years but their use appears to be at odds with classical spectroscopy. These instruments use NIR energy in the 800 - 1100 nm range for transmission measurements and normally operate over an absorption range from 1.0 - 3.0A. The reason for this apparent disregard for classical spectroscopy lies in the assumption resulting in the 0.434 figure. The assumption is that the detector is by far the greatest source of noise and in classical spectrometers this is normally true. However advances in electronics have lowered considerably the noise in the silicon diode detectors used in NIR grain analysers and other sources of noise become much more important. The most important of these is path length error which in a cell being filled with grain becomes the "effective" path length. The effective path length is that taken by the "average" photon. Absorption in such cells can be greatly influenced by the packing density of the grain and arrangements must be made to accommodate for this by changes of the cell path length. In the Foss-Electric "Grainspec" the cell can be varied from 6 to 38 mm giving good control over the actual levels of absorptions being measured.

Another important source of noise is "stray light"; of which there are several forms. It can be internal or external to the instrument and it can be energy which is at a different wavelength to the required energy but which passes through the sample to impinge on the detector, or it can be any energy which impinges on the detector without passing through the sample. Path length errors can be minimised by increasing the path length to make measurements at higher absorption values but stray light errors decrease the maximum absorption which can be measured. As there are no definitive equations which can be used to optimise instrument performance, a spreadsheet model was developed in order to study the interaction of these errors

2 SPREADSHEET CALCULATIONS

2.1 Assumptions

The spreadsheet assumes that detector noise is essentially constant as a function of wavelength and that path length error is independent of the absorption. For agricultural samples, which have broad NIR absorption bands, small amounts of stray light which pass into the sample will be absorbed and so the assumption is that only stray light which reaches the detector without passing through the sample has to be considered.

2.2 Calculations

The spreadsheet (QUATTRO PRO for Windows, Borland International, Scotts Valley, USA) was set up using values of 100 for the input energy and values from 80 to 0.01 for the transmitted energy. These provided absorption values from 0.1 to 4.0. A set of five curves for relative error were computed for different values of detector noise set at 0.1% to 0.005% of the input energy. (Figure 1).

2.2.1 Path length errors. Path length errors were computed as follows:-

1) A unit path length was set at 0.434 (the minimum relative error on the detector curves).
2) The required path length to produce each of the given transmitted energies was calculated using the Beer-Lambert law.
3) For a given constant path length error (5% is used in the figures) absorbance errors were calculated at each transmission level.
4) The detector and path length errors were combined using a square root of the sum of squares rule.

2.2.2 Stray light errors. Stray light was assumed to be a constant percentage of the input radiation which added to both the input and the transmitted energy. The effect of this was computed for various levels and, for a given level of stray light, the errors were combined with detector and path length errors.

2.2.3 Final Output. Different values of detector error, path length error and stray light could be tested as these were variables which the operator was allowed to reset.

3 RESULTS AND DISCUSSION

The optimisation of absorption in NIR analysis using calibration equations for measurements at several wavelengths has been studied by Honigs et al.[2] This work assumed that the system was detector noise rather than path length limited. In another study by the same group the effect of path length errors was discussed[3]. Stray light has always been one of the instrumental difficulties in analytical spectroscopy and its importance in NIR analysis has been recognised. Williams and Norris[4] discussed ways of avoiding stray light and more recently Miller[5] has discussed the effect of stray light in causing non-linearities in NIR calibration data. However none of these investigators has attempted to study the interaction of these three sources of noise which are of critical importance to the efficiency of whole grain analysers. This is possible using a dynamic spreadsheet model. The model is based on assumptions, but these can be tested by altering the model. The model is flexible and could be extended to cover other sources of noise. Results from the model are shown graphically in figures 2 to 7.

3.1 Path length

The classical model is considerably affected when path length errors are taken into account and added to the detector noise. In whole grain analysers, such as the "Grainspec" the physical path length is around 18 mm for the measurement of wheat protein and moisture of whole grains. Errors in the effective path length of the randomly filled cell can be quite large and the effect of a 5% error is shown in Figure 2. The minimum error is now dependent on noise and absorption so for the given values of detector noise the minimum occurred between about 1.3 and 3.0A. This explains why whole grain analysers use absorption in this range. Actual path length errors will be reduced by using the maximum possible path length but stray light errors must also be taken into account.

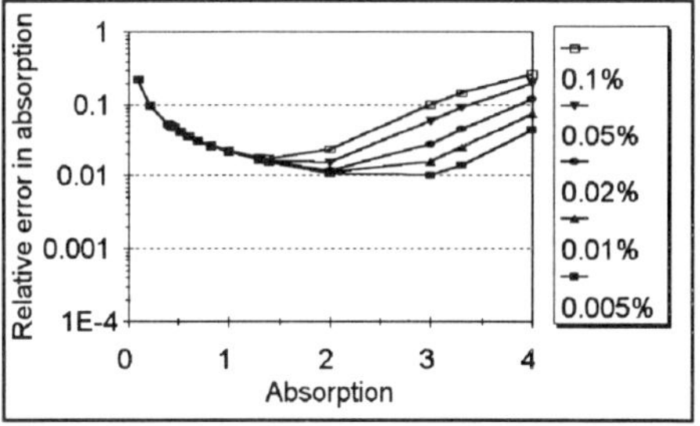

Figure 2. *The effect of combining detector noise and path length error on the relative error in absorption measurements.*

3.2 Stray light

Because the stray light always reaches the detector, it introduces a ceiling above which absorption measurements cannot be made. The non-linearities introduced by stray light mean that very serious errors would be encountered before this ceiling is reached which will depend on the amount of stray light and the absorption level being measured.

Figure 3 shows the relative error which would be caused by levels of stray light

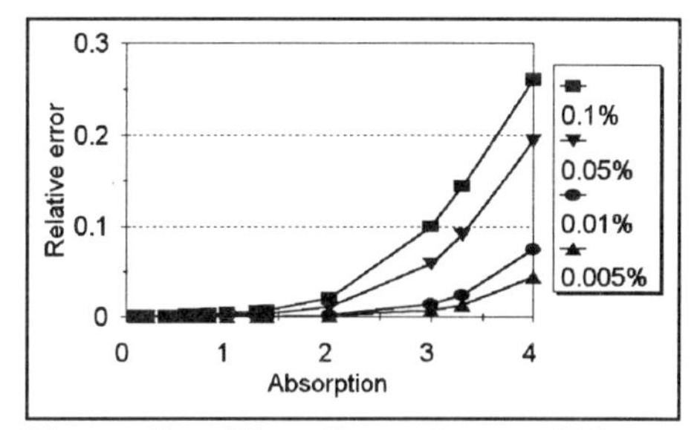

Figure 3. *The effect of stray light on relative absorption errors.*

from 0.1 to 0.005%. When this was added to detector and path length errors the minimum relative error was moved towards lower absorptions. This is demonstrated in Figure 4 when the model used a stray light level of 0.01%, with a path length error of 5% and the previous range of detector noise levels.

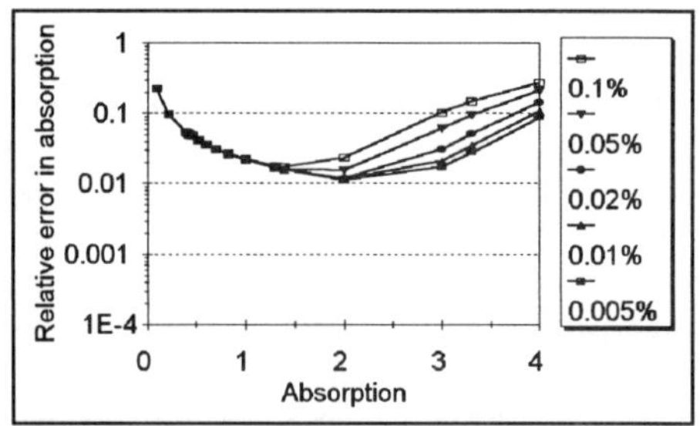

Figure 4. *Combination of detector noise, path length error and stray light.*

3.3 Illustration

The application of the spreadsheet can be illustrated by considering the effect of fitting a new detector with a lower detector noise level of 0.001% while maintaining a path length error of 5% and a stray light level of 0.01%. For these values the model indicates (Figure 5) that although the detector + path length minimum has moved towards higher absorption, the stray light removes this improvement (Figure 6) and no increase in path length would be recommended. In order to utilise the improved detector the stray light must be reduced. Figure 7 shows that the minimum relative error would occur at 3.0A when the stray light was reduced to 0.005%

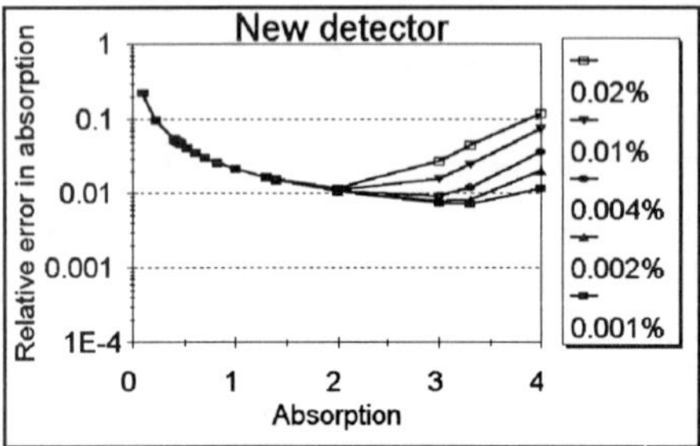

Figure 5. *Combination of detector noise plus path length error for an improved detector.*

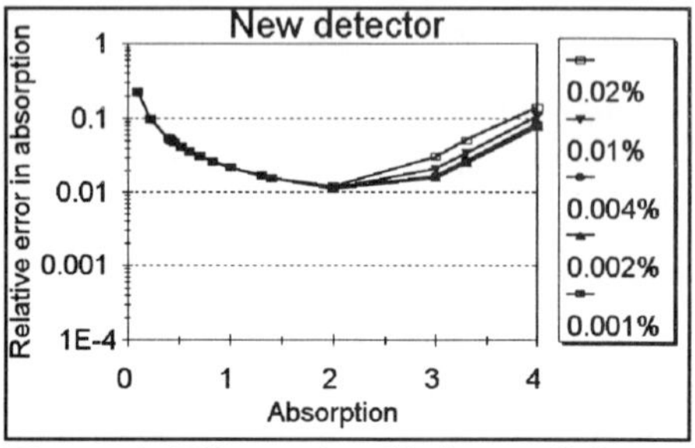

Figure 6. *Effect of added stray light on the relative absorption errors for the improved detector.*

However, experience with these instruments has shown that the assumption that the maximum possible path length will give the lowest relative error in path length (and hence the lowest absorption error) is not always true. This is thought to be related to the fact that the filling of the cell tends to be a chaotic process. Under chaotic conditions some path lengths are more prone to filling errors than slightly smaller ones. Nevertheless, this model does

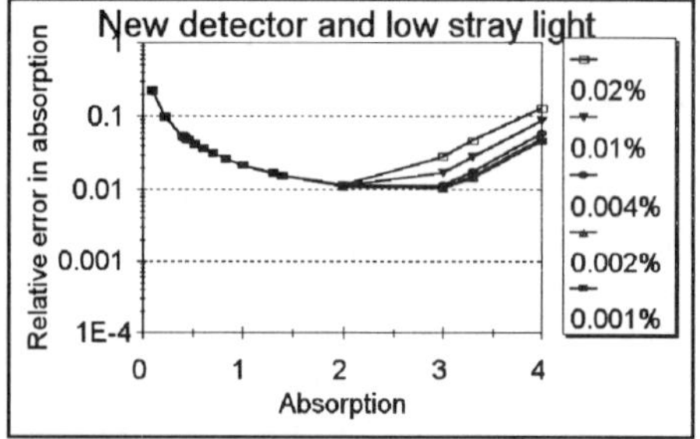

Figure 7. *Shift of minimum error to higher absorptions by reducing the stray light level.*

provide a useful guide when determining instrument conditions for new crops.

3. 4 Conclusions

The model is useful both for demonstrating the relative importance of the various errors and testing results from instruments to see if they perform to specification. It satisfactorily accounts for the performance of current instruments and will be used in the future when designing new systems.

References

1. H. Willard, L. Merritt, J. Dean and F. Settle, 'Instrumental Methods of Analysis', Wadsworth, Belmont, 1981, 6th ed., p.73.
2. D.E. Honigs, G.M. Hieftje and T. Hirschfeld, *Anal. Spectrosc.*,1985, **39**, 253.
3. T. Hirschfeld, D.E. Honigs and G.M. Hieftje, *Anal. Spectrosc.*,1985, **39**, 430.
4. P.C. Williams and K.H. Norris, 'Near-Infrared Technology in the Agricultural and Food Industries', American Association of Cereal Chemists, St. Paul, 1987, p. 140.
5. C.E. Miller, *N.I.R news*, 1993, **4(6)**, 3.

The New Possibilities of Luminescence Spectroscopy of Microscopic Matter: Characterization of Single Hydrocarbon Fluid Inclusions by Fluorescence Excitation–Emission Microspectroscopy

Jan Kihle

INSTIUTT FOR ENERGITEKNIKK, DEPARTMENT OF RESERVOIR AND EXPLORATION TECHNOLOGY, SECTION OF PETROLEUM GEOLOGY, PB. 40, N-2007 KJELLER, NORWAY

1 INTRODUCTION

Fluid inclusions are small (normally < 20 micron size) sealed cavities within mineral species having trapped fluids in geological time when the fluid phase was in equilibrium with the growing mineral crystal. Hydrocarbon fluid inclusions from North Sea reservoir sand stones are commonly more than 30 million years old and display large variations in chemical composition. The combined use of the non-destructive analytical methods of microthermometry and fluorescence micro-spectroscopy gives a unique possibility of deciphering paleo-temperatures, paleo-pressures and timing of entrapment as well as the discrimination of different geological events of fluid trapping — information of great importance to petroleum geologists and engineers with respect to fluid migration mapping and reservoir basin development modeling[1,2].

1.1 Fluorescence of Hydrocarbon Fluid Inclusions

Conventional microscopic luminescence spectroscopy in hydrocarbon fluid inclusion research has hitherto been limited to fluorescence *emission* spectroscopy of approximately 400 nm wavelength and above[3,4]. We have expanded the use of well established luminescence techniques of excitation spectroscopy, synchronous excitation-emission scanning spectroscopy and phosphorescence spectroscopy into the micro-scale domain by using a commercially available luminescence spectrometer adapted to a microscope via Suprasil grade fiber optic cables (Figure 1). These novel techniques were developed to pursue non-destructive fingerprinting of individual hydrocarbon fluid inclusions by use of the quantitative parameters of *optimum excitation wavelength* and *Stokes' shift*.

2 RESULTS

We have tested the technique of fluorescence excitation-emission micro-spectrometry (FLEEMS) on different hydrocarbon fluid inclusion samples from the Norwegian continental shelf. This new micro-technology readily allows unique luminescence information to be extracted for discrimination of hydrocarbon fluids with otherwise seemingly identical fluorescence color or emission behavior (Figures 2ab). Preliminary data indicate that there is a positive correlation of hydrocarbon maturation[5] and Stokes' shifts as illustrated in Figure 3. Hydrocarbon fluid inclusions high in short chain aliphates and 1,2,3-ring PAH's typical for well matured hydrocarbons (condensates), show the larger Stokes' shifts in Figure 3. It might also be possible to discriminate hydrocarbons originating from different sources by the use of such correlation diagrams. Entrapment of hydrocarbon fluid inclusions can also be made experimentally. Knowing the chemical

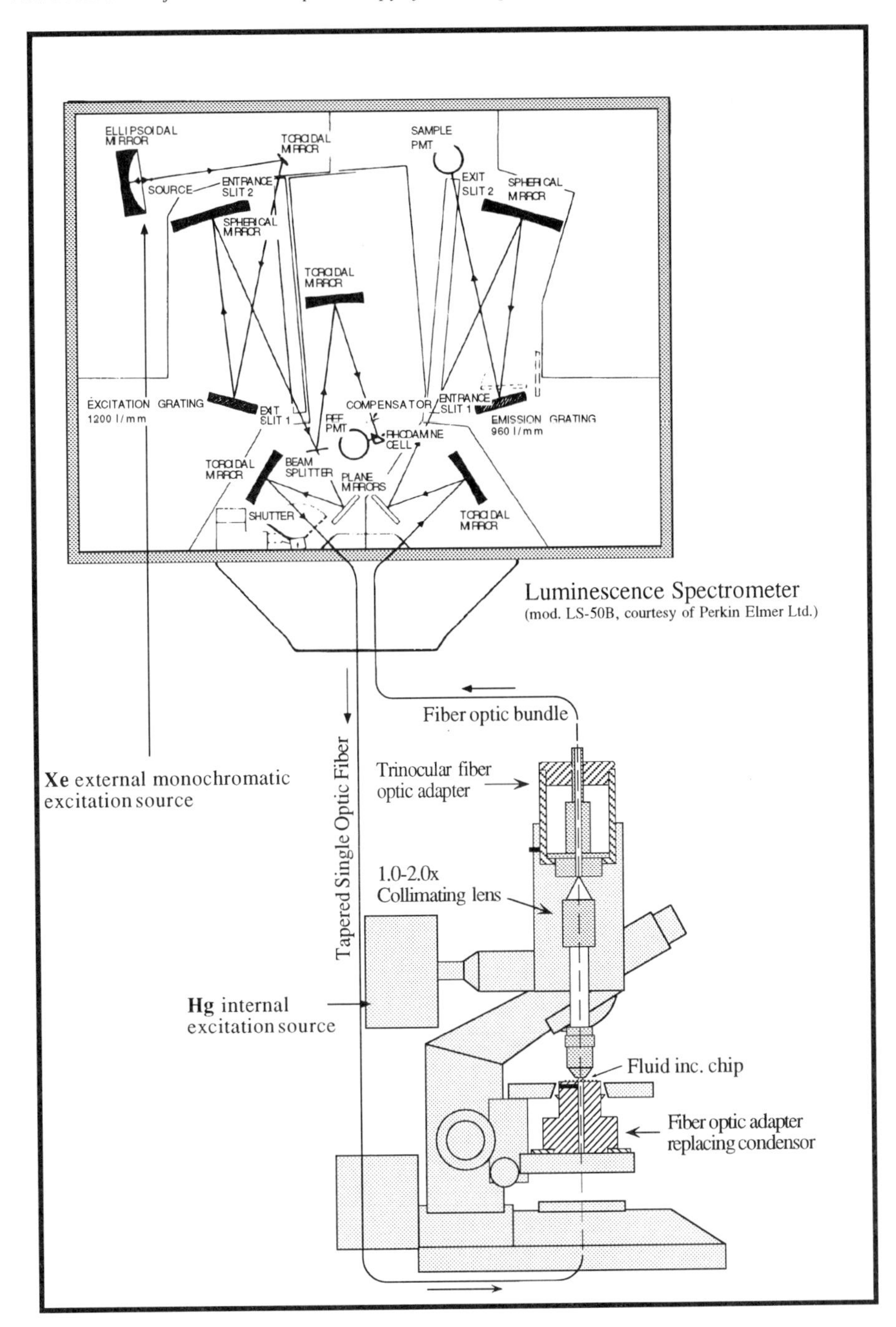

Figure 1 *Instrumentation used in FLEEMS spectroscopy*

composition of synthetic hydrocarbon fluid inclusions[6] FLEEMS spectrum peaks will be related to prominent groups of hydrocarbon molecules and comparisons with natural fluid inclusion FLEEMS spectra can be performed.

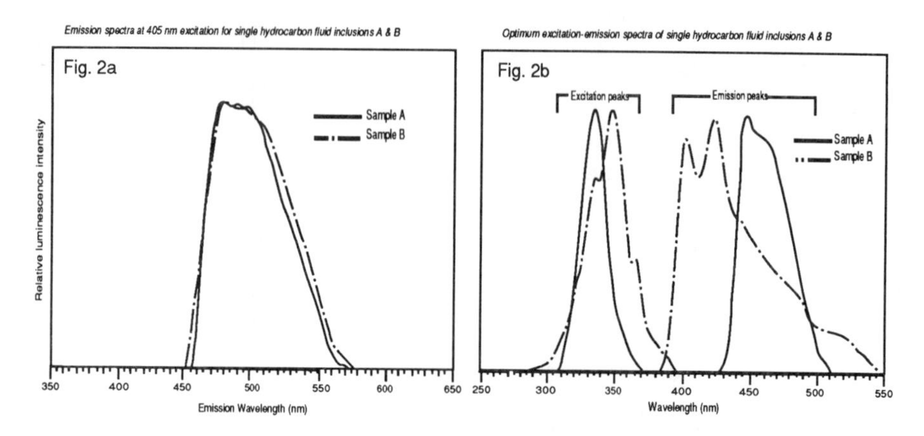

Figure 2 *(a) Emission spectra at 405 nm excitation for single fluid inclusions A and B (b) Optimum excitation-emission spectra for same inclusions as in (a)*

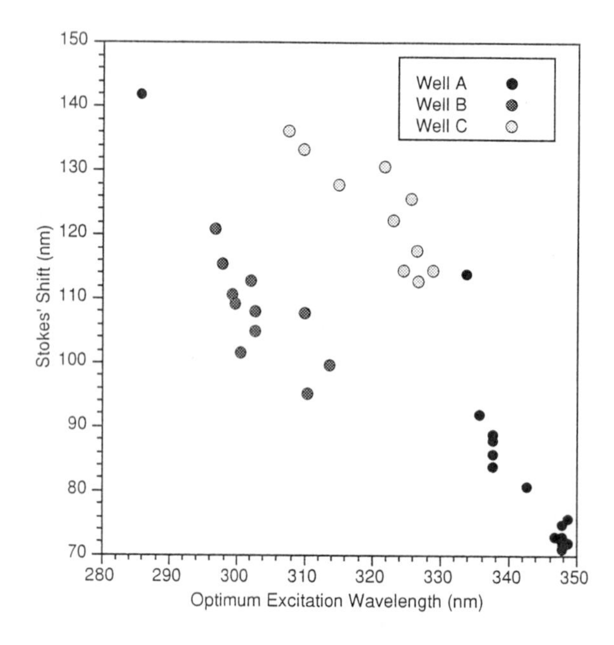

Figure 3 *Stokes' shift as a function of wavelength.*

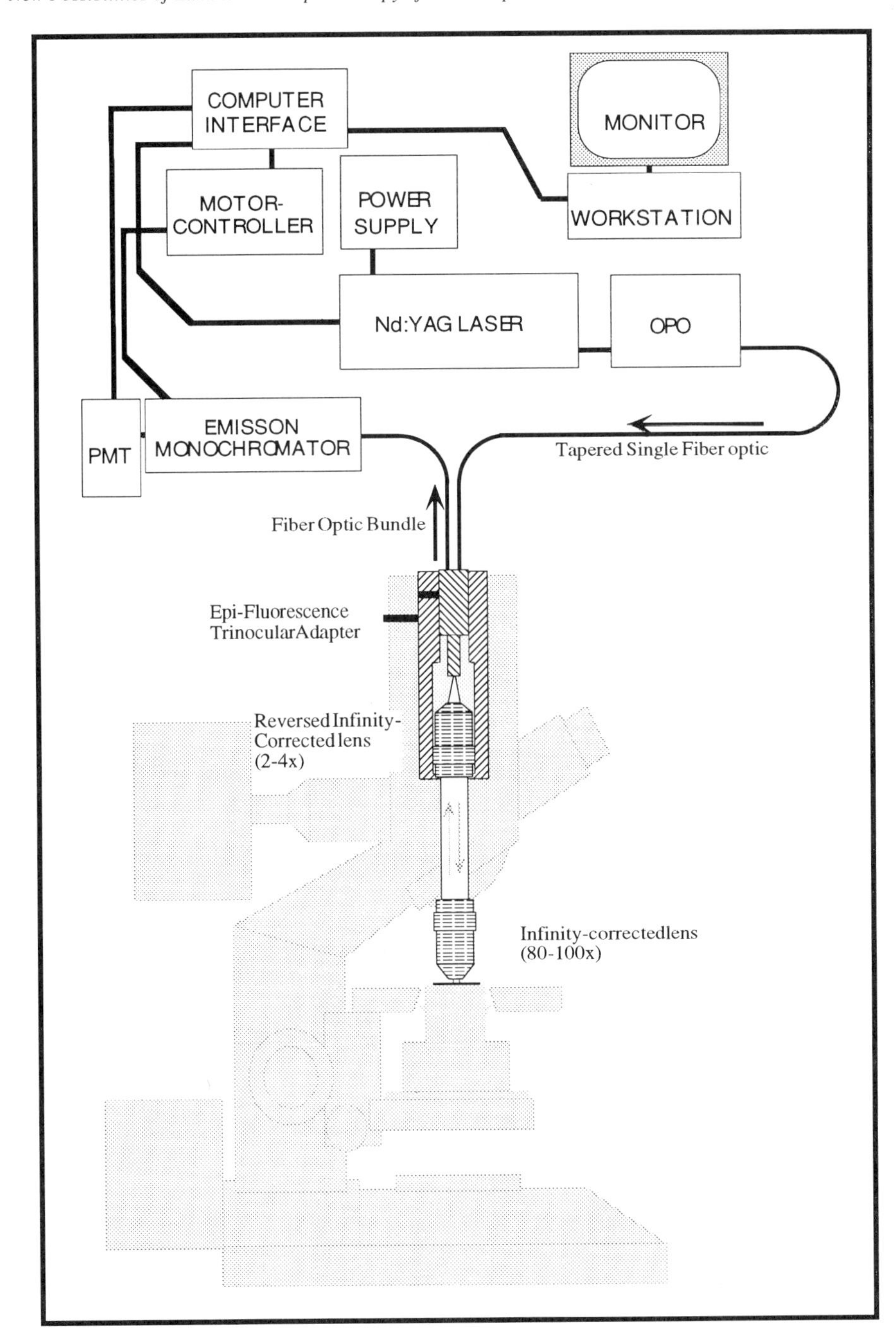

Figure 4 *Instrumentation for Epi-FLEEMS spectroscopy*

3 DISCUSSION

FLEEMS is applicable not only to fluid inclusion research but also to any transparent or translucent mineral or material of approximately 10 microns size or larger. Spectroscopic information in the range of 240 nm to 800 nm is accessible with a high wavelength accuracy (± 1.0 nm) and reproducibility (± 0.5 nm) and signal-noise ratios are better than previously available for microscopic applications.

With the advent of commercial all-solid-state tunable laser systems (OPO = Optical Parametric Oscillators) based on β-Barium Borate crystals, continuously variable excitation wavelengths from 200 nm to more than 2000 nm are available having output linewiths below 0.01 nm and output energies higher than 20 mJ at any wavelength. Such laser-based systems will allow FLEEMS spectroscopy of any microscopic matter equal to appr. 2 micron size or larger (Figure 4). These technologies (Figures 1 & 4) can readily be adapted to most brands of infinity-corrected microscopes in a modular way without any irreversible microscope customization by the incorporation of two dedicated adapters, one replacing the condenser lens system, the other substituting for the trinocular C-mount adapter commonly used in combination with video cameras, or by using the epi-fluorescence fiber optic adapter illustrated in Figure 4.

This study was supported by the Royal Norwegian Science Foundation (NFR) project # 440.90/064.

References

1. McLimans, R.K., (1987) The applications of fluid inclusions to migration of oil and diagenesis in petroleum reservoirs. *Applied Geochem.*, **2,** 585-603.2. Karlsen, D.A., Nedkvitne, T., Larter, S.R. & Bjørlykke, K., (1993) Hydrocarbon composition of authigenic inclusions; applications to elucidation of petroleum reservoir filling history. *Geochim. Cosmochim. Acta*, **57,** 3641-3661.
3. Barker, C.E. & Kopp, O.C. (Eds.) (1991) Luminescence microscopy and spectroscopy: Qualitative and quantitative applications. *Society for Sedimentary Geology* (SEPM) (USA) short course **# 25**.
4. Guilhaumou, N., Szydlowskii, N. & Pradier, B. (1990) Characterization of hydrocarbon fluid inclusions by infra-red and fluorescence microspectrometry. *Min. Mag.*, **54,** 311-324.
5. Tissot, B.P. & Welte, D.H. (1984) Petroleum formation and occurence. SpringerVerlag, 669pp.
6. Kihle, J. & Johansen, H., (1994) Low temperature isothermal trapping of hydrocarbon fluid inclusions in synthetic crystals of KH_2PO_4. *Geochim. Cosmochim. Acta*, **58,** 1193-1202.

Spectroscopic Imaging of Polyethylene

Stewart F. Parker,[1,2] Choon Chai,[1] and Clive Baker[1]

[1] BP RESEARCH CENTRE, CHERTSEY ROAD, SUNBURY-ON-THAMES, MIDDLESEX TW16 7LN, UK

[2] PRESENT ADDRESS: ISIS DIVISION, RUTHERFORD APPLETON LABORATORY, CHILTON, DIDCOT, OXON OX11 0QX, UK

1. INTRODUCTION

Environmental stress cracking in polyethylene[1,2] is a cause of failure in medium density polyethylene used for gas and water pipes. The tendency to stress cracking is assessed on a relative scale by the Bell Telephone Test. This involves notching a section of polymer, bending it and immersing it in a bath of surfactant held at constant temperature. The time taken for half the specimens to fail is then recorded. An understanding of how the test causes the samples to fail potentially offers a way to understand environmental stress cracking in polyethylene in a wider sense.

2. EXPERIMENTAL

Infrared images of the samples were recorded using an Inframetrics 600 thermal imaging camera. Images were recorded both with and without a narrow bandpass filter (1116 cm^{-1} centre, 30 cm^{-1} full width at half maximum from Coating and Filter Design). The infrared source was a laboratory hotplate at a temperature of ~300°C. The samples were suspended approximately 30 cm from the hotplate, close to the infrared camera.

The samples were mounted onto the motorised stage of a Nicplan infrared microscope coupled to a Nicolet 730 Fourier transform infrared spectrometer. The maps were obtained by recording the infrared spectra (32 scans at 4 cm^{-1} resolution with a 100 μm spot size) from a 19 x 7 grid using 100 μm steps to give a total of 133 spectra. The maps were generated by plotting the intensity of the 1150 cm^{-1} band of Igepal as a function of position.

Confocal laser microscopy images were recorded on a Zeiss LSM laser scanning optical microscope (LSOM). For these experiments the 488 nm laser was used and the images were recorded using both the normal viewing mode and the laser induced fluorescence mode.

The polymer sample examined was standard pipe-grade polyethylene that had undergone the Bell Telephone Test. The sample was taken from a region that had <u>not</u> failed.

3. RESULTS AND DISCUSSION

LSOM[3,4] is a recently developed branch of optical microscopy that eliminates the problem of limited depth of field with a dramatic improvement in image quality. Figure 1 shows an LSOM image using the reflected light, Figure 1a, and the fluorescence, Figure 1b, modes. The light mode shows the tip of the notch and the propagation of a crack. The fluorescence mode confirms this observation and demonstrates the presence of a chemical in and near the crack. However, it does not give any indication of the nature of the emitting species.

The advent of Fourier transform infrared spectrometers has made possible the routine use of infrared microscopy[5]. The spatial resolution is ultimately limited by the diffraction limit of the radiation, thus for infrared the limit is ~10 μm, although in practice this is rarely acheived because of the very small apertures needed. To enable spectra of reasonable

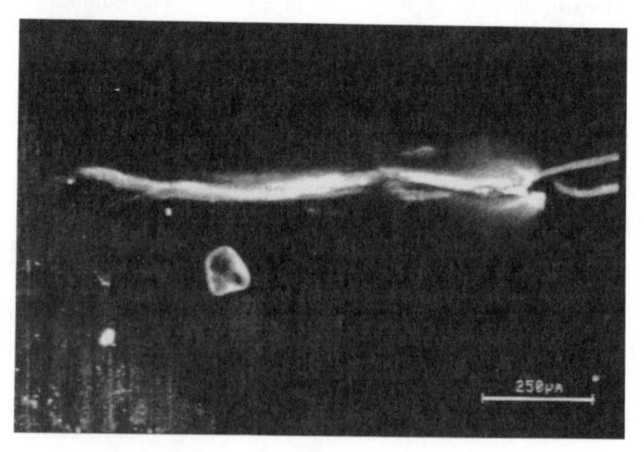

Figure 1 : *LSOM images of the crack region in the reflected light mode (upper) and in the fluorescence mode (lower).*

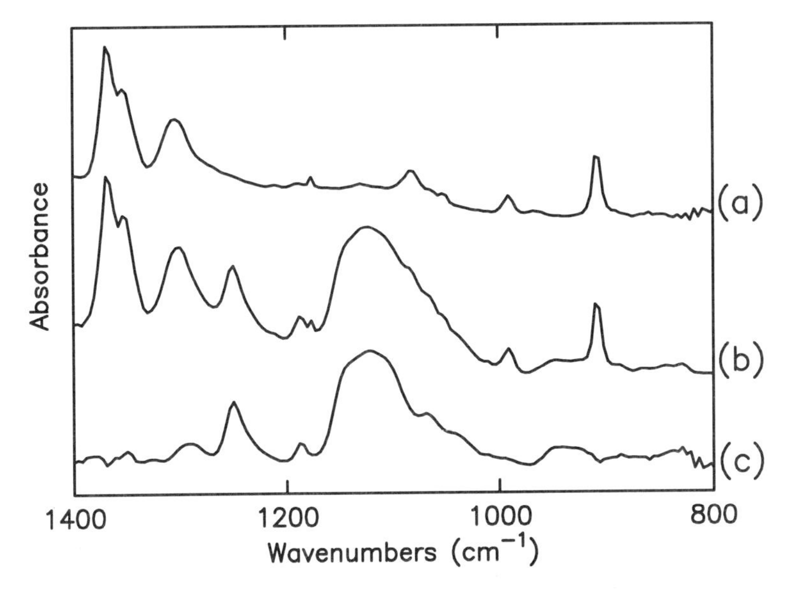

Figure 2: *Infrared spectra of polyethylene after the Bell Telephone test, (a) away from the crack, (b) in the crack region, (c) the difference spectrum (b) - (a).*

quality to be obtained in a few minutes, the diameter of sampled area was increased to 100 μm. Figure 2, shows the spectra recorded (a) away from the crack, (b) in the crack region and (c) the difference spectrum. Comparison with Figure 3 shows unambiguously the difference spectrum is the surfactant used in the test, Igepal CO-720.

A functional group map[6] based on the 1150 cm^{-1} band of Igepal is shown in Figure 4. The similarity with Figure 1 is apparent. The major difference is that the map shows a region of high absorption in the notch region. This is because the spot size is greater than the width of the crack in this region so a contribution from the surrounding area is also recorded.

The two techniques together give a great deal of information about the location of the stress inducing agent but neither technique alone is sufficient. A better method would be one that combines the visual information of the LSOM image with the molecular specificity of the infrared map.

This was achieved by combining an infrared imaging camera with a narrow bandpass filter. The spectra of Igepal and the narrow bandpass filter are shown in Figure 4, upper and lower respectively. It can be seen that the intense C-O stretching band of Igepal at 1150 cm^{-1} and the transmission window of the filter exactly match.

An infrared image recorded using the infrared imaging camera and the filter is shown in Figure 5. (The colour scale runs from white for no absorption to black for total absorption). The sample shows a region of intense absorption along the length of the crack,

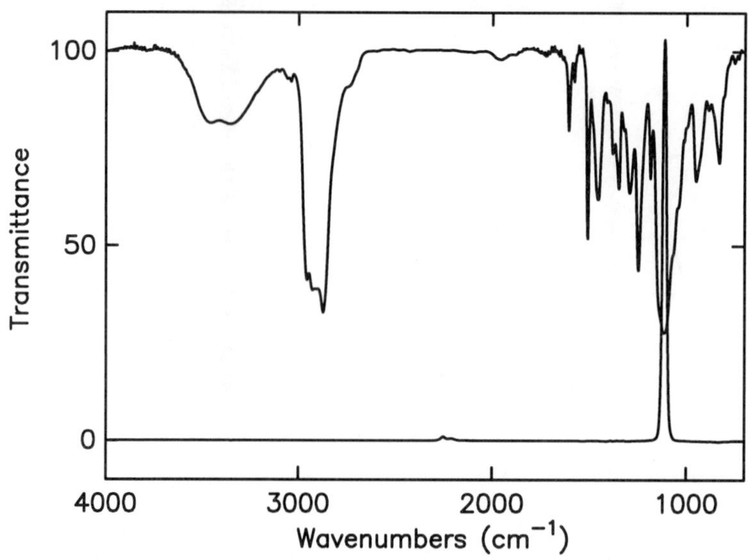

Figure 3: *Transmission infrared spectra of Igepal CO-720 (upper) and the narrow bandpass filter (lower).*

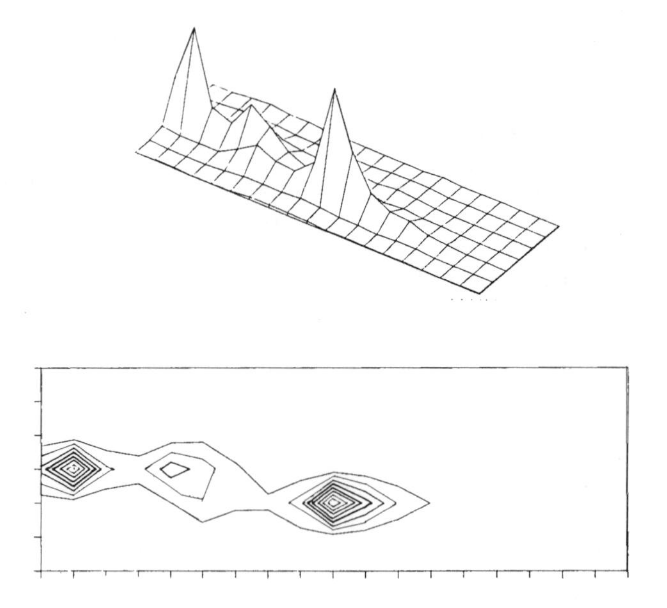

Figure 4: *Infrared map of the cracked region. Upper: axonometric projection, lower: contour plot.*

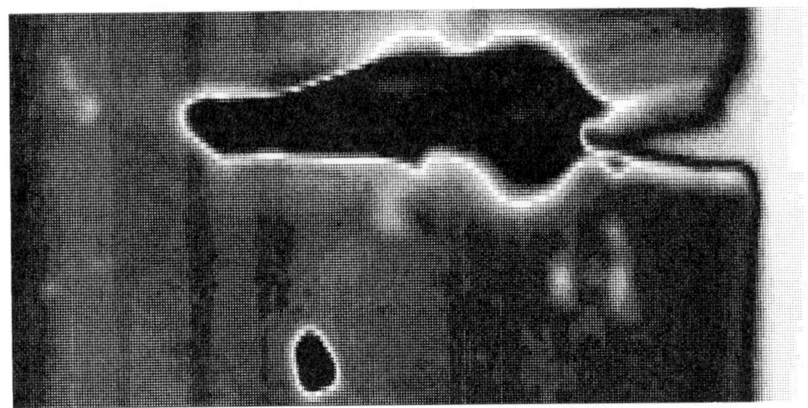

Figure 5: *Infrared image using the bandpass filter of the crack region.*

mirroring what was seen in the fluorescence image, Figure 1b.

Previous work[1] had shown that Igepal could penetrate polyethylene; the authors speculated that the Igepal was concentrated in the interspherulitic region but could not determine the location. The work reported here shows unambiguously, by all three techniques, that the surfactant is concentrated along the direction of the crack propagation. This suggests a model whereby the stress crack agent penetrates the polymer in areas of stress, further increasing the stress and allowing more penetration. This feedback provides a simple mechanism for crack propagation.

4. CONCLUSIONS

LSOM and infrared microscopy of the same sample, showed that the surfactant is concentrated along the direction of crack propagation. However, LSOM does not provide any molecular information while the infrared point-by-point mapping technique is time consuming and collects relatively few points. To overcome this problem a new technique of rapid infrared imaging with spectroscopic analysis was developed. The technique uses a standard broadband infrared camera with a narrow bandpass filter to select the spectral region of interest. Using this method an image can be acquired in a few minutes and readily allows the distribution of surfactant to be mapped.

References

1 P. D. Frayer, P. P-L. Tong and W. W. Dreher, Polymer Eng. and Sci. 17 (1977) 27.
2 A. Lustiger and R.L.Markham, Polymer 24 (1983) 1647.
3 C.J.R.Sheppard and T.Wilson, Optik 55 (1980) 331.
4 A.Hall, M.Brown and V.Howard, Proceedings of the Royal Microscopical Society, 26 (1991) 63.
5 J.E.Katon, G.E.Pacey and J.F.O'Keefe, Anal. Chem. 58 (1986) 465A.
6 M.A.Harthcock and S.C.Atkin, Appl. Spec. 42 (1988) 449.

Ozone Measurements with Star-pointing Spectrometers

H. K. Roscoe, W. H. Taylor, J. D. Evans, E. K. Strong,[1] D. J. Fish,[1]
R. A. Freshwater,[1] and R. L. Jones[1]

BRITISH ANTARCTIC SURVEY, MADINGLEY ROAD, CAMBRIDGE CB3 0ET, UK

[1] CAMBRIDGE CENTRE FOR ATMOSPHERIC SCIENCE, DEPARTMENT OF CHEMISTRY, CAMBRIDGE UNIVERSITY, LENSFIELD ROAD, CAMBRIDGE CB2 1EW, UK

1 INTRODUCTION

Ozone in the stratosphere was first detected by Huggins from observations of the spectrum of Sirius. Later, stars were used to measure polar ozone in the winter dark, but these measurements suffered significant problems which led to contested results.

These problems can now be avoided by using cooled 2D CCD arrays as detectors. We have made two instruments which exploit the advantages of these detectors. Like earlier ground-based UV-visible spectrometers with array detectors, they have the potential of simultaneously measuring O_3, NO_2, NO_3 and OClO.[1] However, at night these earlier sensors could only observe the Moon.[2] Our new instruments have significantly extended the range of sources and elevation angles at night, and so of measurement opportunities, by having sufficient sensitivity to observe stars and planets. This is particularly useful for measurements near the poles in winter when it is dark, when conditions leading to possible ozone depletion occur. Near the poles outside winter, and at mid-latitudes, it also has particular potential for measurements of OClO, because OClO amounts increase significantly during the night;[2] of NO_3, because NO_3 only exists at night;[3] and of the night-time changes in NO_2, which enable N_2O_5 to be deduced.

The new instruments measure ozone at visible wavelengths where depths of absorption are less than in the UV but signal levels are greater and less light is scattered out of the beam when the star is close to the horizon. This is particularly important at the latitudes of the UK and of Antarctica, where Sirius, the brightest star in the sky, is too close to the horizon for UV measurements of ozone with a good signal-to-noise ratio.

The first of these new instruments was configured for simplicity in manned field measurements, and was deployed outside in winter in Northern Sweden in 1991. The second is fully automated for long-term deployment in Antarctica.

2 EARLY MEASUREMENTS

The earliest work on atmospheric ozone[4,5] used the spectra of Sirius and its variation with elevation angle to help identify the constituent and to demonstrate that it was in the Earth's atmosphere. Fowler and Strutt[5] showed that the same UV absorption lines existed in spectra of the sun. However, ozone was not measured in this way - these early measurements with stars were qualitative, not quantitative. These early techniques used photographic plates to avoid spectral noise from atmospheric scintillation (twinkling) by measuring simultaneously at all wavelengths. The intensity variations are then equal at each wavelength, unlike more recent detection systems using a photomultiplier and scanning the wavelengths.

Later quantitative measurements using stars[6,7,8] also benefited from this advantage of photographic plates. The signals were measured after exposure by microdensitometers, allowing the spectra to be manipulated. However, these techniques used a prism without collimator or entrance slit. Although this resulted in simple optical systems without a telescope, it also restricted the collecting aperture (and hence signal) to that of the prism, and meant that the image position depended on both incidence angle and wavelength. Hence the same system could not be used for measuring a distributed source, nor could light from the sky near to the star be measured and subtracted (twilight and aurora caused fogging of the plates used by Hamilton[7]); and the star had to be very accurately tracked for the duration of the exposure otherwise it degraded the resolution. It is conspicuous that these early papers contain no examples of Langley plots or fits of laboratory to measured spectra, to show the quality of the data. Hamilton's results were severely criticised.[7]

These difficulties were later avoided by Wardle et al,[9] who used a 10" telescope with a grating spectrometer and entrance slit. The apparatus retained immunity to scintillation by measuring the ratio of intensities of signals at two wavelengths simultaneously, using two independent photomultipliers. Because of its good range of elevation angles above 10° they chose to observe ϵ-Orion, a weak but very blue star. Below 10° elevation, attenuation of UV light by scattering from air molecules becomes severe. However, the apparatus retained the sensitivity to aerosol of techniques which only used one pair of wavelengths.

3 ADVANTAGES OF ARRAY DETECTORS

By contrast to spectrometers which scan in wavelength and use a single detector, array detectors prevent scintillation noise and allow measurements to be made when cloud cover varies. Coupled with the differential analysis technique (see below), the stability of modern array detectors also allows measurements of small depths of absorption. This allows measurements in the visible band of ozone rather than the UV, which eliminates problems of scattered light.

In the novel arrangement in our new instruments, the image at the entrance slit of the spectrometer contains the blur circle of the star at its centre and light from the sky on either side. The image at the detector thus contains the spectrum of the star in a central stripe, with the spectrum of the sky on either side. Hence we use an array detector with pixels in two dimensions so that light from the sky adjacent to the star can be measured simultaneously and later subtracted. This new design feature is important at twilight, in hazy conditions near city lights, or near the poles during aurora. In the example in Figure 1, the auroral peak of green light at 558 nm can be clearly seen in the sky spectrum.

All of these advantages could have been realised in the 1930s with film as the detector if the correct design choices had been made.

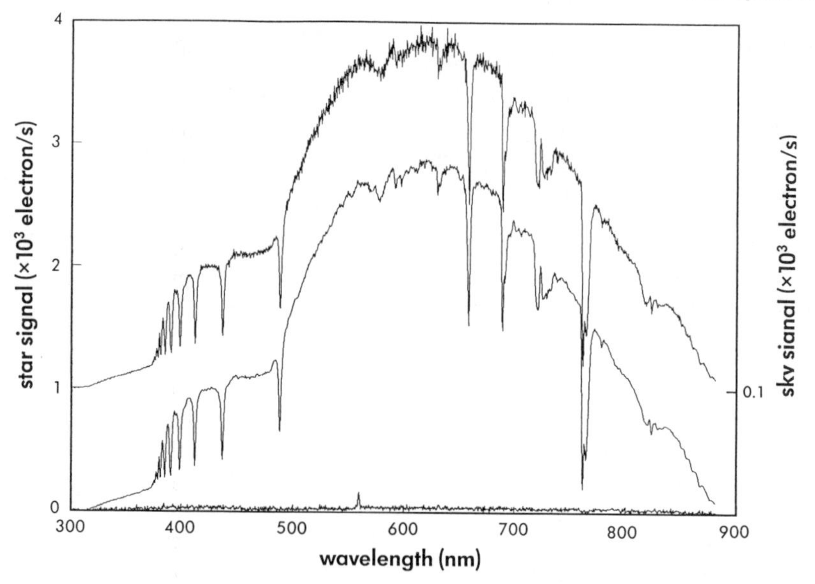

Figure 1. *A raw spectrum of Vega after 30 minutes' integration of signal from a single stripe of detector pixels (upper trace), and after dividing by inter-pixel variability (middle trace). By using a 2-D detector array, the spectrum from the sky adjacent to Vega (lower trace, x10) can be obtained simultaneously and subtracted from the star spectrum.*

4 THE FIRST INSTRUMENT

This instrument is described more fully elsewhere.[10] Briefly, it consists of an amateur astronomy telescope of diameter 12", with a spectrometer mounted at the focus. The mirrors are of modest quality, sufficient because the spectrometer slit width is equivalent to 4 minutes of arc. The telescope is housed in a low-profile weatherproof structure rather than a dome which would offer a higher profile to strong winds when closed. Field operation is simplified by a small TV camera with tracking software, mounted adjacent to the spectrometer. This allows accurate tracking with poor polar alignment, restores the image to the spectrometer slit if strong winds displace the telescope, and avoids repeated adjustments when observing planets. Simplicity is maintained by our use of a small spectrometer with no moving parts.

Wavelength calibration is determined from lines in the measured spectrum due to absorption in the atmosphere of the star or of the Sun (reflected from the Moon or Jupiter). Stability of the wavelength calibration is only necessary for the duration of an integration (maximum 30 minutes), so that temperature control of the spectrometer is not critical.

Although the ratiometric nature of the measurement should cancel absorption lines in the atmosphere of the star, scattered light causes Fraunhofer lines in the cooler stars to remain a problem. Balmer lines in the hotter stars miss the spectral region of visible O_3 absorption but interfere with absorption by NO_2 and OClO. Nevertheless the hotter stars (Sirius, Vega) are still potentially more useful for measurements of NO_2 and OClO because they give so much more blue light. The instrument functions with stars of magnitude M3, but useful signals are only obtained from the brighter planets and from stars brighter than M1.[10]

By using a cooled detector designed for the highest quality astronomical imaging, dark current is small enough that shot noise on the photon flux at the detector dominates, despite the modest telescope diameter. The readout noise is negligible with signals above 10 electron/s with our exposure time.

The differential analysis procedure (see below) is insensitive to smooth variations in responsivity along the detector, variations between adjacent pixels are important. We measured these variations to be 2% peak-to-peak between single pixels, as shown by the noisy trace in Figure 1. Figure 2 shows the improvement in quality of a spectrum after division by inter-pixel variability.

It is important to measure this inter-pixel variability because absorption by stratospheric constituents is less than 3%. Variability was measured with a tungsten lamp, which has an inherently smooth spectral emission. Successive lamp measurements showed that its limit to stability from ratios of single pixels is about 1.5×10^{-3}. Because we observe signals summed over about 7 pixels perpendicular to the wavelength axis, the practical limit to observation of optical depth is about 5×10^{-4}.

In Figure 2, the intensity units, electrons accumulated by the detector, include grating and detector response functions. Absolute calibrations of the spectra are unnecessary because of the differential analysis procedure.

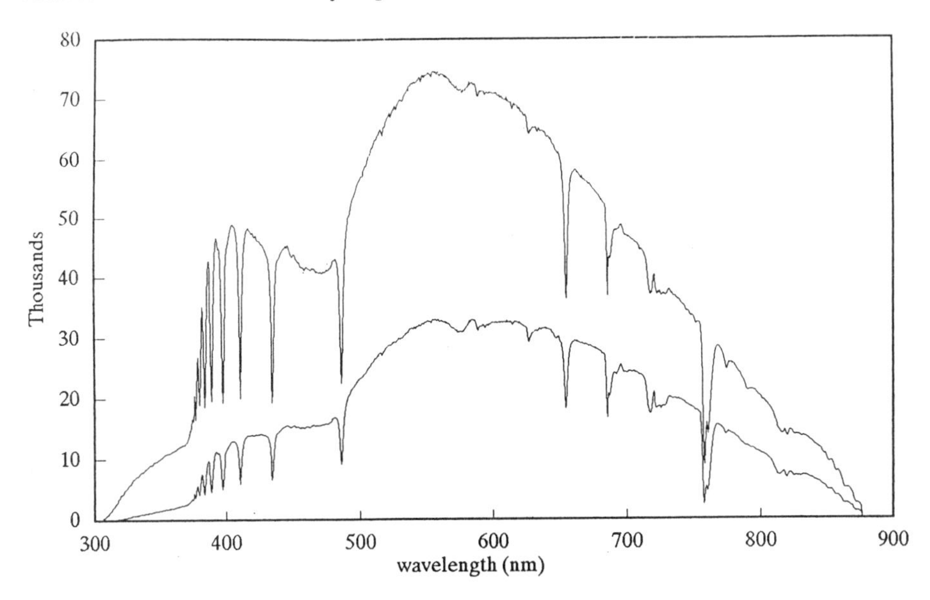

Figure 2. *Spectra of Vega measured during the night of 1 to 2 March 1992 at 69°N, at elevations of 50° (upper trace) and 17° (lower trace), averaged over about 2 hours each. At 500 nm, the signal-to-noise ratio exceeds 3000 in each of these spectra.*

5 THE SECOND INSTRUMENT

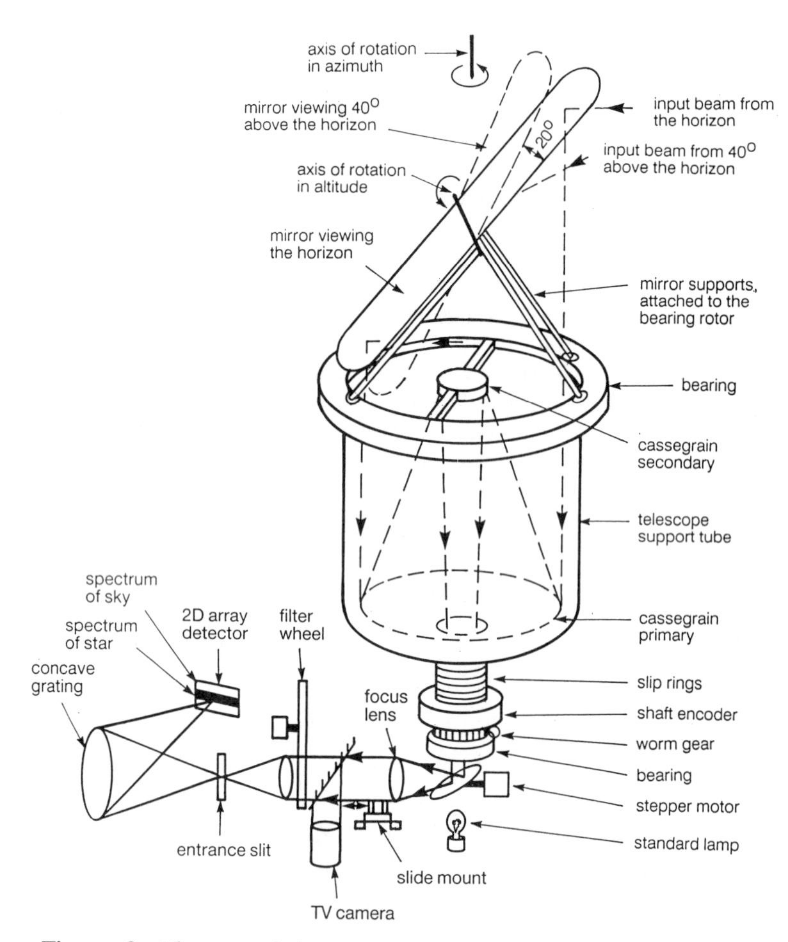

Figure 3. *The second instrument's pointing mirror and telescope assembly, which directs light from the star to a stationary spectrometer and detector. The focused beam exits through hollow components via relay optics to the spectrometer.*

This instrument is designed for long-term deployment in Antarctica. It combines the general principles of the first instrument, described above, with the novel viewing geometry suggested by Roscoe.[10] This geometry incorporates an altitude-azimuth pointing mirror, so that the spectrometer does not move, thereby helping to eliminate cables which flex. Such a pointing system cannot have a co-aligned telescope to sight on the star because the input mirror moves by half the altitude angle of the input beam. Instead, a beam splitter directs light to a TV camera, which incorporates tracking software as in the first instrument.

Another new feature is the slip-rings to provide power and sense position in the altitude direction. This further avoids flexing of cables, thereby improving reliability.

It also dispenses with the need for a stop in the azimuth rotation, which would restrict all-round viewing

The final important development was to place a glazed enclosure around the mirror to ensure protection from rain and snow. This was made of UV-transmitting perspex, from flat plates to avoid changing the focal length of the telescope.

Other features of the optics, shown in Figure 3, include a filter wheel, and a mirror to switch the input beam to different views, by-passing the telescope. One view is to the zenith sky which allows the instrument the full capability of the SAOZ spectrometers,[11] but with improved signal-to-noise ratio because of its more sensitive detector. The input mirror, telescope, and optics plate form a rigidly-coupled unit. The unit is supported from the base of the housing but is not attached elsewhere except for shipping purposes. The optics plate is made from cast aluminium alloy for increased stability.

The instrument is housed in an environmental box with removable insulated panels, sealed against snow by Silicon-rubber gaskets. Internal components function at -40°C but performance is optimised by heating the box to 0°C. The upper panel includes a channel through which interior air is blown when above +10°C, to provide cooling in sunlight.

The system is automated by means of a desk-top PC indoors and a more robust PC in the spectrometer system. The computers communicate via fibre-optic links to reduce the possibility of RF interference from the radar at the Antarctic site, and to reduce problems of static in such a location with no proper ground.

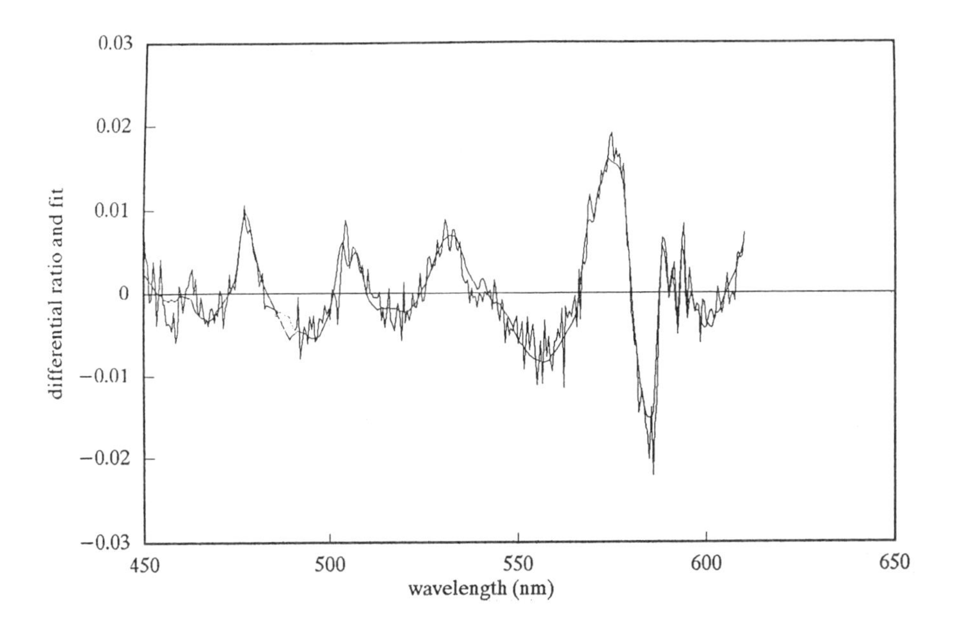

Figure 4. *A differential spectrum calculated from laboratory cross-sections of O_3, H_2O and O_4 (smooth line), after a least-squares fit to the differential of the log of the ratio of two spectra of Vega measured on 23 February 1992 at 69°N by the first instrument (noisy line). From the fit, the vertical column of ozone was 400 DU.*

6 SOME SAMPLE RESULTS

Sample spectra from the first instrument were acquired during a field deployment at Abisko in Northern Sweden in the winter of 1991/92, as part of the European Arctic Stratospheric Ozone Experiment (EASOE). The telescope was mounted on a pier from which the sky across the lake could be viewed to within 3° of the horizon.
Unfortunately, Sirius, the brightest star, is not visible at 69°N, and Vega does not set, so that its range of slant paths in the atmosphere is small. Figure 4 shows a sample of our results for ozone, which are described more fully elsewhere[14].

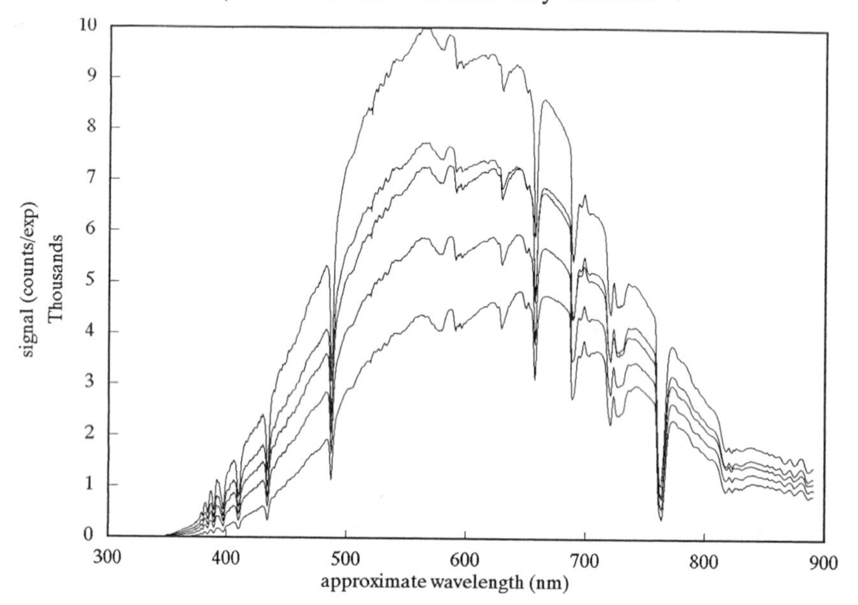

Figure 5. *Spectra of Sirius observed from Cambridge on 7 February 1994 at elevations of 21, 20, 16, 13 and 9°, the most signal being at the highest elevation. Each spectrum is an average of 27 exposures, and the signal-to-noise ratio of the highest is 2500/pixel.*

The analysis shown in Figure 4 starts with two spectra of the same star, one low in the sky and the other high in the sky, as in Figure 2. The log of their ratio contains small but highly structured features due to atmospheric constituents. It also contains features which vary smoothly with wavelength, due to scattering by air molecules and aerosol, which are eliminated by a high-pass filter in wavelength space. This standard procedure results in a so-called differential spectrum[13].
Use of the log of the ratio means that the result is scaled so that no absolute calibration of intensities is needed. Differential laboratory spectra are then least-squares fitted to the measured differential spectrum, resulting in the difference in amounts of constituents between the different slant paths of the two measurements. From the angles at which the spectra were measured, the vertical amounts of constituents are deduced.

The second instrument has been making measurements in the winter from the Met tower at BAS HQ in Cambridge, as part of its tests. Figures 5 and 6 show sample spectra, and Figure 7 shows some analyses.

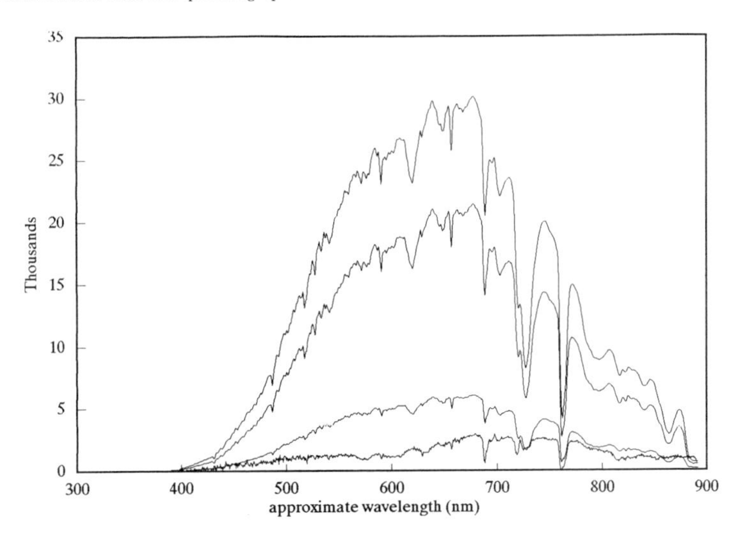

Figure 6. *Spectra of Jupiter observed from Cambridge on 8 April 1994. The last spectrum shown here is noisier because it was measured across sunrise - a large statistical noise is present on the large sky signal, which must be subtracted from that of Jupiter+sky to produce the small spectrum shown.*

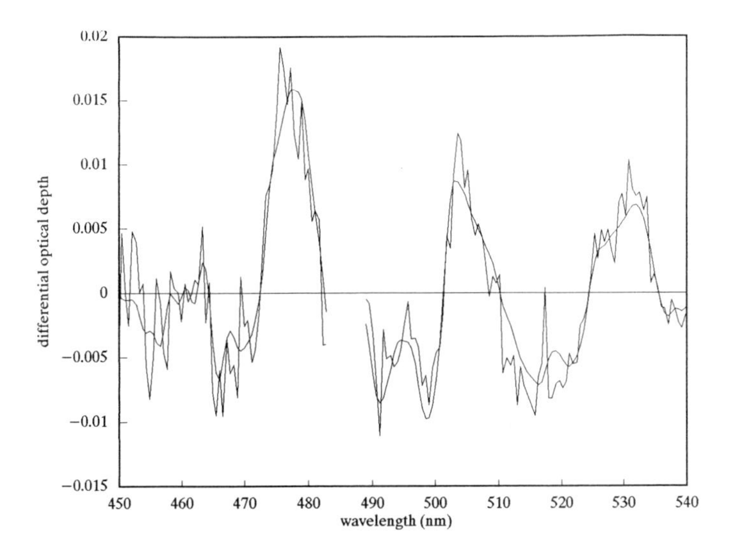

Figure 7. *Differential optical depth from observations of Sirius by the second instrument from Cambridge on 7 February 1994 (noisy line), and the calculated optical depth (smoother line). At 500 nm, the statistical noise from the photon flux is* $\pm 0.9x10^{-3}$, *comparable to the differences between observation and calculation. From the fit, the vertical column of ozone was 382 DU.*

Acknowledgements

Development of the first instrument was supported by the CEC STEP programme. DJF was supported by a NERC studentship.

References

1. G. H. Mount et al, *J. Geophys. Res*, 1987, **92**, 8320-8328.
2. S. Solomon et al, *J. Geophys. Res*, 1987, **92,** 8329-8338.
3. S. Solomon et al, *J. Geophys. Res*, 1989, **94**, 11041-11048.
4. W. Huggins and Mrs Huggins, *Proc. Roy. Soc. London A*, 1890, **48**, 216-217.
5. A. Fowler and Hon.R.J.Strutt, *Proc. Roy. Soc. London A*, 1917, **93**, 577-586.
6. D. Chalonge and E.Vassy, *Rev. Opt. Théo. Instrum,* 1934, **13**, 113-126.
7. R. A. Hamilton, *Quart. J. Roy. Met. Soc*, 1939, **65**, 210-214.
8. D. Chalonge, *Sci. Proc. Intern. Assoc. Meteorol., Rome 1954 (Butterworth)*, 1956, 163-164.
9. D. I. Wardle, C.D. Walshaw, T.W. Wormell, *Nature*, 1963, **199**, 1177-1178.
10. H. K. Roscoe et al, in press, *Applied Optics*, 1994.
11. H. K. Roscoe, *Proc. First Europ. Workshop on Polar Strat. Ozone Res. (ed. J.A. Pyle & N.R.P. Harris), ISBN 2-87263-060-0*, 1991, 91-94.
12. J-P. Pommereau and F. Goutail, *Geophys. Res. Lett*, 1988, **15,** 895-897.
13. M. Q. Syed and A. W. Harrison, *Can. J. Phys*, 1980, **58,** 788-802.
14. D. J. Fish et al, in press, *Geophys. Res. Lett. (EASOE special issue)*, 1994.

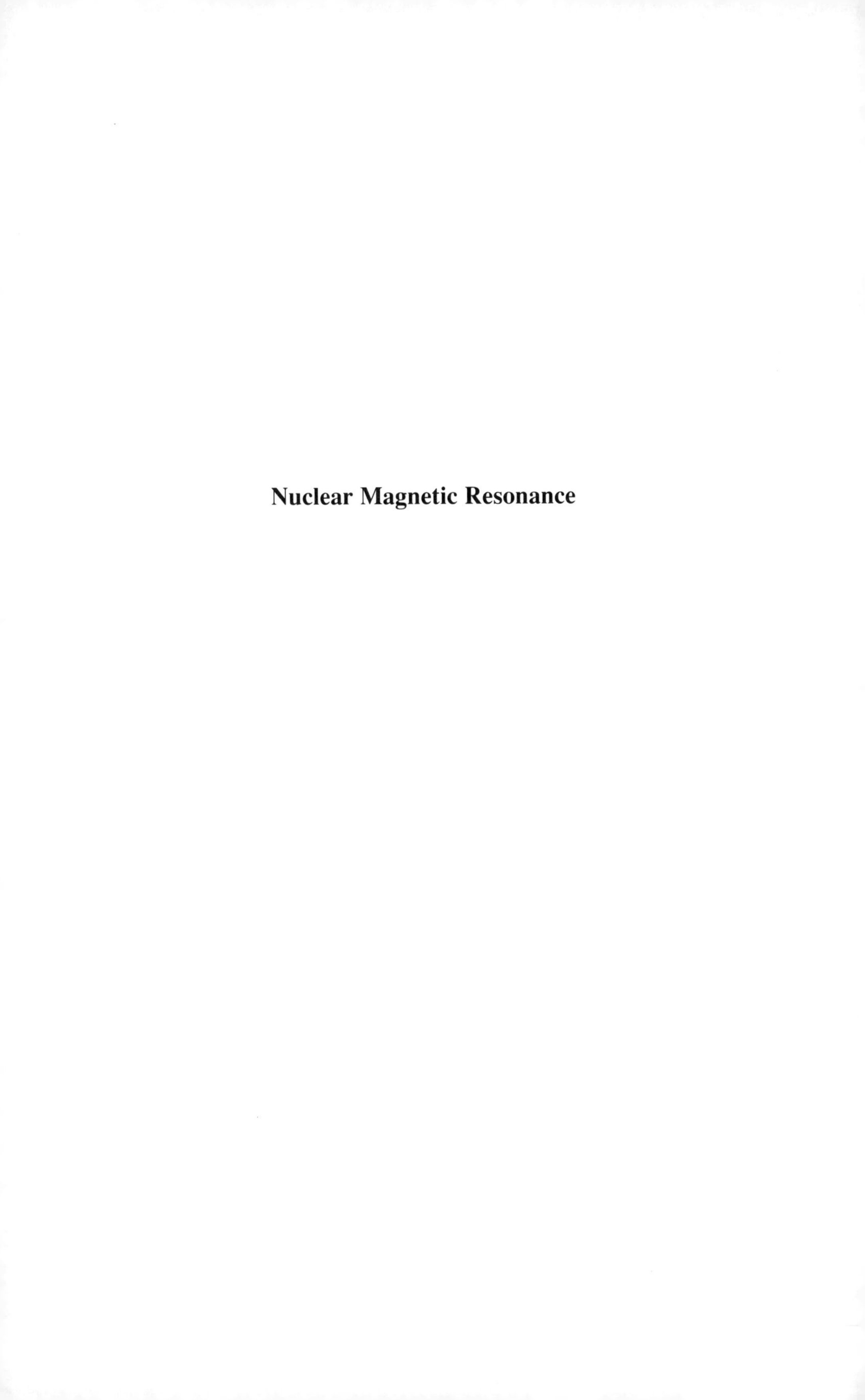

Nuclear Magnetic Resonance

Recent Advances in High-resolution Chemical NMR Spectroscopy of Solids

Robin K. Harris

DEPARTMENT OF CHEMISTRY, UNIVERSITY OF DURHAM, SOUTH ROAD, DURHAM DH1 3LE, UK

1 INTRODUCTION

High-resolution NMR spectroscopy of solids can now be considered to have come of age. In 1994 it was 18 years since Schaefer and Stejskal first demonstrated the potential of combining cross-polarization (CP), high-power proton decoupling and magic-angle spinning (MAS) for ^{13}C spectra. These methods are now ubiquitous for a whole range of nuclei, though sometimes MAS on its own suffices (e.g. in the absence of protons). Occasionally, when there is considerable molecular motion or where there is a low concentration of hydrogen atoms in a sample, MAS will also produce high-resolution proton spectra (and analogously for ^{19}F) of solids. However, more commonly for ^{1}H and ^{19}F it is necessary to combine MAS with multiple-pulse sequences such as MREV8, giving CRAMPS (Combined Rotation and Multiple Pulse Spectroscopy). All these techniques are now available commercially. Any solid material can be studied, including both crystalline and amorphous solids (gels, rubbers, glasses etc.). The sample may be homogeneous or heterogeneous. To illustrate the wide variety of samples that can be examined, the following lists esoteric cases which have been submitted since 1986 to a commercial research service operated from the University of Durham:

a. Fossil graptolites from the Ordovician period.

b. A stick of celery.

c. Cement from a barge which sank in the River Thames 150 years ago.

d. Human atheroma from atheroschlerotic plaques.

e. Grass from an experimental agricultural plot.

f. Whole pharmaceutical pills.

g. Silk from Queen Caroline's bed.

However, in this article it is intended to concentrate on more "normal" chemical systems. In fact only essentially crystalline samples will be discussed. They are normally studied as microcrystalline powders.

Since the basic techniques are now well-known and well-established, they will not be discussed further here. Recent technical considerations have consisted partly in the increasing use of a wide variety of pulse sequences, including those used for two-dimensional spectra, and partly in topics which may be classified under the general heading of spectral analysis. The latter becomes necessary whenever spectra are governed by more than the isotropic chemical shifts. All NMR interactions (shielding, indirect coupling, dipolar coupling and quadrupolar coupling) are, of course, tensors. That is to say they are orientation-dependent. Whereas the first efforts in high-resolution solid-state NMR are normally directed to removing the orientation dependence (leaving only isotropic chemical shifts and isotropic indirect coupling constants), more refined work is inevitably aimed at reintroducing the full interactions in such a way as to simplify their measurement. This frequently involves the interplay of the various tensors and, in particular, the use of spinning sidebands which result whenever the MAS rate is insufficient to give full averaging. Measurement of dipolar interactions is especially desirable, since this leads to molecular geometry information.

Another facet of growing importance (paralleling the earlier development of solution-state NMR) is the multinuclear nature of the technique: most elements of the Periodic Table are accessible. However, it is frequently of value to isotopically enrich samples, especially in cases such as ^{2}H, ^{13}C, ^{15}N, ^{17}O and ^{29}Si.

In this article, a few selected topics will be chosen for brief discussion, involving only spin-$\frac{1}{2}$ spectra.

2 SPECTRAL EDITING

As for solution-state NMR, there are a number of pulse sequences which can assist assignment of spectra. One such experiment is that of selective polarization inversion[1] (Figure 1). Following cross-polarization, to ^{13}C say, the phase of ^{1}H irradiation is reversed, which means that ^{13}C magnetization will follow suit at a rate dependent on (^{13}C, ^{1}H) dipolar interactions. For a relatively rigid organic molecule, this process will occur twice as fast for CH_2 as for CH groups. Choice of a polarization inversion time of ca. 25 μs suffices to null CH_2 signals. As an example,[2] Figure 2 shows ^{13}C spectra for form I of cortisone acetate (**A**). If 50 μs is chosen, the CH peaks are nulled and the CH_2 signals inverted.

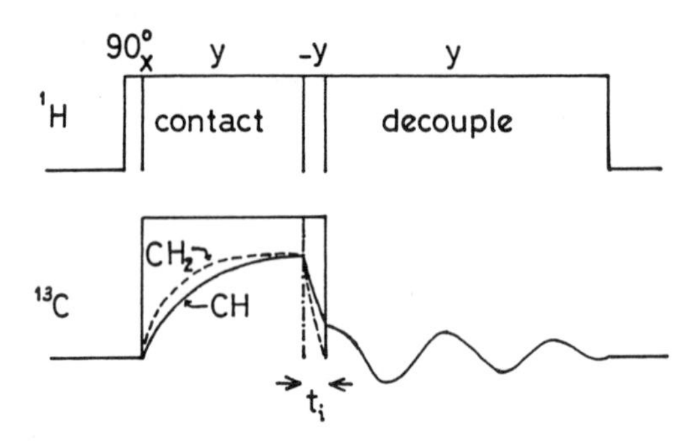

Figure 1 *The selective polarization inversion pulse sequence.*[1]

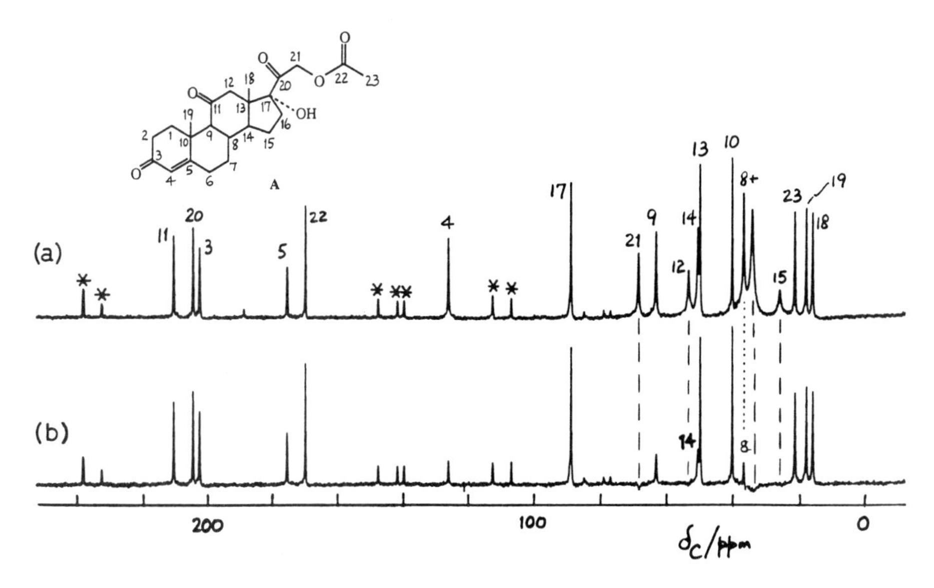

Figure 2 *(a) Normal and (b) methylene-suppressed 75.4 MHz ^{13}C CPMAS proton-decoupled spectra*[2] *of solid cortisone acetate (form I), obtained using the selective polarization inversion pulse sequence of Figure 1 with polarization-inversion times of (a) zero and (b) 25 μs. Peaks marked by an asterisk in (a) are spinning sidebands.*

3 ROTATIONAL RESONANCE

For compounds with two non-equivalent nuclei of the same isotope, it is possible to selectively reintroduce dipolar coupling in a MAS experiment by setting the MAS rate equal to the chemical shift difference between the nuclei. The result consists of powder patterns with widths equal to $D/2\sqrt{2}$ (ignoring anisotropy in J), where D is the dipolar coupling constant. Figure 3 shows an example[3] - the ^{15}N resonance of doubly-labelled 5-methyl-2-diazobenzenesulphonic acid (**B**) (as the hydrochloride). Fast MAS gives linewidths of 12 Hz, whereas under rotational resonance the width of each resonance is 320 ± 4 Hz, which is consistent with a N-N distance of 1.110 ± 0.004Å.

4 INTERPLAY OF DIPOLAR AND QUADRUPOLAR COUPLING

When quadrupolar energies are significant with respect to Zeeman energies, the wave functions become of mixed character. This allows dipolar coupling to spin-$\frac{1}{2}$ nuclei to become secular and results in second-order energies which are not entirely averaged by MAS. In consequence, spin-$\frac{1}{2}$ spectra may contain extra shifts and/or splittings. Figure 4 shows[4] a case for the (^{13}C, $^{35/37}Cl$) spin pair which, at moderate to high magnetic fields, gives doublet splittings in ^{13}C spectra. Such splittings can be used to assign resonances, and also to derive dipolar coupling constants (D) if quadrupolar coupling constants (χ) are known, or v.v. In the case illustrated, the splittings can be accurately reproduced using χ = -67.8 MHz and C,Cl distances 1.77Å for C-2 and 2.74 Å for C-1, assuming axial symmetry for the electric field gradient (along the C-Cl bond).

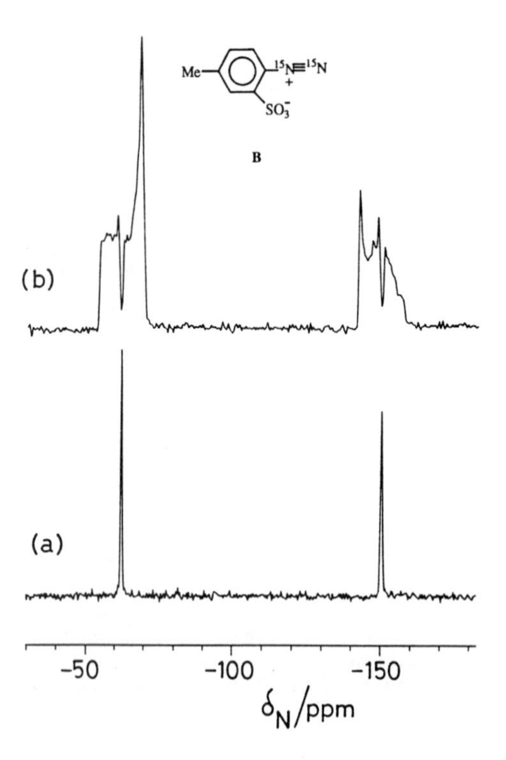

Figure 3 *Nitrogen-15 CPMAS proton-decoupled NMR spectra[3] at 20.3 MHz of solid 5-methyl-2-diazobenzenesulphonic acid (as the hydrochloride), doubly enriched with ^{15}N: (a) with fast MAS, and (b) with MAS rate 1786 Hz, equal to the chemical shift difference between the ^{15}N signals (rotational resonance condition).*

5 SPINNING SIDEBAND ANALYSIS

Spinning sidebands, separated from centrebands by multiples of the spinning rate, span the static spectrum. Under proton decoupling, this normally implies a dependence of the intensities of any spinning sideband manifold solely on the principal components of the relevant shielding tensor. Therefore computer analysis can be implemented to obtain the principal components. These are usually alternatively listed as the isotropic value, σ_{iso}, the anisotropy, ζ, and the asymmetry, η, defined as follows:

$$\sigma_{iso} = \tfrac{1}{3}(\sigma_{XX} + \sigma_{YY} + \sigma_{ZZ})$$

$$\zeta = \sigma_{ZZ} - \sigma_{iso}$$

$$\eta = (\sigma_{XX} - \sigma_{YY})/\zeta$$

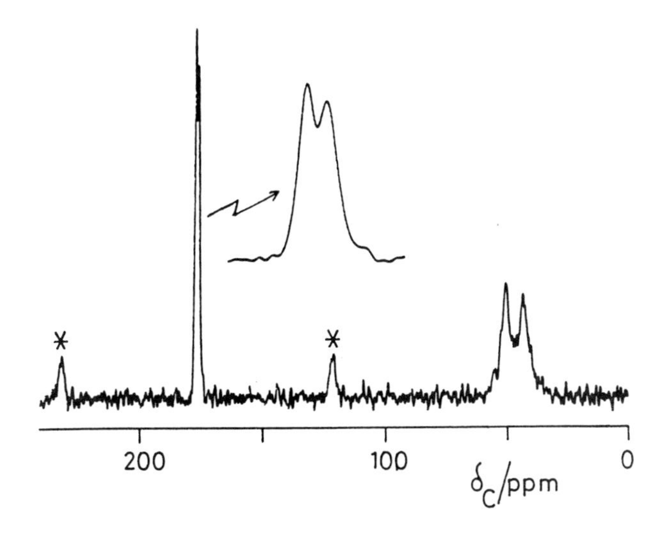

Figure 4 *Carbon-13 CPMAS proton-decoupled NMR spectrum*[4] *at 75.4 MHz of solid sodium monochloroacetate, illustrating residual splittings arising from* (^{13}C, $^{35/37}Cl$) *dipolar coupling. Peaks marked ssb are spinning sidebands.*

Whereas σ_{iso} is comparable to the value obtained by solution-state NMR, data on ζ and η are normally unobtainable for solutions. Figure 5 shows an example[5] of spinning sideband analysis for the ^{119}Sn nucleus in dimethylbis(2-pyridinethiolato-N-oxide)tin(IV) (**C**), resulting in δ_{iso} = -214 ppm, ζ = 570 ppm, η = 0.49. Such fundamental spectroscopic parameters can be of value for structure elucidation, and are sensitive tests of theory.

6. INTERPLAY OF SHIELDING, DIPOLAR COUPLING AND INDIRECT COUPLING

When heteronuclear coupling is involved as well as shielding anisotropy, analysis of spinning sideband patterns yields substantial data provided an isotropic indirect (**J**) coupling exists to differentiate shielding and dipolar interactions. Figure 6 shows[6] the ^{31}P spectrum of solid triphenylphosphine dimanganese nonacarbonyl, $Mn_2(CO)_9PPh_3$. The centreband and each spinning sideband is split into six peaks by coupling to ^{55}Mn ($I = \frac{5}{2}$). Detailed analysis of the sideband manifolds of each multiplet peak shows that the shielding anisotropy, $\sigma_{ZZ} - \sigma_{iso}$, is 111 ppm, with an asymmetry of zero. The multiplet splittings themselves are influenced by the quadrupolar effect discussed above, but it can be readily shown that $|J_{iso}|$ = 297 Hz, χ = 41.6 MHz and the anisotropy in J, ΔJ = 1027 Hz (assuming the more reasonable of two alternatives). Theoretical understanding of ΔJ is still lacking, partly because so few reliable data are available to date.

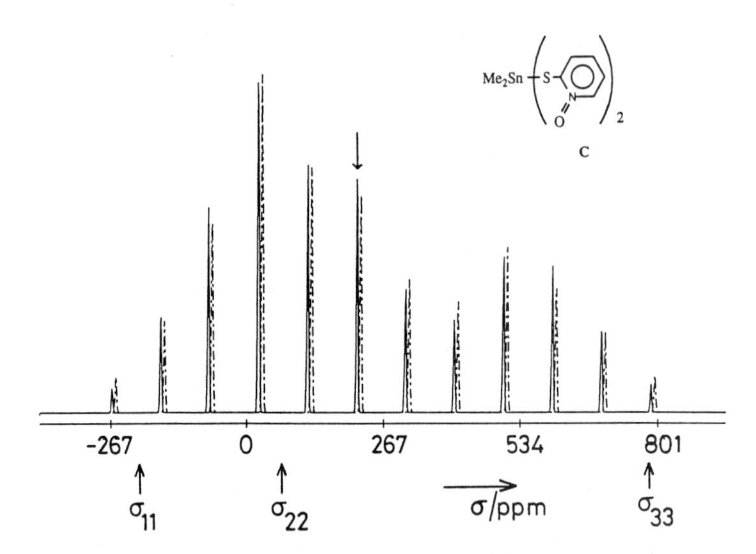

Figure 5 *Tin-119 CPMAS proton-decoupled NMR spectrum[5] at 111.9 MHz of solid **C**, illustrating spinning sideband analysis. The solid line is (schematically) the observed spectrum and the dashed line (slightly offset for convenience) the computer-fitted spectrum. The arrow indicates the centreband.*

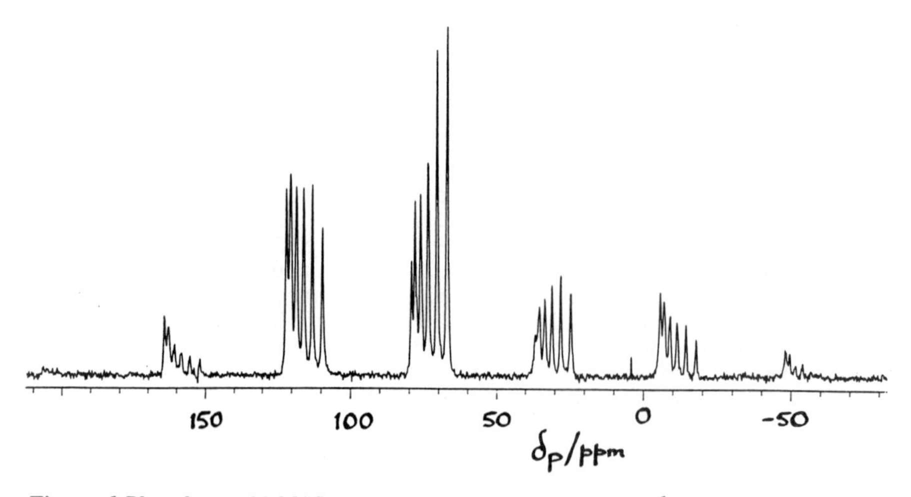

Figure 6 *Phosphorus-31 MAS proton-decoupled NMR spectrum[6] at 121.4 MHz of solid $(OC)_5MnMn(CO)_4PPh_3$, illustrating the unsymmetrical multiplet splittings and the variation in spinning sideband intensities of the multiplet components.*

7 TWO-DIMENSIONAL SPECTRA - EXSY

As for solution-state spectra, bandshapes for solid-state NMR can yield information on chemical exchange processes such as internal rotation. Two-dimensional spectra can be vital in understanding such processes. For example, it is now known that many tin compounds with trigonal bipyramidal coordination can undergo rotation about the axial bonds in the solid state. This occurs for three-dimensional coordination polymers of the type $[(Me_3E)_4M(CN)_6]_\infty$, where E = Sn or Pb and M = Fe, Os or Ru. The methyl groups are equatorial, with nitrogen atoms (of cyanide groups which link E to the metal M) axial. However, the asymmetric unit contains two such Me_3Sn or Me_3Pb groups, so the low-temperature ^{13}C spectra show six CH_3 resonances. The EXSY experiment,[7] illustrated[8] in Figure 7, is essential to determining which three signals belong to the same Me_3E group. Once this is established, bandshape calculations on the one-dimensional spectra lead to data for the thermodynamic parameters of the exchange process.

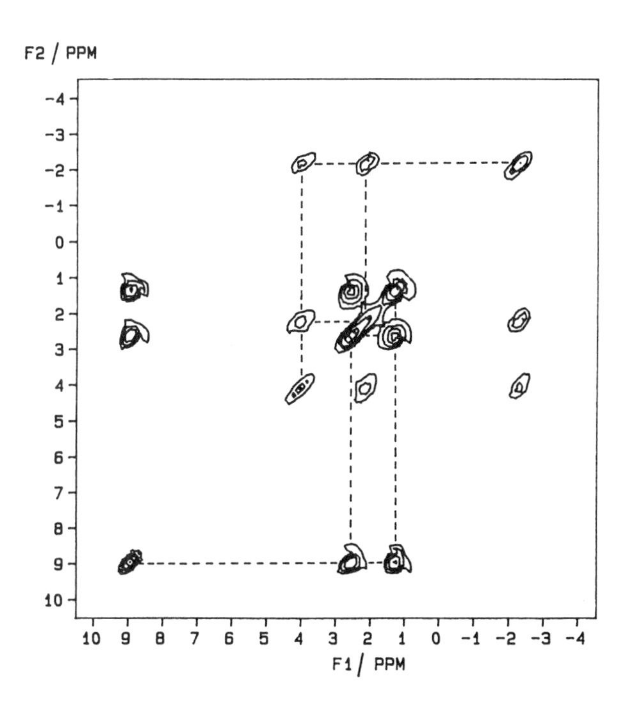

Figure 7 *Carbon-13 proton-decoupled two-dimensional exchange spectrum[8] (EXSY) at 75.4 MHz, obtained with CPMAS, of $[(Me_3Sn)_4Ru(CN)_6]_\infty$. The off-diagonal signals link methyls belonging to the same Me_3Sn group as indicated by the dashed lines.*

8 TWO-DIMENSIONAL SPECTRA - HETCOR

Cross-polarization between, say, ^1H and ^{29}Si, occurs via dipolar coupling and therefore is only feasible when the two nuclei are relatively close in space and the molecular fragments are rather rigid. It is often important to know which protons are active in the CP process, though such questions are only meaningful when MAS suffices to give relatively high-resolution proton spectra. Such a situation occurs for the layered system sodium octosilicate, $Na_2O.8SiO_2.9H_2O$ (at least when rather dry), which is of unknown structure. The proton spectrum contains signals at δ 3.6 (water) and δ 15.9 (strongly hydrogen-bonded OH). The ^{29}Si spectrum has resonances from silicon sites linked by three siloxane bridges (Q^3, δ -100.1) and four siloxane bridges (Q^4, δ -110.9). The selective cross-polarization two-dimensional[9] spectrum shown[10] in Figure 8 indicates clearly that it is almost entirely the H-bonded hydrogen that results in CP to all silicon sites. Such experiments assist in building a structural model of octosilicate.

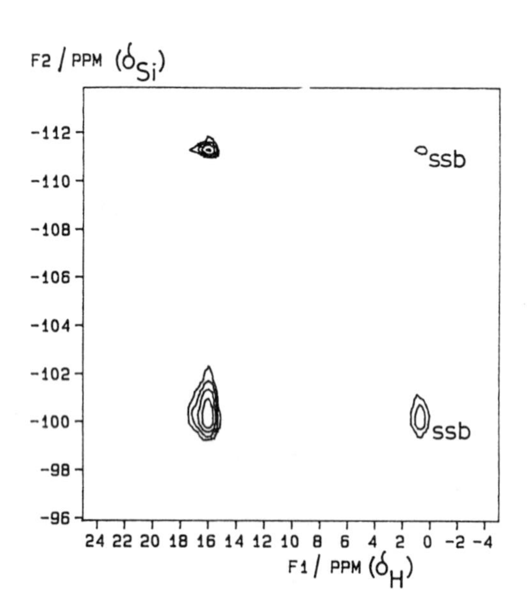

Figure 8 *Two-dimensional (^{29}Si, ^1H) heteronuclear correlation (HETCOR) spectrum[10] of dried sodium octosilicate, obtained with CPMAS at 7.05 T. Peaks marked ssb are spinning sidebands.*

9 ISOTOPIC ENRICHMENT - BORON CARBIDE

The structure of B_4C has long been a puzzle. Although X-ray studies show it consists of icosahedral units and linear three-atom groups, it has not been clear where the carbon atoms are. Carbon-13 MAS NMR provides the answer, but because of the long relaxation time for ^{13}C, isotopic enrichment is highly desirable. Figure 9 shows[11] the spectrum, which proves that B_4C is actually $[B_{11}C][CBC]$. Most of the carbon in the icosahedral unit is probably at the polar sites, but the smaller peak at δ = 101.3 may arise from equatorial sites. At the time we were recording such spectra, similar results were reported by Kirkpatrick et al.[12]

10 CONCLUSION

Modern high-resolution NMR of solids can involve a variety of sophisticated techniques which yield detailed information not only about solid-state chemical structure and geometry but also about fundamental spectroscopic parameters, including their orientation-dependence.

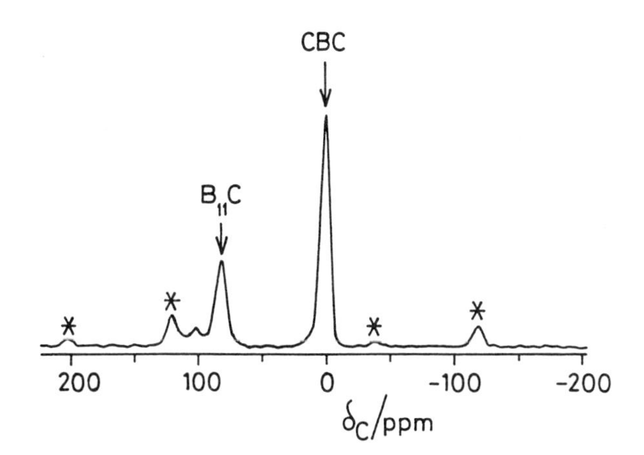

Figure 9 *Carbon-13 MAS spectrum[11] at 75.4 MHz of boron carbide, B_4C, enriched in ^{13}C. Spinning sidebands are indicated by asterisks.*

11 ACKNOWLEDGEMENTS

I thank the many research colleagues who have collaborated in the work reported here, and the SERC for financial support, including use of the National Solid-state NMR Service based at Durham.

References

1. X. Wu, S. Zhang and X. Wu, *J. Magn. Reson.*, 1988, **77**, 343.
2. D.C. Apperley, R.K. Harris and R.R. Yeung, unpublished work.
3. R. Challoner and R.K. Harris, *Chem. Phys. Letters*.
4. S.H. Alarcón, A.C. Olivieri, S.A. Carss and R.K. Harris, *Angew. Chem. (Internat.)*, in press.
5. R.K. Harris and V.G. Kumar Das, unpublished work.
6. R. Gobetto, R.K. Harris and D.C. Apperley, *J. Magn. Reson.*, 1992, **96**, 119.
7. N.M. Szeverenyi, M.J. Sullivan and G.E. Maciel, *J. Magn. Reson.*, 1982, **47**, 462.
8. D.C. Apperley, N.A. Davies, R.K. Harris, S. Eller, P. Schwarz and R.D. Fischer, *J.C.S. Chem. Comm.*, 1992, 740.
9. A.J. Vega, *J. Amer. Chem. Soc.*, 1988, **110**, 1049.
10. G.G. Almond, R.K. Harris and P. Graham, *J.C.S. Chem. Comm.*, 1994, 851.
11. D.C. Apperley, R.K. Harris, R.J. Oscroft and D.P. Thompson, unpublished work.
12. R.J. Kirkpatrick, T. Aselage, B.L. Phillips and B. Montez, AIP Conf. Proc. **231** "Boron-rich Solids", Eds. D. Emin, T.L. Aselage, A.C. Switendick, B. Morosin and C.L. Beckel, 1990, 261.

SFC/NMR On-line Coupling

Klaus Albert and Ulrich Braumann

INSTITUT FÜR ORGANISCHE CHEMIE, AUF DER MORGENSTELLE 18, D-72076 TÜBINGEN, GERMANY

1. INTRODUCTION

The direct *on-line* coupling of High Performance Liquid Chromatography (HPLC) and Nuclear Magnetic Resonance (NMR) Spectroscopy has proved to solve analytical problems in pharmaceutical, polymer and biomedical research (1-6). Despite all progress in the performance of the HPLC-NMR interface (e. g. the probe), the solvent signals and the signals of the impurities associated with the commonly used HPLC eluents remain problems with this hyphenated technique. Up to this point, the manufacturers of the HPLC eluents guarantee the UV-purity of these solvents, whereas the absence of ^{1}H-NMR-Signals of impurities is not checked.

One possibility to eliminate these problems is to resort to a chromatographic separation technique which has the equivalent separation power of HPLC but does not use proton containing eluents. These theoretical requirements are ideally fit by Supercritical Fluid Chromatography (SFC) with supercritical carbon dioxide as eluent. The feasibility of direct SFC-NMR coupling remains a question. Therefore, first attempts were performed to directly couple a SFC instrument together with an NMR spectrometer using a pressure-proof NMR detection cell.

2. SFC PROBE

For direct HPLC-NMR coupling, the measurement configuration consists of a vertically fixed, non-rotating glass tube with internal diameters of 2, 3 or 4 mm, tapering at both sides to the outer diameter of the inlet and outlet Polyetheretherketone (PEEK) tubing. The NMR detection coil is directly fixed to the glass cell. The entire measurement device is located within a glass dewar, in which a thermocouple is inserted, enabling exact temperature control of the measurement region. This measurement configuration was modified for SFC-NMR-application (7). The glass-tube was substituted with a sapphire tube (o.d. 5mm, i. d. 3mm), resulting in a detector volume of 120 μl. In addition, the PEEK tubing was replaced by Titan tubing.

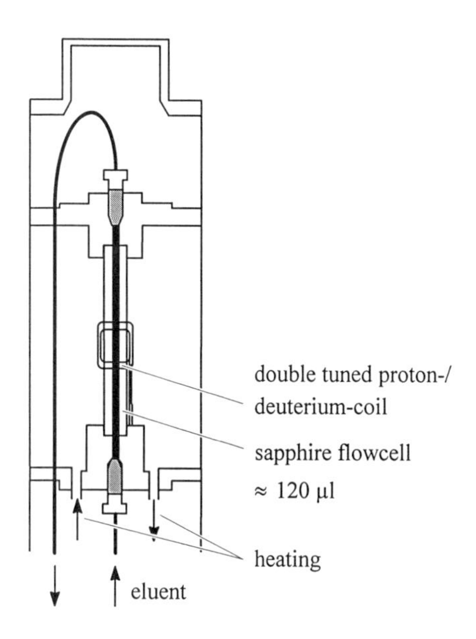

Fig. 1: Schematic diagram of the SFC-NMR probe.

3. SFC-NMR ON-LINE COUPLING

A Hewlett Packard Supercritical Fluid Chromatograph G1205A was located 1.5 m from the 9.4 T magnet of a Bruker ARX 400. The outlet of a pressure stable DAAD UV detector of the SFC instrument and the inlet of the SFC probe were connected by stainless steel capillaries (i. d. 0.2 mm). To maintain supercritical conditions the sapphire cell within the probe was kept at a temperature of 321 K using preheated air.

3.1 ^{1}H NMR spectra under supercritical conditions

The effect of flow rate on ^{1}H NMR signals in the liquid state has been well studied (4), however, to our knowledge, no investigations have been performed considering flowing supercritical eluents. The basic design of the sapphire measurement configuration exhibited resolution values similar to those of conventional HPLC probe.

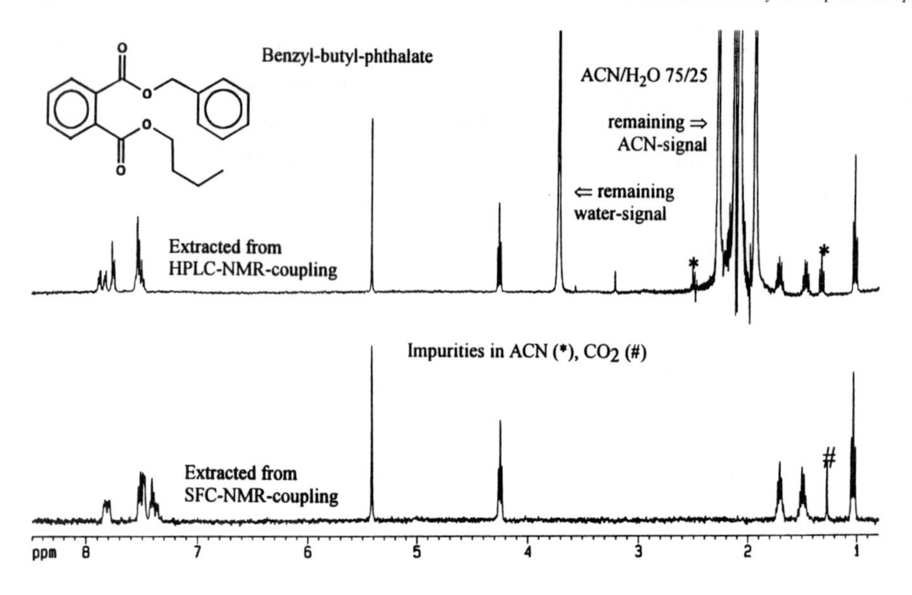

Fig. 2: ¹H-NMR spectra (400 MHz) of benzyl-butyl-phthalate.
a.) under HPLC conditions
b.) under SFC conditions

This behaviour is demonstrated in figure 2, showing the comparison of ^{1}H-NMR-spectra of benzyl-butyl-phthalate under HPLC conditions in the liquid state (flow rate 0,3 ml/min, pressure 1 bar, temperature 303 K) and under SFC conditions in the supercritical state (flow rate 2,0 ml/min, pressure 165 bar, temperature 321 K).

Taking the signals of the butyl chain as an indicator, it is evident that no degradation of the NMR resolution in the supercritical state can be observed. Because NMR signal line widths are reciprocal to the spin-spin relaxation time, T_2, it is evident that T_2 does not change dramatically in the supercritical state. This is not the case with respect to the spin-lattice relaxation time T_1.

Figure 3 shows the stacked plot of the determination of ^{1}H-T_1 values of butyl-benzyl-phthalate in the supercritical state (250 bar, 321 K), using the inversion recovery method. Values in the order of 20 s result for the aromatic proton signals, whereas those on the order of 8-9 s indicate the ester, methylene and methyl proton signals. These values are two to three times higher than the corresponding values in the liquid state at the same temperature. This phenomenon can be explained by the decreased viscosity in the supercritical compared to the liquid state.

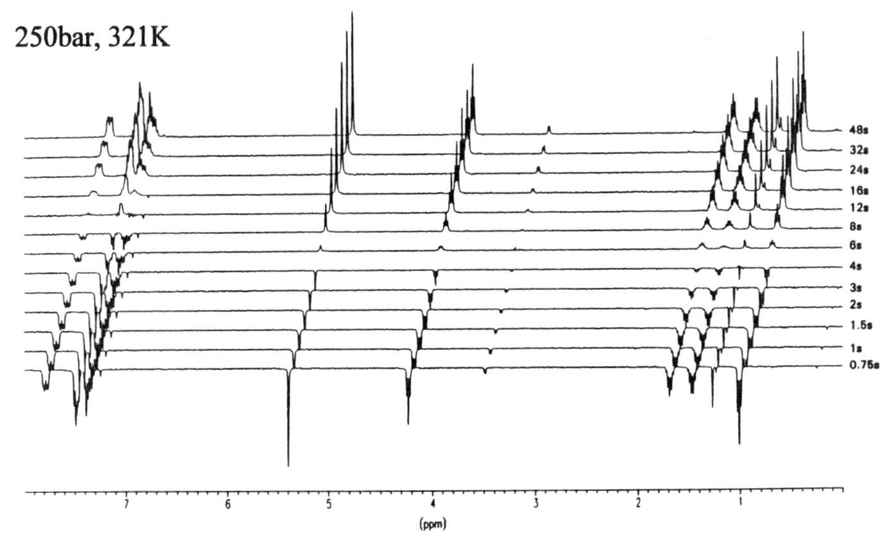

Fig. 3: Inversion-recovery spectra (400 MHz) of benzyl-butyl-phthalate in the supercritical state (250 bar, 321 K).

Separations in the supercritical state are usually performed using a pressure gradient. Figure 4 shows the signal of chloroform in the supercritical CO_2, recorded in a time increment of 12.19 s with a pressure gradient from 90 bar to 244 bar at 325 K. It is evident that due to the increasing density with increasing pressure the ^1H NMR signals shift to higher field. This behavior necessitates small pressure changes during the acquisition of ^1H NMR spectra, otherwise a signal line broadening, that is dependent onthe pressure gradient, would result. Taking into account a pressure gradient of 115 bar to 180 bar within 20 min, actually used during a separation, up to 16 scans can be co-added introducing a negligible line broadening.

The drift of the signal, as it can be seen in the contourplot, is in a first estimation proportional to the calculated density of the CO_2. The two steps are introduced by imperfect temperature control during the first experiments.

3.2 *On-line* SFC-NMR separations

This is proved in Figure 5, showing the continuous-flow ^1H NMR spectra of a plastifier separation in the supercritical state. The separation was performed with an analytical RP 18 column (LiChrosphere RP Select B 5μ, 250 x 4.6 mm).

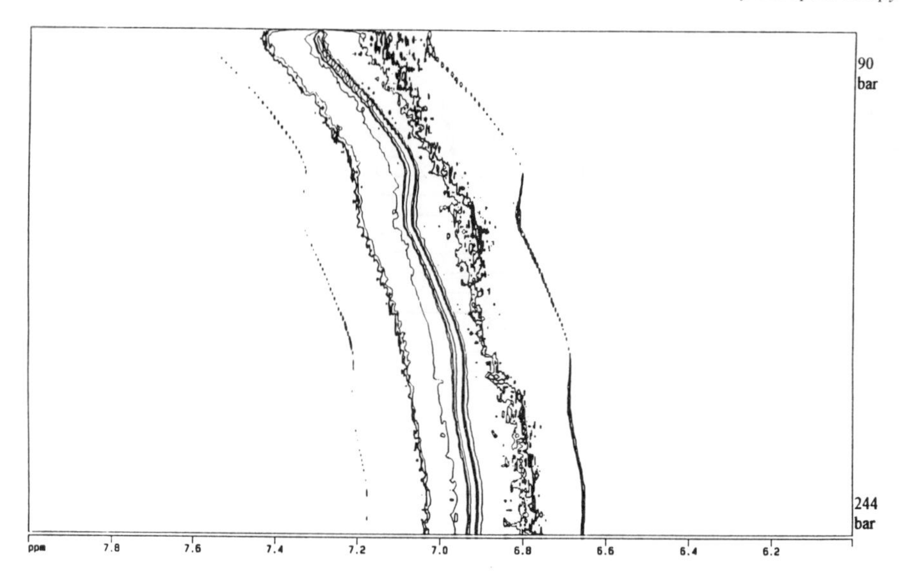

Fig. 4: Contour plot (400 MHz) of the signal of chloroform in the supercritical state at pressure gradient from 90 to 244 bar.

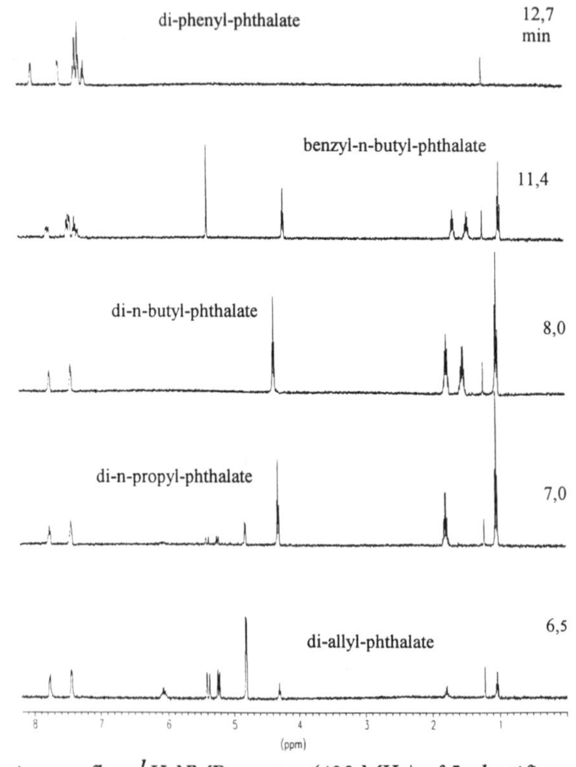

Fig. 5: Continuous-flow ^{1}H-NMR spectra (400 MHz) of 5 plastifiers recorded during a separation in the supercritical state.

The resolution of all five spectra enables the determination of coupling constants down to a value of 1.4 Hz.

3.3 *Stopped-flow* measurements

Signal assignment in the routine NMR spectroscopy is usually facilitated by the application of 2-D assignment techniques such as homonuclear $^1H^1H$ correlated spectroscopy ($^1H^1H$ COSY).

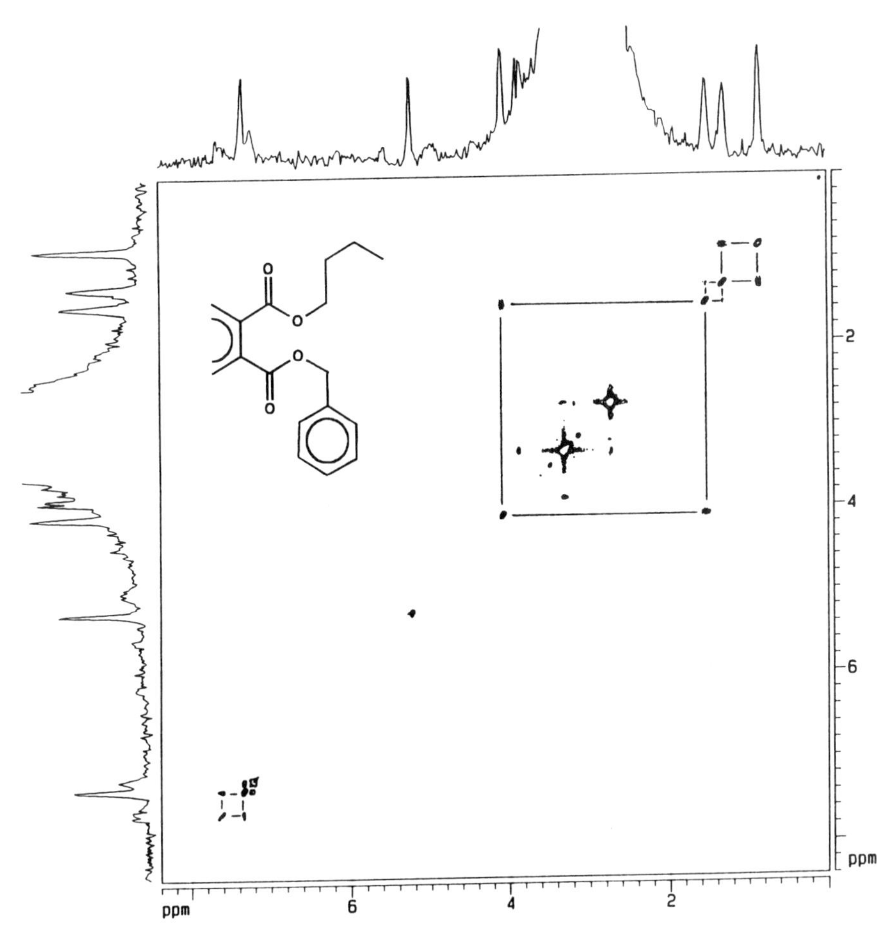

Fig 6: Contour plot (400 MHz) of a ^1H homo nuclear shift correlated experiment (COSY) of benzyl-butyl-phthalate in the supercritical CO_2 with 10% methanol. Pressure 154 bar, temperature 321 K.

Figure 6 shows the contour plot of one constituent of the phthalate separation. Here the dead volume betwen the UV-detector and the SFC flow cell was determined before separation. After an adequate delay after the occurrence of the UV maximum of benzyl-butyl-phthalate in the UV-detector, the SFC separation was stopped and the 2D acquisition was started. The pressure proved to be stable for several hours, which was sufficient for the acquisition of the 2-D COSY spectrum. Despite the intense signal of undeuterated methanol, which was used as a modifier in the SFC separation, the ^{1}H-NMR signal connectivities are clearly visible. The corresponding cross peaks in the aromatic system and the aliphatic part of the butyl chain are indicated in the contour plot. This example proves that even under the exotic supercritical fluid conditions, elaborate 2D assignment techniques can be used.

4. SUMMARY

The direct coupling of supercritical fluid chromatography and Proton Nuclear Magnetic Resonance spectroscopy is feasible. The advantage of NMR detection in the supercritical state is the absence of ^{1}H NMR signals of solvents and impurities, the drawbacks of this hyphenated technique are the pressure dependence of NMR signals and the increased spin lattice relaxation times. Despite these problems the spectral quality obtained together with the capability to acquire 2-D NMR spectra may bring SFC-NMR-coupling to an established hyphenated technique. Furthermore the enormous application power of supercritical fluid extraction in combination with ^{1}H NMR detection will lead to new structural elucidation pathways in the search for new, natural occurring compounds which may be used in medical therapy.

Acknowledgement

The authors gratefully acknowledge the help of Li-Hong Tseng in spectrometer set up. They are indebted to Bruker company (Rheinstetten, Germany) for technical support and to Hewlett Packard (Waldbronn, Germany) for the loan of an SFC instrument.

References

1 H. C. Dorn, *Anal. Chem.* 1984, **56**, 747.

2 D. A. Laude, Jr. and C. L. Wilkins, Trends *Anal. Chem.* 1986, **9**, 231.

3 K. Albert and E. Bayer, Trends *Anal. Chem.* 1988, **7**, 288.

4 K. Albert, Habilitationsschrift, Tübingen University, 1988.

5 K. Albert and E. Bayer, 'HPLC Detection Newer Methods', (G. Patonay Dd.), VCH Publishers New York, 1992, p. 197.

6 M. Spraul, M. Hofmann, P. Dvortsak, J. K. Nicholson, I. D. Wilson, *Anal. Chem.* 1993, **65**, 327.

7 K. Albert, U. Braumnn, L.-H. Tseng, G. Nicholson, E. Bayer, M. Spraul, M. Hofmann, C. Dowle and M. Chippendale, *Anal. Chem.* 1994, **66**, in press.

Structural Determinations of Organic Compounds Found in the Environment by NMR Spectroscopy and Mass Spectrometry

S. T. Belt, D. A. Cooke, S. J. Hird, E. J. Wraige, P. Donkin,[1] and S. J. Rowland

DEPARTMENT OF ENVIRONMENTAL SCIENCES, UNIVERSITY OF PLYMOUTH, PLYMOUTH, DEVON PL4 8AA, UK

[1] PLYMOUTH MARINE LABORATORY, PROSPECT PLACE, WEST HOE, PLYMOUTH, DEVON PL1 3DH, UK

1 INTRODUCTION

This paper describes the application of NMR spectroscopy and mass spectrometry to the structural determination of some organic compounds found in the environment.

The first of these is a simple C_{11} branched hydrocarbon (4-propyloctane) that has been synthesised in the laboratory. This compound is a constituent member of a much larger set of unresolved hydrocarbon mixtures (UCMs) which are found in refined and crude oils.[1,2] UCMs represent a high environmental burden of pollutant hydrocarbons but are not generally toxic (however few tests on low molecular weight UCMs have been made).[3]

The second example represents the characterisation of a commonly occurring highly branched isoprenoid C_{25} diene obtained directly from a sediment, and represents an important study of this type. Since the origins of compounds of this classification have recently been shown to come from diatomaceous algae,[4] their importance as palaeoenvironmental biomarkers is likely to increase.

2 EXPERIMENTAL

2.1 Toxicity Roles of 4-propyloctane

Evaluating relevant toxicity data for individual hydrocarbons found in pollutants remains a major challenge largely due to the unresolved nature of the hydrocarbon mixtures by current analytical techniques. Increased resolution of these unresolved complex mixtures (UCMs) of hydrocarbons can be achieved via chemical modification such as oxidation. Although complete resolution is not yet achievable via this method, the identification of some components can be used to trace and identify some of the original unresolved hydrocarbons. UCMs are known to be toxic to simple marine bivalves including mussels (*M.Edulis*) when UCMs are oxidised, and some correlation between individual components and toxicity has been established.[5] Straight chain n-alkanes appear to diminish in toxicity approximately between decane and undecane (ie C_{10} - C_{11}). However, since branched isomers of compounds of this molecular weight are likely to be more soluble than their straight chain analogues, their identification in UCMs suggests that toxicological roles should also be investigated

2.1.1 *Synthesis and Toxicity of 4-propyloctane.* 4-propyloctane was synthesised by initially reacting 1-bromopropane with ethylpentanoate via a Grignard reaction to yield 4-

propyloctan-4-ol. This was dehydrated with P_2O_5 to give a mixture of alkene isomers which were subsequently hydrogenated using H_2 and a hexane suspension of $PtO_2.H_2O$. Argentation TLC yielded the pure product which was subjected to MS and NMR analysis. The successful synthesis of a $C_{11}H_{24}$ hydrocarbon was verified via the following data. The ^{13}C NMR spectrum of 4-propyloctane shows 8 aliphatic carbon resonances as expected from the symmetry of the molecule. Analysis via the DEPT sequence reveals the presence of 2 methyl, 5 methylene and a single methine carbon as required. In the ^1H spectrum, each methyl group appears as a triplet while the methylene and methine protons are unresolved due to overlap. Noticeably, the lowest field carbon resonance arises from the methine carbon at the branch point. The MS of 4-propyloctane shows the molecular ion ($M^{+\cdot}$ = 156) and fragments corresponding to α-cleavage about the branch point ($C_4H_9^+$ (m/z 57) and $C_3H_7^+$ (m/z 43)).

Following the successful synthesis and characterisation of 4-propyloctane, early studies carried out with mussels have demonstrated that this compound is toxic during the first 48 hours of exposure (as evidenced from clearance rates, CR). We also have some preliminary findings that the bioaccumulation-toxicity relationships may show marked dependency on exposure time for 4-propyloctane.

2.2 Structure of 2,6,10,14-tetramethyl-7-(3-methylpent-4-enyl)pentadec-5-ene: a Highly Branched C_{25} Sedimentary Isoprenoid Biomarker

C_{20}, C_{25} and C_{30} hydrocarbons with a range of degrees of unsaturation are commonly found in marine sediments.[6] These highly branched compounds may be envisaged to be oligomers of isoprene units although the biosynthetic origin is unknown. Until recently, interest in highly branched isoprenoid hydrocarbons (HBIs) has been largely academic though it has been postulated that their occurrence may be useful guides to the presence of other organic matter in sediments (ie as biomarkers).[7,8] Support for this has recently been demonstrated and it has been shown that these compounds can be synthesised by a number of diatomaceous algae.[4] Despite the widespread occurrence of HBIs and the associated interest in them as biomarkers, a detailed structural characterisation of these compounds remains to be achieved. To date, characterisation has been established largely by mass spectrometry in order to determine degrees of unsaturation (though this has involved ambiguity on occasion due to the possibility of cyclic compounds) coupled with hydrogenation to the parent alkanes (comparison with synthesised compounds) to establish the basic carbon skeleton.[7,9,10] Other chemical modifications including micro-ozonolysis have also been employed for some examples. The application of NMR spectroscopy to the structural characterisation of HBIs has essentially been limited to a recent study of a previously unreported compound.[11] Of the more frequently reported HBIs, although their occurrence is widespread, their low concentrations have precluded any detailed analysis via NMR spectroscopy. In this current study, a complete structural characterisation via ^1H, ^{13}C NMR and MS of a HBI diene with a common C_{25} skeleton is described.

2.2.1 *Isolation and Identification of a HBI C_{25} diene from Caspian Sea sediments.* A colourless oil was isolated from recent benthic sediments of the Caspian Sea by solvent extraction with $MeOH/CH_2Cl_2$, column chromatography (hexane/silica) and argentation TLC. Identification of a C_{25} diene initially came from observation of a molecular ion (M^+) at m/z = 348. At this stage, the observed fragmentation pathway was consistent with several possible isomeric structures. Hydrogenation of this oil gave an alkane which had identical retention indices (on 3 different GC phases) to previously synthesised 2,6,10,14-

tetramethyl-7-(3-methylpentyl)pentadecane,[9] which enabled the basic skeleton of the diene, and in particular the location of the T-branch, to be established.

2.2.2 *NMR analysis of the HBI C_{25} diene.* A simple permutation treatment for a diene within an established C_{25} framework reveals that there are over 400 different structural permutations for the positions of the double bonds. Although many of these could be eliminated, a unique structural identification was not achievable from the MS data alone. It was therefore decided to consider a complementary method for analysis. A bulk extraction from the sediment enabled approximately 5 mg of diene to be isolated. This proved to be sufficient for both ^{1}H and ^{13}C NMR analysis and allowed the structure given in Figure 1 to be elucidated.

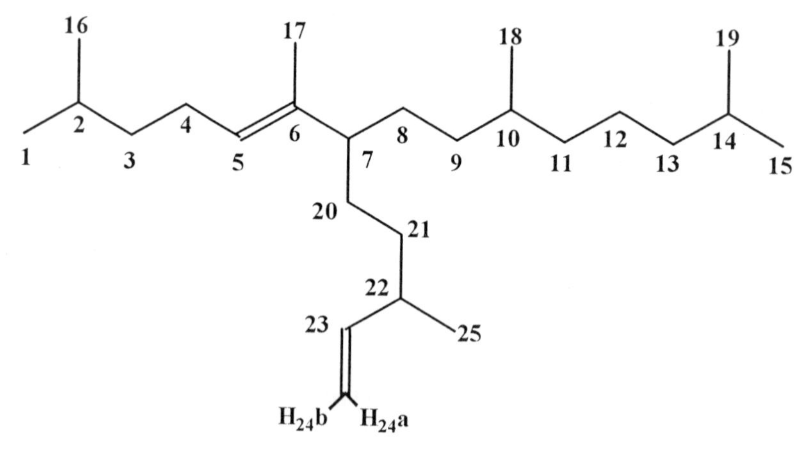

Figure 1. *Structure of the Highly Branched Isoprenoid C_{25} Diene*, 1

The ^{13}C NMR spectrum of the diene shows 25 individual carbon resonances which can be divided into 4 alkenic and 21 aliphatic type carbons as expected for a C_{25} diene with low symmetry. Analysis via the DEPT sequence establishes the number of methyl, methylene and methine carbons to be 7, 10 and 7 respectively with one additional quaternary carbon which is alkenic in nature. The observation of a methylenic sp^2 carbon established the presence of a terminal alkene in the molecule. Further, the location of this terminal double bond could be assigned by consideration of the ^{1}H spectrum. Three resonances at 5.65, 4.93 and 4.89 ppm arise from a vinyl functionality. The first of these which appears as a 7 line multiplet is particularly diagnostic and arises due to H-23 coupling to H-24a, H-24b and H-22. The expected 8-line multiplet (ddd) is not resolved at 270 MHz. Significantly, this region of the spectrum can be modelled extremely well by comparison with the spectrum of 3-methylpent-1-ene. A consideration of the C_{25} skeleton enables the location of this double bond to be established without ambiguity. In particular, a single possibility for a vinyl group with only one associated allylic proton. The chemical shift of the allylic proton H-22 and its coupling to H-23 was established via decoupling experiments. In addition, the irradiation of H-22 resulted in the transformation of a doublet at 0.95 ppm to a singlet allowing an assignment for H-25. The remaining methyl groups in the molecule appear as doublets due to methine proton couplings except for a singlet at 1.42 ppm indicating the location of the second double bond to be at a branch point (H-17). This observation is also consistent with the presence of a quaternary alkenic carbon described earlier.

The remaining alkenic proton appears as a triplet (5.08 ppm). Since there is only a single proton resonance for this double bond, this coupling must arise from an allylic methylene group. By irradiating at 5.08 ppm, the position of the allylic resonance could be identified by observing the transformation of a quartet at 1.99 ppm to a triplet. A reciprocal decoupling experiment carried out on H-4 resulted in the observation of a singlet for the H-5 alkene resonance. The observation of a second double bond with a single proton coupling to an allylic methylene group can be satisfied by 6 other structural isomers to the one given above and are given in Figure 2.

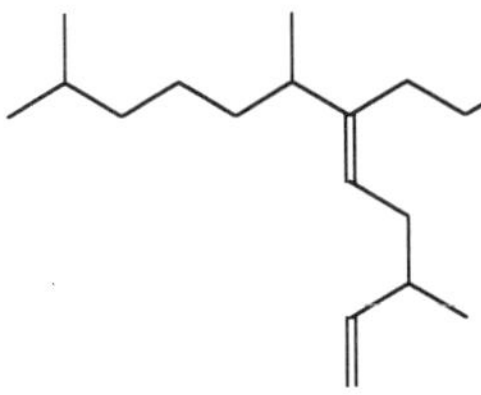

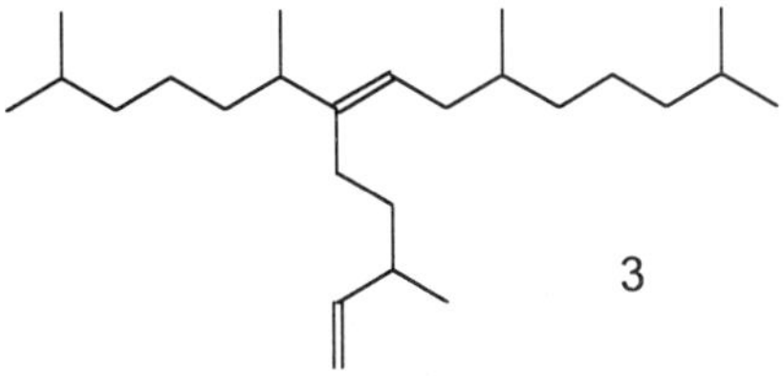

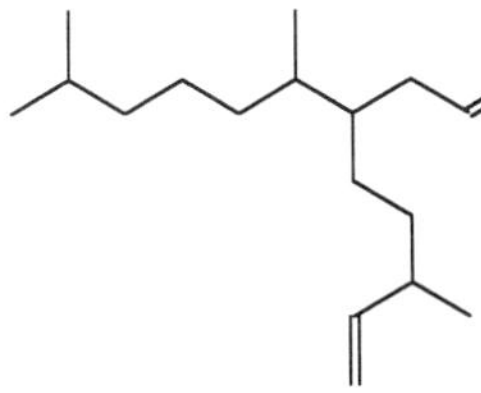

Figure 2. *Some structural isomers of the C_{25} diene*

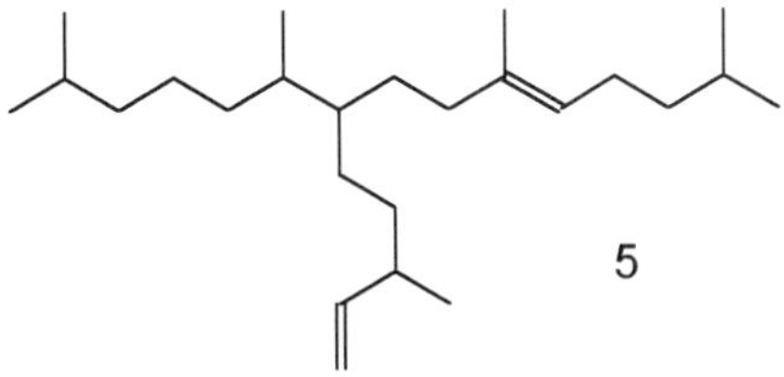

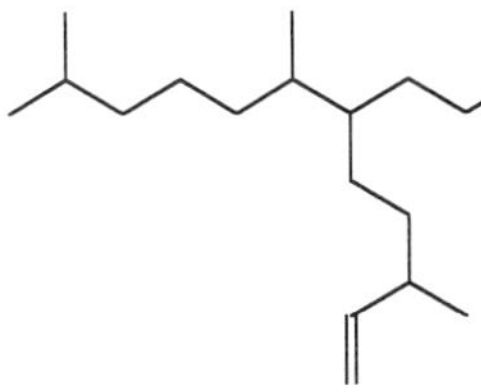

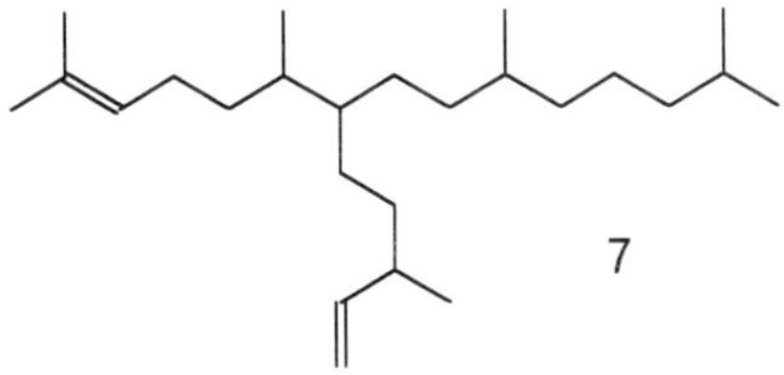

Isomer 1 can be verified as a single assignment according to the following criteria: (a) One (and only one) methyl singlet is observed which allows structures 2,3,6 and 7 to be eliminated. (b) A quartet for an allylic proton eliminates structures 2,3,4 and 7. Structure 5 would exhibit an allylic quartet but there would also be an allylic triplet. This is not observed. (c) Integration of the allylic methine and methylene resonances reveals 4 protons of these types. This can only be satisfied by structure 1. Decoupling observations can also be used to support these verifications.

The remaining allylic proton is located at the T-branch and appears as a multiplet at 1.8 ppm with couplings to H-20 and H-8. Accordingly, this resonance is unchanged by any of the decoupling experiments described. The corresponding ^{13}C nucleus resonates at 49.3 ppm as identified from a ^{1}H - ^{13}C COSY spectrum. This chemical shift is ca. 10 ppm to lower field compared with the other sp^3 hybridised carbons and is in accord with the findings of Summons *et al* for a related structure.[11]

2.2.3 *Analysis of the HBI C_{25} diene by mass spectrometry*

The structure of the HBI C_{25} diene 1 has also been verified by mass spectrometry. This analysis has been carried out on the parent diene, a chemically modified adduct following epoxidation and from ozonolysis products. These reactions together with the observed characteristic fragmentation pathways are shown in Figure 3.

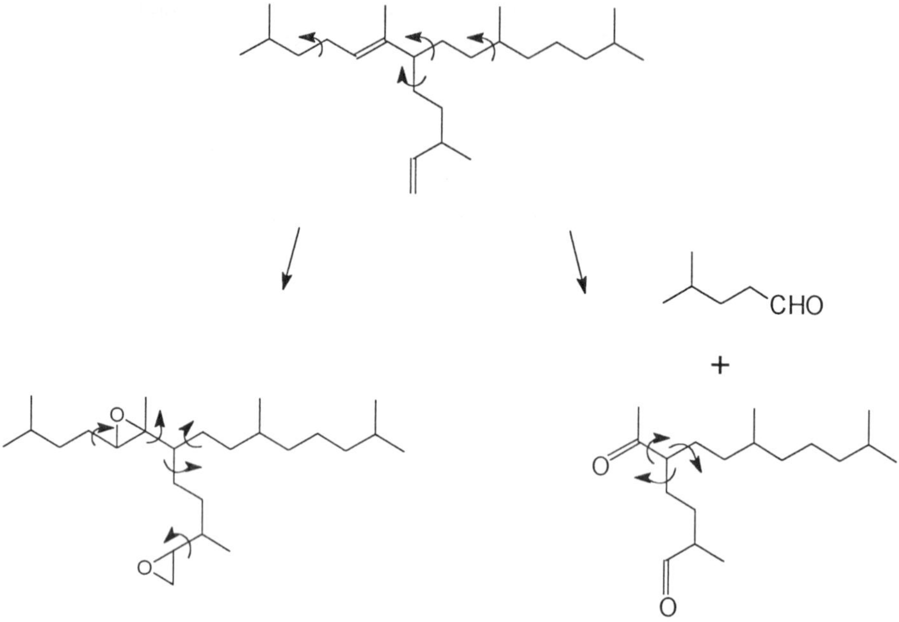

Figure 3 *Results from mass spectral analysis showing fragmentation pathways for 1 and its epoxidation (m-chloroperbenzoic acid) and ozonolysis products*

3 CONCLUSION

NMR spectroscopy has been used in conjunction with mass spectrometry to characterise the structures of 2 organic compounds found in the environment. The first of these, 4-propyloctane, which is thought to be a component in unresolved complex mixtures of hydrocarbons, has been independently synthesised and characterised. Preliminary toxicity studies demonstrate that mussels (*M. Edulis*) exhibit a time dependent clearance rate (CR) to this compound which emphasises the importance of exposure times when evaluating EC_{50} parameters. In contrast to this, the second example of a structural characterisation by these methods has been carried out on the highly branched isoprenoid C_{25} diene 1, which was obtained directly from the environment. This demonstrates that studies of this type can be carried out despite the relatively low concentrations of these compounds. Preliminary studies on more highly unsaturated C_{25} polyenes indicate that there may be common positions for the double bonds as described for 1.
We acknowledge Dr. Roger Evens for obtaining the NMR spectra , Mr R Srodzinski for help with the MS and M-Scan Ltd for the sediment samples.

References

1. S. Thompson and G. Eglinton, *Marine Pollution Bulletin*, 1978, **9**, 133.
2. M. A. Gough and S. J. Rowland, *Energy and Fuels*, 1991, **5**, 869.
3. J. W. Farrington, A. C. Davies, N. M. Frew and A. Knap, *Marine Pollution Bulletin*, 1988, **19**, 372.
4. J. K. Volkman, S. M. Barrett and G. A. Dunstan, *Org. Geochem.*, 1994, **21**, 407.
5. P. Donkin, J. Widdows, S. V. Evans, C. M. Worral and M. Carr, *Aquatic Toxicology*, 1989, **14**, 277.
6. S. J. Rowland and J. N. Robson, *Mar. Env. Res.*, 1990, **30**, 191.
7. D. A. Yon, G. Ryback and J. R. Maxwell, *Tetrahedron Lett.*, 1982, **23**, 2143.
8. R. Dunlop and P. Jefferies, *Org. Geochem.*, 1985, **8**, 313.
9. J. N. Robson and S. J. Rowland, *Nature*, 1986, **324**, 561.
10. J. N. Robson and S. J. Rowland, *Tetrahedron Lett.*, 1988, **29**, 3837.
11. R. E. Summons, R. A. Barrow, R. J. Capon, J. M. Hope and C. Stranger, *Aust. J. Chem.*, 1993, **46**, 907.

Solid State Proton NMR and Dynamic Mechanical Analysis Studies of Polymer Latex Blends

Roger Ibbett,[1] Martin James,[1,3] Doug Hourston,[2,4] and Ian Aucott[2]

[1] COURTAULDS RESEARCH AND TECHNOLOGY, PO BOX 111, COVENTRY CV6 5RS, UK

[2] THE POLYMER CENTRE, UNIVERSITY OF LANCASTER, BAILRIGG, LANCASTER LA1 4YA, UK

[3] PRESENT ADDRESS: PROCTOR AND GAMBLE LTD., EGHAM, SURREY, UK

[4] PRESENT ADDRESS: THE UNIVERSITY OF LOUGHBOROUGH, LOUGHBOROUGH, LEICESTERSHIRE, UK

1. INTRODUCTION

Polymer latexes are a class of materials which have wide ranging uses in many industrial fields. Their film forming properties mean they have important applications as ingredients in coatings formulations and in the manufacture of structural ingredients.

However, the emulsion polymerisation techniques required for the synthesis of latex particles with bi-component morphologies may be demanding[1]. In many cases it would be ideal if two single component latexes could be mixed together in the desired proportions and then directly used to make a solid. Such blending techniques have been developed to a large degree[2], and it has been shown that the resulting materials have properties which are similar to those made, for example, using shell/core particles.

This current study is aimed at characterising a series of solid latex blends, and at the same time assessing the usefulness of solid state proton NMR and Dynamic Mechanical Thermal Analysis techniques[3,4]. These should provide complementary information. Highly controlled polymer morphologies were obtained, which gave an unparalleled opportunity to consider some of the physics associated with experimental measurement[5].

2. EXPERIMENTAL

2.1 Synthesis

A Poly(ethylacrylate) latex was manufactured by emulsion polymerisation, with particles of a mean diameter of 100 nanometres [PEA(100)]. Poly(methylmethacrylate) and polystyrene latexes were similarly made, with 100 and 500 nanometre mean diameters [PMMA(100),PMMA(500),PS(100),PS(500)]. Pairs of latexes were mixed together in the following weight proportions, and solid blend were made by freeze/thaw water removal.

Table 1 *Solid Latex Blend Compositions (w/w)*

components	*compositions*						
PMMA(100):PEA(100)	100:0,	80:20,	70:30,	50:50,	30:70,	20:80,	0:100
PMMA(500):PEA(100)	"	"	"	"	"	"	"
PS(100):PEA(100)	"	"	"	"	"	"	"
PS(500):PEA(100)	"	"	"	"	"	"	"

2.2 Dynamic Mechanical Thermal Analysis

The solids were hot pressed into bars of approximately 2x1x0.2 centimetre dimensions, suitable for torsional clamp geometry. A Polymer Labs DMTA was used to measure tan (δ) over a temperature range from -30 to +150°C.

2.3 Solid State Proton NMR Relaxation Analysis

The freeze/thaw dried solids were measured using a modified Bruker SXP pulsed NMR spectrometer, at the ambient probe temperature of 30°C. T1 spin-lattice relaxation behaviour was determined using an inversion recovery pulse sequence and $T_{1\rho}$ rotating frame relaxation behaviour was determined using a variable spin-lock pulse sequence. Fitting of single or multiexponential relaxation times was carried out using a non-linear regression analysis computer program[5]

3. RESULTS

Transmission electron images of osmium tetroxide stained sections of typical solid latex blends are shown in figure 1.The rubbery component (PEA) forms the continuous phase. The glassy component (PMMA or PS) forms the discrete particulate phase.

The temperature at which the rubbery (PEA) component give a maximum in tan(δ) is known as its glass transition temperature ($T_{g(r)}$). The variations in $T_{g(r)}$ with blend composition are shown in figure 2 for the different latex pairs. The glass transition temperature of the glassy component ($T_{g(g)}$) did not change significantly with composition.

The extent of pair compatibility or interfacial mixing can be correlated with the height of tan(δ) midway between $T_{g(r)}$ and $T_{g(g)}$[6]. The variation with component composition is shown in figure 3.

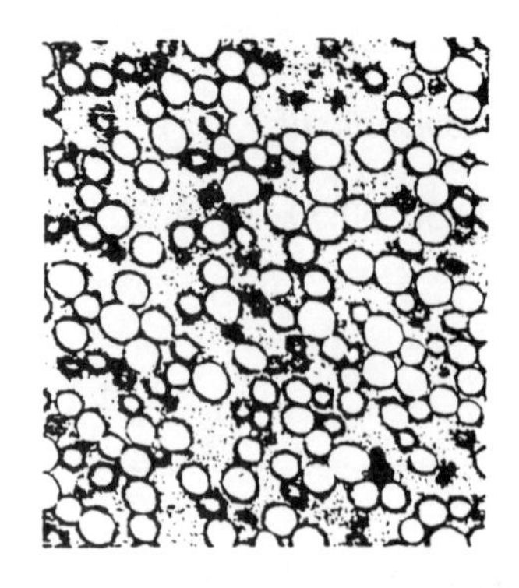

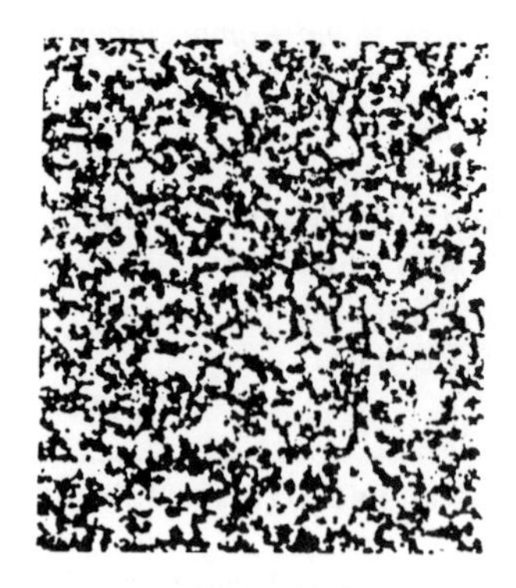

Figure 1 *TEM images of latex blends (*10,000), left; 50:50 glassy(100): rubbery(100), right; 50:50 glassy(500):rubbery(100).*

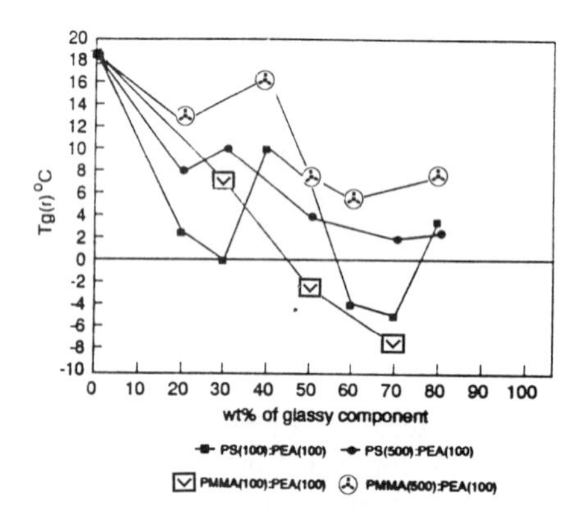

Figure 2 *DMTA Tg of rubber component of solid latex blends*

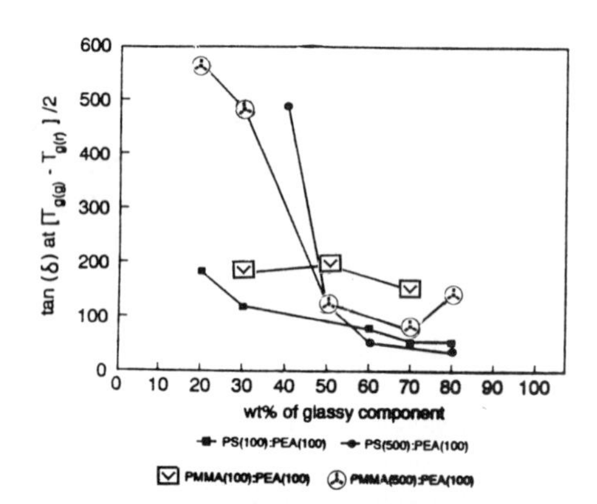

Figure 3 *DMTA tan(δ) at mid point between rubbery and glassy Tg's in solid latex blends.*

Solid state proton NMR relaxation is a measure of average molecular motion. Spin diffusion will operate across component boundaries and single exponential T_1 behaviour will be observed provided that diffusional distances are larger than the domain sizes. Figure 4 demonstrates the change in spin diffusion averaged T_1 with component composition for all latex pairs.

Proton $T_{1\rho}$ relaxation is faster than T_1 relaxation, so spin diffusion will occur over much shorter distances. Figures 5 and 6 show the variations in $T_{1\rho}$ with composition for a latex pairs, and it was found that bi-exponential behaviour was observed in all cases. The separate $T_{1\rho}$ times for each component were determined by computer fitting.

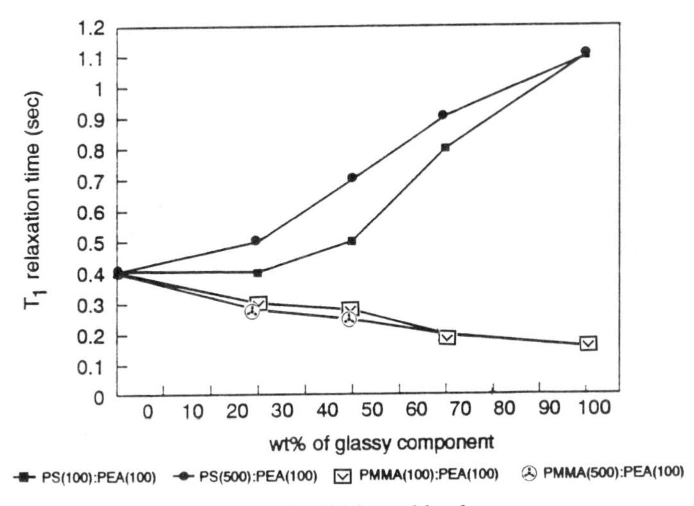

Figure 4 *Proton T_1 NMR analysis of solid latex blends.*

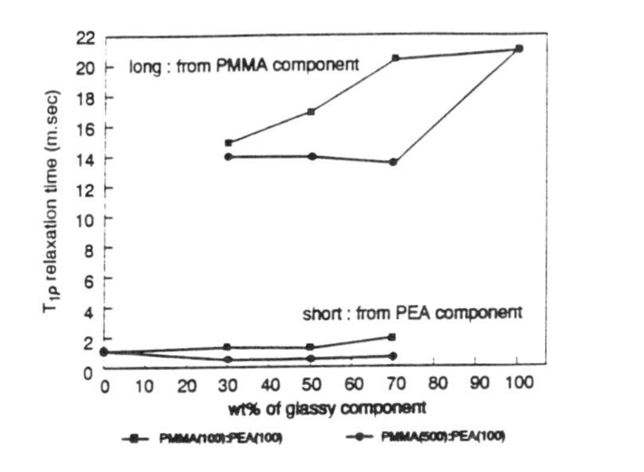

Figure 5 *Proton $T_{1\rho}$ NMR Analysis of PMMA:PEA solid latex blends*

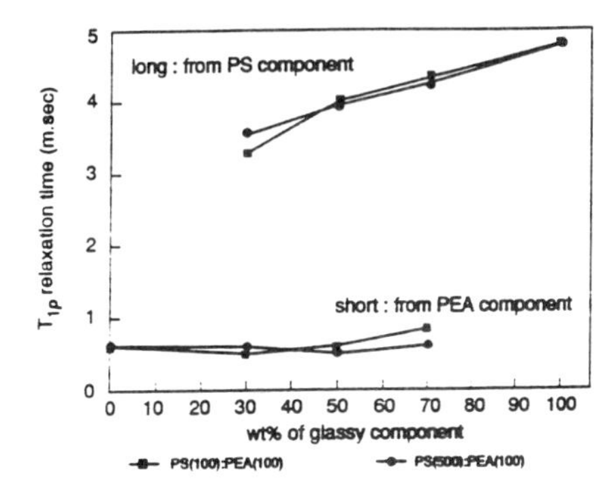

Figure 6 *Proton $T_{1\rho}$ NMR Analysis of PS:PEA solid latex blends*

4. DISCUSSION

Both the NMR and DMTA data indicated some form of interaction between the components of the latex blends. It was decided to carry out relaxation simulations using a one dimensional model, to investigate whether the NMR observed interactions might be the result of spin diffusion. T_1 and $T_{1\rho}$ simulations are shown for PMMA:PEA and PS:PEA in figures 7 and 8. The simulations show that T_1 can be averaged to a single exponential, despite the large domain sizes. Although the simulation only refers to a lamellar type structure, it seems reasonable to assume that the experimental spherical geometries would also give rise to similar averaging. The simulations also reveal that spin diffusion can influence the separate $T_{1\rho}$'s of the blend components, especially in the interfacial regions.

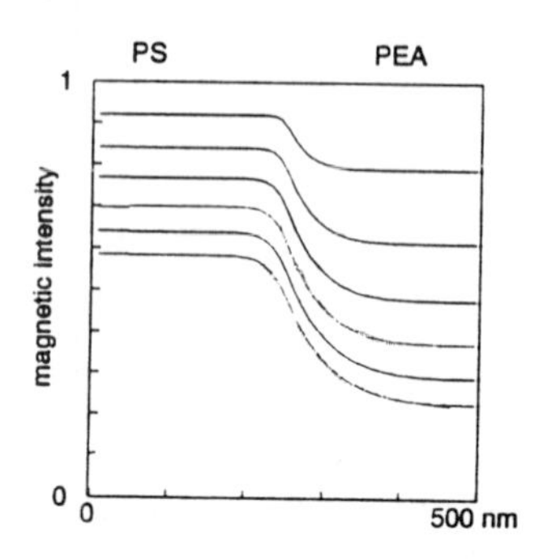

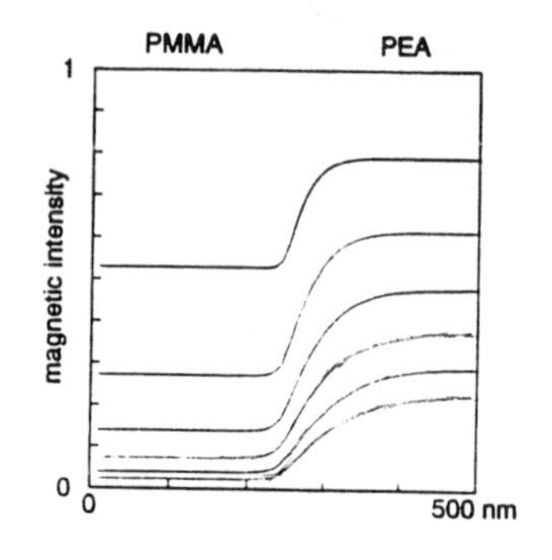

Figure 7 *Proton T_1 relaxation/spin diffusion simulation; One dimensional models of: left; PS:PEA 50:50 blend, right; PMMA:PEA 50:50 blend, sum of half widths = 500 nm, time step = 0.1 sec*

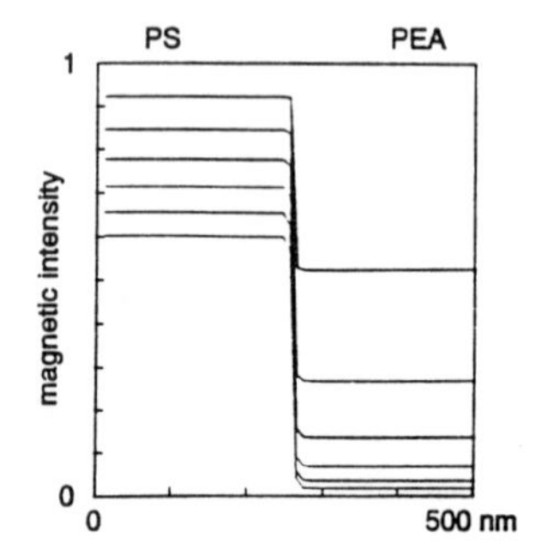

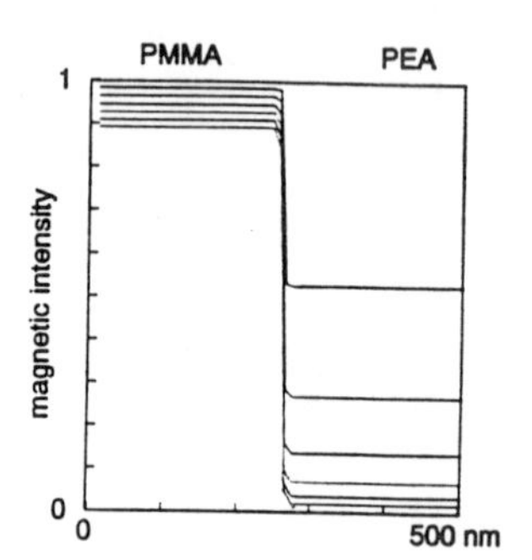

Figure 8 *Proton $T_{1\rho}$ relaxation/spin diffusion simulation; One dimensional models of: left; PS:PEA 50:50 blend, right; PMMA:PEA 50:50 blend, sum of half widths = 500 nm, time step = 0.4 msec*

The variations in $T_{g(r)}$ may indicate a long range mechanical interaction between the separate blend components. The extent of the observed effect may depend on the level of compatibility between the polymer types, component proportions, domain sizes and domain geometries. Alternatively, changes in $T_{g(r)}$ and tan(δ) may suggest physical interfacial mixing. The height of tan(δ) mid way between $T_{g(r)}$ and $T_{g(g)}$ may increase as the proportion of a mixed interface becomes larger. There is some indication that the effect is greater for blends with larger size glassy particles, especially when their proportion falls below 50% w/w.

References

1. J. W. Goodwin, J. Hearn, C. C. Ho, *Br Polymer J*, 1973, **5**, 347.
2. I Aucott, PhD Thesis, in progress.
3. R. E. Wetton and P. J. Corish, *Polymer Testing,* 1989, **8**, 303.
4. K. J. Packer and I. J. Colquhoun, *Br Polymer J.*, 1987, **19**, 151.
5. A. M. Kenwright, K. J. Packer, B. J. Say, *J. Mag. Res.* 1986, **69**, 426.
6. D. A. Greenhill, PhD Thesis, University of Lancaster, 1990.

Biological Applications

Speciation of Selenium in Human Serum by Size Exclusion Chromatography and Inductively Coupled Plasma Mass Spectrometry

Tonya M. Bricker and R. S. Houk

AMES LABORATORY-US DEPARTMENT OF ENERGY, DEPARTMENT OF CHEMISTRY, IOWA STATE UNIVERSITY, AMES, IA 50011, USA

1 INTRODUCTION

Since its development, inductively coupled plasma mass spectrometry (ICP-MS) has been a widely used analytical technique. ICP-MS offers low detection limits, easy determination of isotope ratios, and simple mass spectra from analyte elements. ICP-MS has been successfully employed for geological, environmental, biological, metallurgical, food, medical, and industrial applications.[1-3] One specific use important to many areas of study involves elemental speciation with ICP-MS as an element specific detector interfaced to liquid chromatography.[4] Elemental speciation information is important and cannot be obtained by atomic spectrometric methods alone which measure only the total concentration of the element present. This study describes the speciation of selenium in human serum by size exclusion chromatography (SEC) and detection by ICP-MS.

1.1 ICP-MS Overview

The inductively coupled plasma is an electrodeless discharge sustained by radio frequency power coupled through a load coil. The plasma is supported inside a torch made of concentric quartz tubes (see Figure 1). The sample is introduced as an aerosol through the central tube in a flow of argon. The temperature of the ICP is in the range of 6000 K to 8000 K.[1] This high temperature allows the introduced sample to be vaporized, atomized, and ionized. Most elements are ionized in the plasma with 90 to 100% efficiency.[1,2] However, certain elements with high ionization energies are ionized to a lesser degree. For example, selenium has an ionization energy of 9.75 eV and is only 33% ionized.

Due to the atmospheric pressure conditions in the plasma, the ions must be extracted into a vacuum system before mass analysis. Figure 1 shows a typical interface for the extraction of ions. The ions are sampled by a sampling orifice of about 1.0 mm diameter positioned in the normal analytical zone (see Figure 1). Ions and neutral gas flow into the first stage of the vacuum system where a supersonic jet is formed. The center portion of the jet flows through a second orifice called the skimmer. Behind the skimmer, several ion lenses focus the ions into a quadrupole mass analyzer .[1,2]

For most ICP-MS experiments a conventional pneumatic nebulizer is used to introduce the liquid sample into the plasma. The liquid sample is introduced through a narrow tube and shattered into droplets at the end by interaction with a flow of argon gas. To avoid solvent overload of the plasma, a spray chamber is typically employed to remove the large droplets. With this type of nebulizer all but 1-3% of the sample is lost to the drain. When coupled to liquid chromatography, the spray chamber also causes band

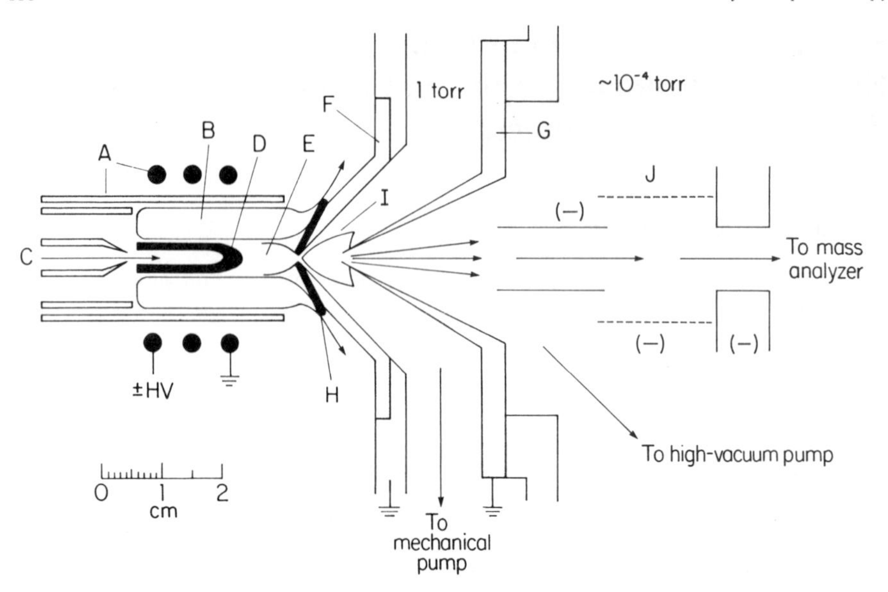

Figure 1. ICP and ion sampling interface. A = torch and load coil (HV = high voltage), B = induction region of ICP, C = a solution aerosol being injected into axial channel, D = initial radiation zone, E = normal analytical zone, F = nickel cone with sampling orifice in tip, G = skimmer cone, H = boundary layer of ICP gas deflected outside sampling orifice, I = expanding jet of C gas sampled from ICP, and J = ion lens elements. Reproduced from reference (1) with permission.

broadening due to its large dead volume. For chromatographic separations, the direct injection nebulizer (DIN) is a more attractive sample introduction system for ICP-MS. The design and construction of the DIN have been described elsewhere.[5] The DIN is essentially composed of a fused silica capillary inside a ceramic support rod with a metal tip at its end. The liquid is introduced through the capillary while an argon gas flow is introduced into the ceramic rod allowing a spray of droplets to be created pneumatically. The DIN is directly inserted inside the ICP torch and therefore all of the sample reaches the ICP.

1.2 Speciation of Selenium

Forms of selenium that occur in living organisms include the low-molecular-weight compounds (SeO_3^{-2}, SeO_4^{-2}, $(CH_3)_3Se^+$), selenocysteine, selenohomocystine, selenomethionine, dimethylselenide, selenotaurine, and the enzymes glutathione peroxidase, formate reductase, and glycine reductase. The nutritional value of selenium and its absorption from the gastrointestinal tract vary with chemical form and the amount of the element ingested.[6] Dietary forms of selenium include selenocysteine and selenomethionine. Fish and whole grains contain high levels of these selenium compounds. The availability of either total selenium or different forms of selenium in the diet and the amount of methionine consumed are only some of the factors that might contribute to the final distribution of selenium in specific selenoproteins (containing selenocysteine) or nonspecific selenoproteins (containing selenomethionine). After association with plasma proteins, selenium is delivered to all tissues including the bones, hair, the erythrocytes, and the leucocytes.[7] Selenium is found at its highest concentration in the kidney, followed by the glandular tissues, including the pancreas, pituitary, and the liver. Many studies have tried to elucidate the incorporation of selenium into the selenoproteins.[8] The major known fates of selenium in animals are incorporation into protein as selenocysteine, incorporation into certain modified tRNAs, and excretion as methylated compounds.[8]

Two plasma selenoproteins, glutathione peroxidase and selenoprotein P, have been characterized. Glutathione peroxidase contains four identical subunits, each containing selenium in the form of a single selenocysteine residue. Selenoprotein P (plasma) is a glycoprotein secreted by the liver containing 7.5 ± 1 selenocysteines per molecule. The functions of these two plasma proteins are unknown. Glutathione peroxidase was considered an important antioxidant. However, the capacity of plasma glutathione peroxidase to remove H_2O_2 under physiological conditions can be questioned, which leaves open the possibility of other functions for this enzyme.[8] Selenoprotein P may play a redox role, although evidence to support this is limited.[9] Another proposed function for Selenoprotein P is the transport of selenium.[10] More work is necessary to elucidate the roles of these selenoproteins.

Metalloproteins in human serum have been separated by size exclusion chromatography SEC-DIN-ICP-MS.[11] In this study, the separation of metalloproteins was carried out by SEC while employing ICP-MS for the determination of selenium.

2 EXPERIMENTAL SECTION

2.2 HPLC-DIN-ICP-MS

The HPLC system was composed of an SSI Model 222D metal-free microflow pump (Scientific Systems, Inc., State College, PA), a Rheodyne 9010 metal-free high-pressure sample injector with a 3 µl PEEK injection loop, and a 2.0-mm-i.d. x 25-cm-long size exclusion column (GPC 300, SynChrom, Inc., Lafayette IN). The outlet of the column was connected to the DIN through a switching valve (9010 Rheodyne) with a narrow bore polysil tube (50 m-i.d. x 5-cm-long, Scientific Glass Engineering, Inc., Austin, TX). The narrow bore connecting capillary minimized extra column band broadening. A polysil tube was also used to connect the sample injector valve to the inlet of the column.

The design and construction of the DIN have been described elsewhere.[5] A 50-µm-i.d. x 40-cm-long fused silica capillary was used to transport the effluent from the column to the plasma. The 50 µm i.d. capillary is less susceptible to plugging with the high salt content in the matrix of human serum than the previous 30 µm i.d. capillary. The width of the annular gap between the inner capillary and the nebulizer tip was ≈25 µm.

The ICP-MS instrument used was the Elan Model 250 (Perkin-Elmer Sciex, Thornhill, ON, Canada). Instrument operating conditions were optimized to provide maximum ion signal for a 10 ppm Se standard solution (PlasmaChem Associates, Inc., Bradley Beach, NJ). The instrument operating parameters are described in ref. 11. Chromatographic conditions are summarized in Table 1.

2.2 Data Acquisition

The ICP-MS device was equipped with the Elan 500 upgraded hardware and software. During the separation of metalloproteins in serum two isotopes of selenium were measured (m/z 78 and 82), along with the calcium isotope at m/z 42. The data were acquired by peak hopping over the three isotopes using a 20 ms dwell time and 1 measurement per peak. Chromatograms were recorded in real time and stored on the hard

Table 1. *Chromatographic conditions*

Separation of metalloproteins	
column	SynChrom, Inc. SynChroPak GPC 300 2-mm-i.d. x 250-mm-long
stationary phase	silica gel (5-μm particles, pore size = 300 Å)
mobile phase	0.1 M Tris/HCl (pH = 6.9)
eluent flow rate	100 μl min^{-1}
injection volume	3 μl
isotopes monitored	m/z = 78, 82, 42

disk of an IBM PS/2 Model 70 computer. These data were then processed as ASCII files in a spreadsheet program. The raw counts were smoothed using Golay smoothing. The peak area was determined by summing all the count rates under each peak. The background was measured while nebulizing only the mobile phase and summing the total counts under the particular chromatographic peaks.

2.3 Reagents and Samples

Deionized water (18 M Ω @ 25°C) obtained from a Barnstead Nanopure-II system (Newton, MA) was used. A 0.1 M tris(hydroxymethyl)aminomethane/ hydrochloric acid (tris/HCl) buffer was used as the mobile phase for the separation. Eluent of high ionic strength, such as 0.1 M NaCl, was not used to avoid plugging of the DIN. The Tris/HCl solution was prepared by dissolving certified ACS grade Tris (Fisher Scientific, Fair Lawn, NJ) in deionized water. The pH was then adjusted to 6.9 by adding concentrated Ultrex II grade HCl (J.T. Baker, Inc., Phillipsburg, NJ). Human serum (NIST SRM 909a-2) was reconstituted in 10 ml of the buffer solution.

3 RESULTS AND DISCUSSION

3.1 SEC of Selenium-Containing Proteins

The determination of selenium compounds is complicated by the low ionization efficiency of selenium in the ICP, its presence in relatively low concentrations in real samples and the abundance of Ar related ions in the region of interest. During the separation of the metalloproteins in human serum, two molecular weight fractions were identified to contain selenium as shown in the chromatograms in Figures 2 and 3 (m/z 78 and 82 respectively). The first peak eluted at 4.6 minutes, followed by a second selenium containing peak at 6 minutes. Sodium elutes from the column at approximately 6.7 minutes and causes a decrease in the background at other masses and a negative peak in the chromatogram. Calcium was also identified at the same elution times as the selenium containing fractions shown in the chromatogram in Figure 4. A large calcium peak elutes later around 10.4 minutes. The size exclusion column was previously calibrated using pure protein standards.[11] This calibration identifies the two selenium containing molecular weight fractions to be 200 and 15 kDa respectively.

Selenium exists as six isotopes including m/z 74 (0.9), 76 (9.0), 77 (7.6), 78 (23.5), 80 (49.8), and 82 (9.2). The natural percent abundance of each isotope is shown

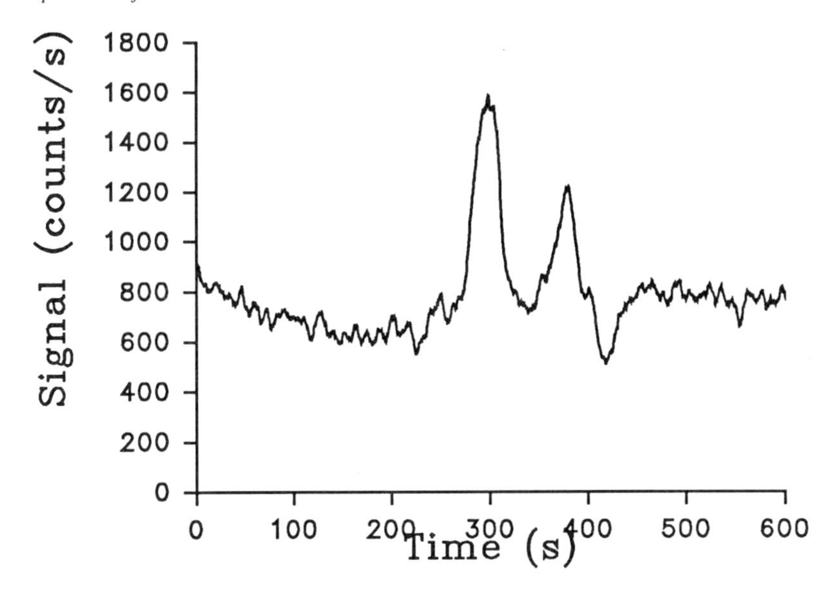

Figure 2. Selected ion chromatogram of metalloproteins in human serum monitoring m/z 78.

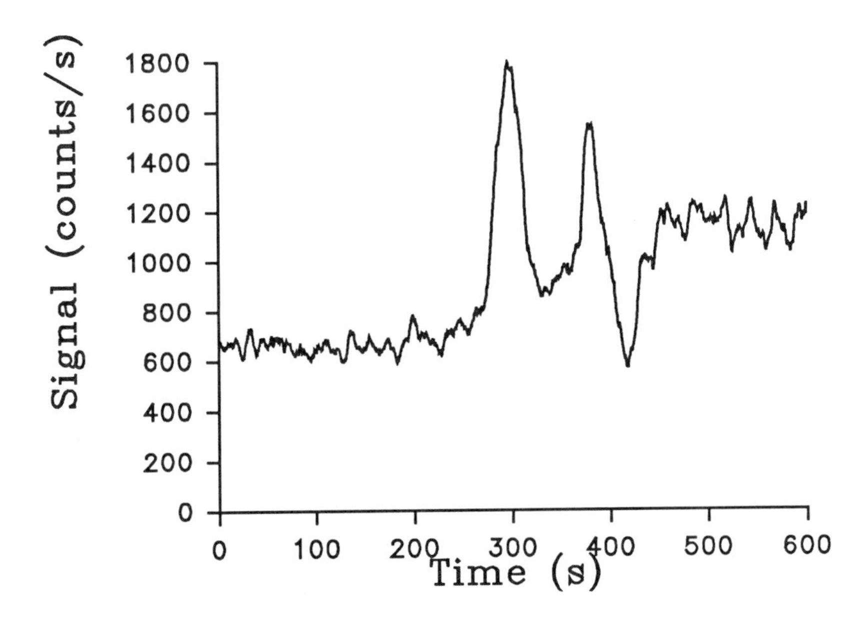

Figure 3. Selected ion chromatogram of metalloproteins in human serum monitoring m/z 82.

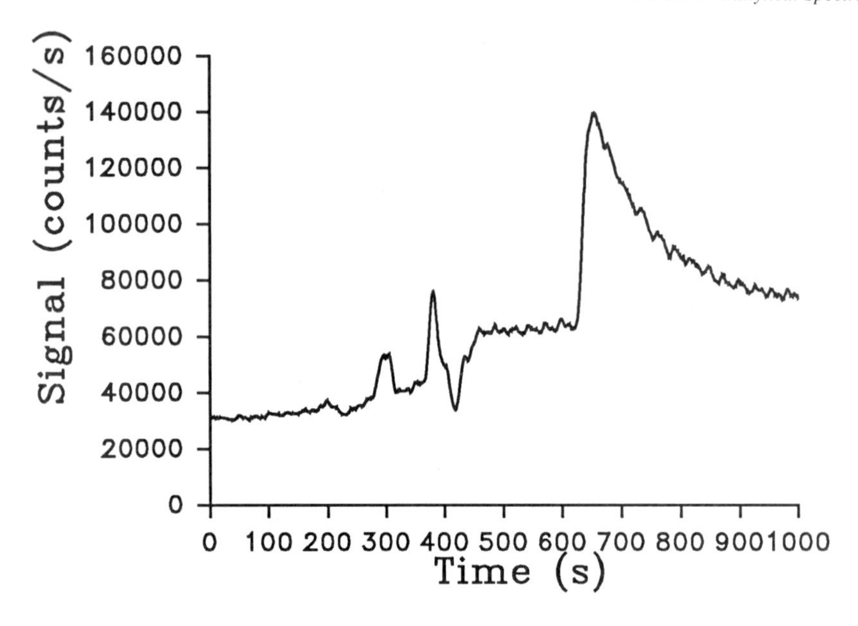

Figure 4. Selected ion chromatogram of metalloproteins in human serum monitoring m/z 42.

in parenthesis. Finding two isotopes of selenium free from mass spectral interference is difficult due to the argon related polyatomic ions in this mass region. Argon dimer ions exist at mass 76 ($^{40}Ar^{36}Ar$), 78 ($^{40}Ar^{38}Ar$), and 80 ($^{40}Ar_2$). ^{36}Ar is more abundant than ^{38}Ar and hence the background at mass 76 is larger than at 78. Due to the chloride in the matrix of human serum, $^{40}Ar^{37}Cl$ eliminates mass 77 as a possible selenium isotope. Other elements in the matrix such as Ca and K can also produce polyatomic interferences when combined with argon. Selenium at mass 74 gives a signal too weak to measure at the low concentration of selenium in human serum. After consideration of these interferences, isotopes 78 and 82 were chosen to monitor selenium during the chromatographic separation. The correct isotope ratio of selenium 78 and 82 was not observed. An attempt was made to subtract the interference from $CaAr^+$ at m/z 82 by measuring another isotope of $CaAr^+$ at m/z 84. However, this data proved inconclusive due to species other than $CaAr^+$ at m/z 84. Hence, $CaAr^+$ does not seem to give a major contribution to m/z 84 or 82. Mass spectral interference from species in the human serum which coelute with the selenium containing molecular weight fractions cause background signal under the selenium peaks. The actual isotope ratio of selenium 78/82 is 2.55. The measured isotope ratio 78/82 (0.64 and 0.80 for peaks 1 and 2 respectively) is much lower than the actual isotope ratio, and therefore signals from selenium and other ions are present in the peaks at m/z 82. Because no other interference-free isotope of selenium exists, it is unclear if there is another ion at m/z 78 from an element that elutes at the same retention time as the selenium containing peaks. In order to accurately determine the amount of selenium in each peak, a second detection method which is free of mass spectral interference and has the capability to measure selenium at approximately 1.5 ppb is necessary.

3.2 Determination of Total Selenium

The total selenium content in the serum was determined by the standard additions method to be 90 ppb using m/z = 78. This value is close to or a little below the usual value of ~100 ppb for total Se in human serum.[12] This result indicates that spectral interference at m/z = 78 is not too bad, otherwise standard additions would yield a result that is too high. . The plot of signal versus concentration of the added selenium spike has a correlation coefficient of 0.9948 and is shown in Figure 5. An approximate concentration in each molecular weight fraction was determined by proportion using the chromatogram obtained at mass 78. By this method, the 200 kDa fraction contains 54 ppb and the 15 kDa fraction contains 36 ppb.

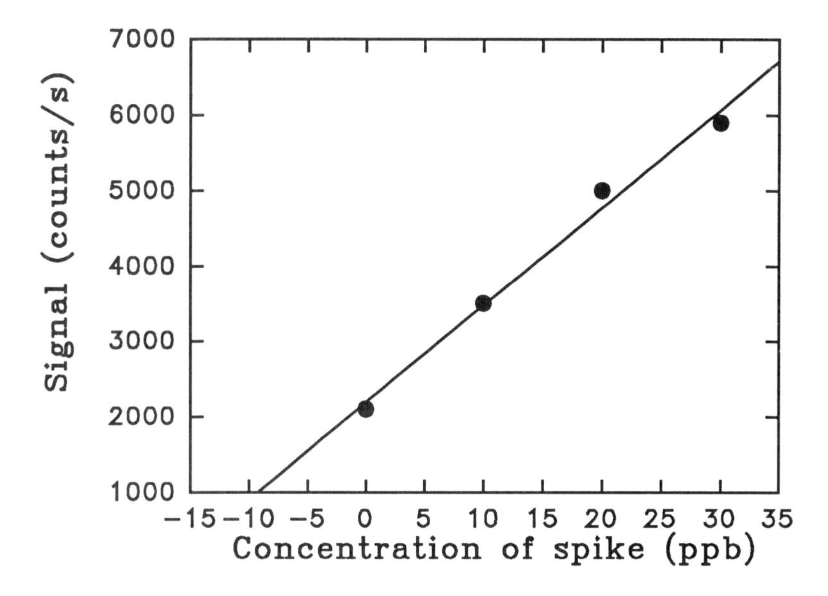

Figure 5. Standard additions plot for selenium in human serum (R = 0.9948).

4. CONCLUSION

Two molecular weight fractions in human serum containing selenium were observed. The total amount of selenium in human serum was determined by the standard additions method to be 90 ppb. Accurate determination of the amount of selenium in each fraction is hindered by mass spectral interferences due to species present in the serum sample which coelute with the selenium containing fractions, but m/z = 78 appears to be a reasonable selection.

ACKNOWLEDGEMENT

Ames Laboratory is operated for the U.S. Department of Energy by Iowa State University under Contract W-7405-Eng-82. This research was supported by the Director for Energy Research, Office of Basic Energy Sciences.

References

1. R. S. Houk, *Anal. Chem.* 1986, **58**, 97A.
2. K. E. Jarvis, A. L. Gray and R. S. Houk, 'Handbook of ICP-MS', Blackie, London, 1992.
3. D. W. Koppenaal, *Anal. Chem.* 1992, **64**, 320R.
4. R. S. Houk and S. J. Jiang, in 'Trace Metal Analysis and Speciation', I. S. Krull, Ed., Elsevier, Amsterdam, 1991, Chap. 5.
5. D. R. Wiederin, F. G. Smith and R. S. Houk, *Anal. Chem.* 1991, **64**, 219.
6. T. M. Florence, in 'Trace Element Speciation: Analytical Methods and Problems', G. E. Batley Ed. CRC Press, Florida, 1989, Chap. 9.
7. E. J. Underwood, 'Trace Elements in Human and Animal Nutrition', Academic Press, New York, 1977, p. 302.
8. R. F. Burk, *FASEB J.* 1991, **5, 2274.**
9. R. F. Burk, R. A. Lawrence, J. M. Lane, *J. Clin. Invest.* 1980, **65**, 1024.
10. M. A. Mastenbacker and A. L. Tappel, *Biochim. Biophys. Acta* 1982, **719**, 147.
11. S. C. K. Shum and R. S. Houk, *Anal. Chem.*, 1993, **65**, 2972.
12. D. E. Nixon, personal communication, 1994.

The Investigation of Varietal Differences Among Sorghum Crop Residues Using Near Infrared Reflectance Spectroscopy

S. J. Lister, M. S. Dhanoa, I. Mueller-Harvey,[1] and J. D. Reed[2]

INSTITUTE OF GRASSLAND AND ENVIRONMENTAL RESEARCH, PLAS GOGERDDAN, ABERYSTWYTH SY23 3EB, UK

[1] DEPARTMENT OF AGRICULTURE, UNIVERSITY OF READING, READING, BERKSHIRE RG6 2AT, UK

[2] UNIVERSITY OF WISCONSIN, 1675 OBSERVATORY ROAD, MADISON, WISCONSIN 53706-1284, USA

1 INTRODUCTION

Sorghum crop residues are an important source of livestock feed in Ethiopia and other African countries. Breeding programmes have resulted in increased sorghum grain yields but there is a need to breed for nutritional quality of the crop residue. Birds are a major crop pest so farmers are being encouraged to plant newer varieties of sorghum which are more bird-resistant than traditional varieties. Unfortunately, these are known to contain higher amounts of polyphenolics[1] which may have adverse effects on livestock production.

Near Infrared Reflectance Spectroscopy (NIRS) is most commonly used to determine the concentration of chemical components in materials such as forage.[2,3] In this conventional use, a large sample set must be available for the calibration process, with the parameter to be predicted already determined by traditional means. However, NIRS can also be used for qualitative analysis. In this study, NIRS was used to perform qualitative analysis using multivariate techniques, with a view to establishing it as a fast screening tool in the study of variety performance and their interaction with the environments in which they are grown.

2 MATERIALS AND METHODS

Sorghum crop residues (Table 1) were obtained from the International Livestock Centre for Africa, Addis Ababa, Ethiopia. The samples were harvested at grain maturity from two matched varietal trials conducted at Melkasa (elevation 1500 m) and Debre Zeit (elevation 1900 m). Separated samples (leaf blade, leaf sheath and stem) were air dried and ground to pass through a 1 mm sieve. The samples were scanned using a NIRSystems 6250 (Perstorp Analytical, Bristol, UK) scanning monochromator over the spectral range from 1100 to 2500 nm. Spectra were stored as the reciprocal logarithm (Ln 1/R) of the reflected energy at 2 nm intervals, giving a total of 700 data points for each sample.

Table 1 *Sorghum Varieties Grown at Sites A (Melkasa) and C (Debre Zeit).*

Variety	*Country of Origin*	*Resistance*
Ikinyaruka	Rwanda	b
Serena	Uganda	b
Seredo	Uganda	b
5D x 135/13/1/31	Uganda	b
X/35:24	Sudan	b
Framida	West Africa	b
ESIP 4	Ethiopia	n
ESIP 7	Ethiopia	n
ESIP 13	-	n
ESIP 17	-	n
ESIP 21	-	n
ESIP 25	-	n
ESIP 40	Sudan	n
ESIP 43	Sudan/Ethiopia	n

b=bird-resistant variety; n=non-bird-resistant

A high level of multi-collinearity is generally found between NIR data points and variation due to particle size can account for as much as 90 per cent of the variance[4] in a set of NIR spectra. In order to highlight chemical characteristics, the variation due to physical effects was reduced by the use of Standard Normal Variate and De-Trend transformations.[5] The SNV transformation centres each spectrum and then scales it by its own standard deviation; correcting any shifts on the y-axis. De-Trending is performed using a second order polynomial in regression analysis where spectral values are the response and wavelength the independent variable. This removes base line curvature caused by variable interactions of moisture and particular effects across the NIR spectral range.

Initially the data was reduced by averaging windows of 10 nm, thus giving 140 data points for each sample and then further condensed by calculating the first 16 principal components (PC). Pairwise plots of the various components were used to obtain a graphical representation and an overview of the inter-relationship between data units representing species. Cluster analysis based on similarities using furthest neighbour criterion was performed to study the emerging clusters. All multivariate analyses were performed using GENSTAT.[6]

3 RESULTS AND DISCUSSION

The first 16 principal components accounted for at least 0.999 of the variation in the data set and 0.778, 0.094, 0.058, 0.026, 0.014 and 0.010 of the variation was associated with the first six components, respectively. Pairwise plots of the main 3 components showed segregation of different plant fractions and different varieties into apparent groups. For illustration, the PC plot of the first 2 components for the different plant fractions (Figure 1) yields distinct groupings for leaf blade, leaf sheath and stem fractions. Leaf blade is clearly defined whereas some overlap of leaf sheath and stem may be a result of incomplete separation.

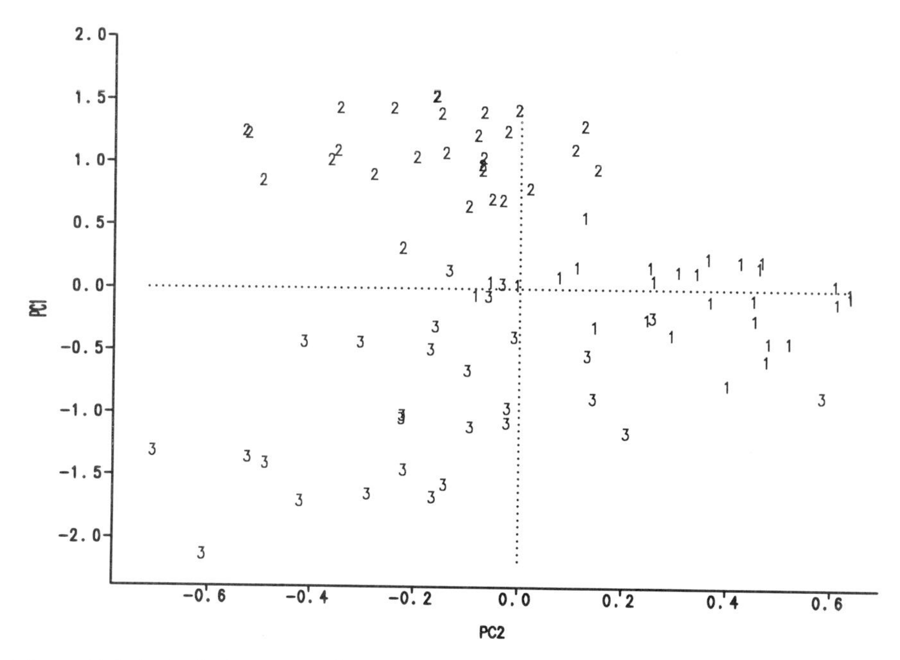

Figure 1 *Principal Component 1 versus Principal Component 2 for Leaf Sheath (1), Leaf Blade (2) and Stem (3) Fractions of Sorghum Crop Residues.*

Within leaf blade the first two components (Figure 2) show clear separation between samples grown at sites A and C and some separation between the bird-resistant and non-bird-resistant varieties. Cluster analysis was conducted using all 16 PC and the groupings thus produced are presented as a dendrogram in Figure 3. This again shows two main groupings, of varieties grown at the two different sites, with the exception of one bird-resistant variety (Framida); the samples grown at sites A and C cluster together, suggesting stability of this variety in different environments.

Leaf sheath fractions also show similar discrimination (Figure 4) into groupings according to site and resistance. Further, to investigate environmental effects on leaf sheath, difference spectra were calculated between the average bird-resistant and average non-bird-resistant samples grown at sites A and C. The resultant difference spectra are

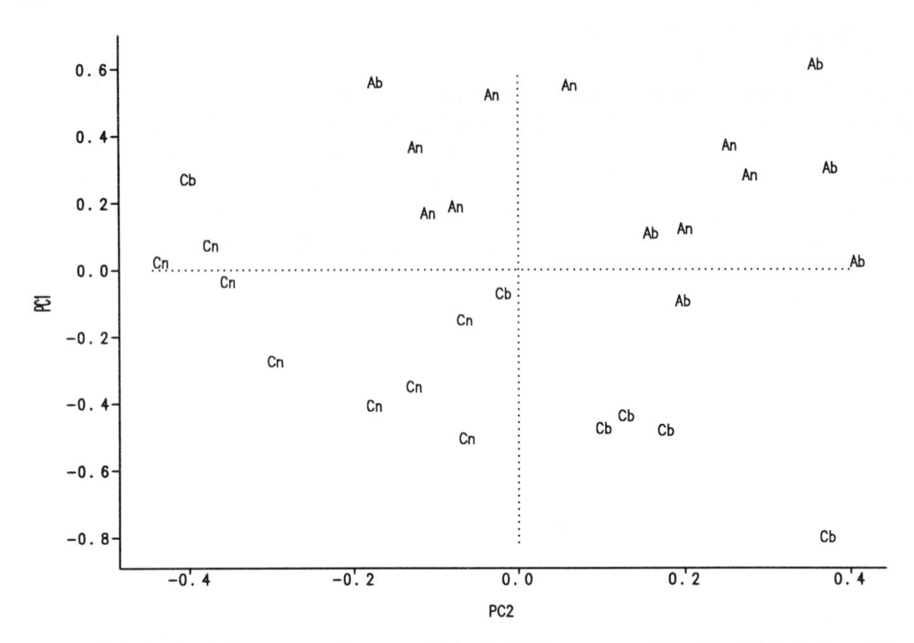

Figure 2 *Principal Component 1 versus Principal Component 2 for Leaf Blade from Sites A and C for Bird-Resistant (b) and Non-Bird-Resistant (n) Varieties.*

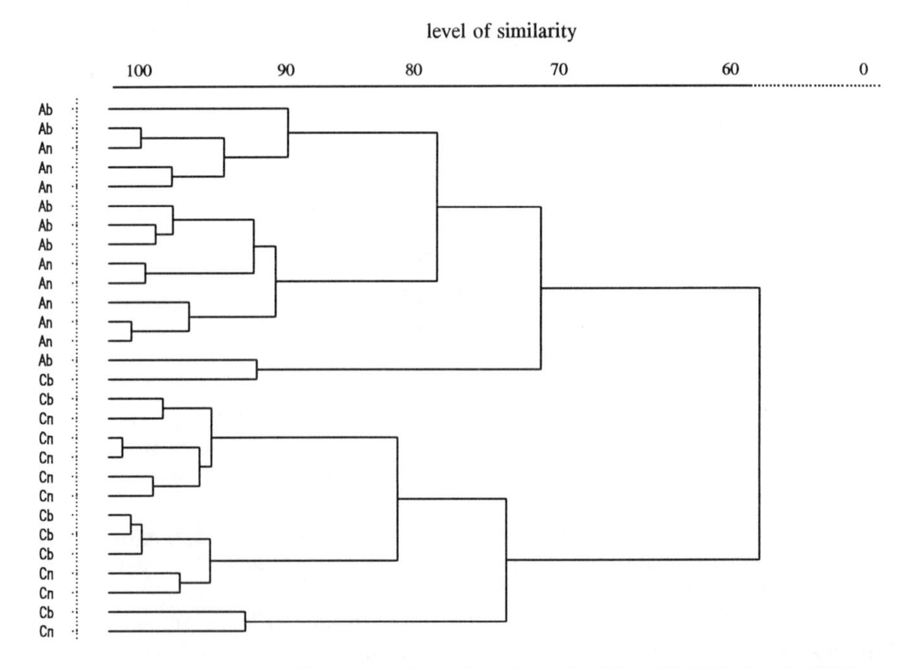

Figure 3 *Dendrogram from Cluster Analysis Based on the First 16 PC's for Leaf Blade from Sites A and C for Bird-Resistant (b) and Non-Bird-Resistant (n) Varieties.*

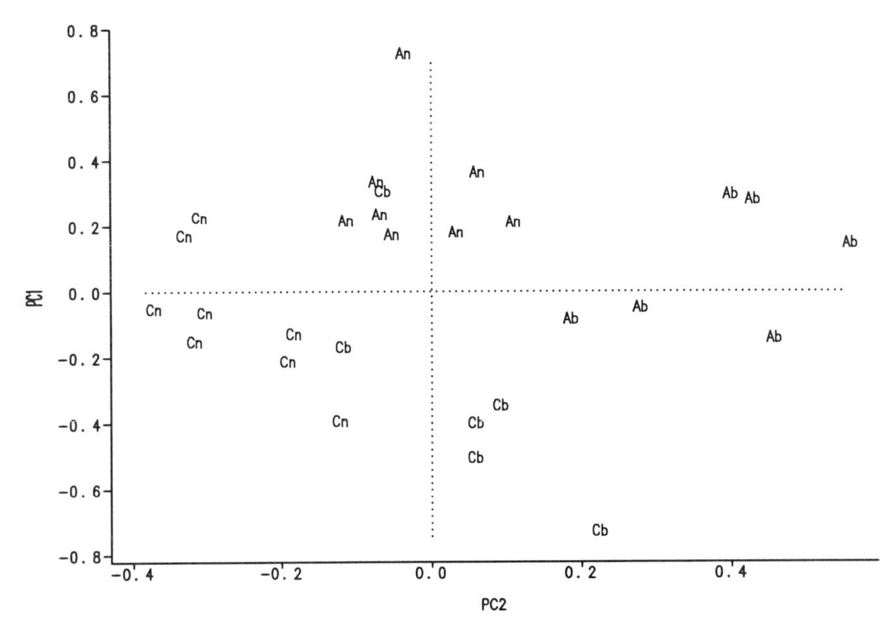

Figure 4 *Principal Component 1 versus Principal Component 2 for Leaf Sheath from Sites A and C for Bird-Resistant (b) and Non-Bird-Resistant (n) Varieties.*

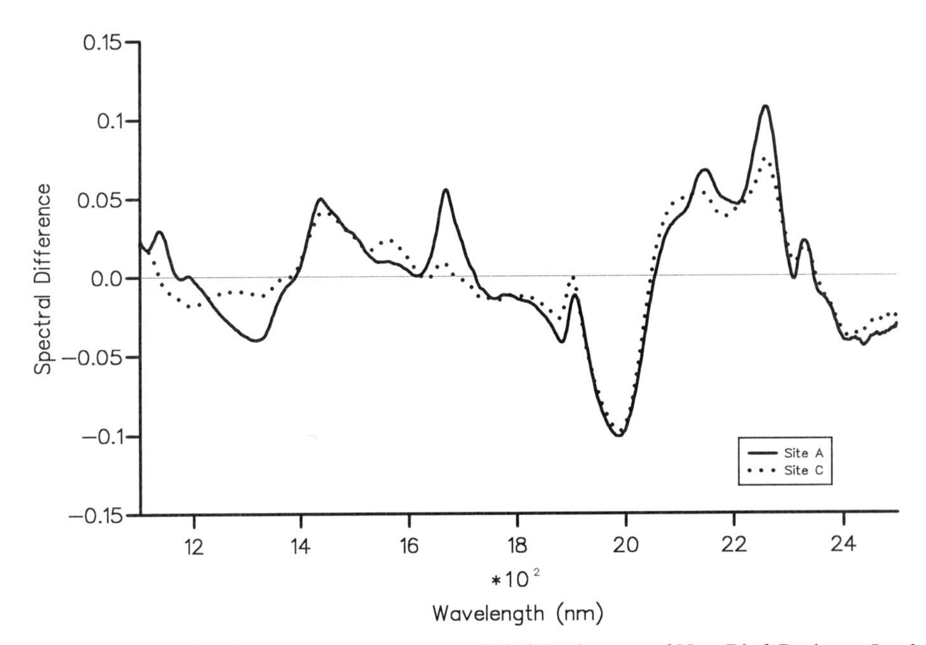

Figure 5 *Spectral Difference between Averaged Bird-Resistant and Non-Bird-Resistant Leaf Sheath Fractions Grown at Sites A and C.*

presented in Figure 5. It is apparent that there are differences between bird-resistant and non-bird-resistant varieties, but these are dependent upon the environment in which they are grown. The difference spectra reveal that more polyphenolics are present in bird-resistant than non-bird-resistant varieties, as indicated by the following regions; the peak centred at 1664 nm arising from the first overtone of the C-H stretching fundamental characteristic of aromatic compounds,[7] and the combination bands in the 2260 nm region which are associated with phenolic compounds.[8]

4 CONCLUSIONS

Environmental effects tend to be greater compared to varietal differences. In this study most varieties seem to show a strong environment x genotype interaction. Differences between bird-resistant and non-bird-resistant varieties were evident. NIRS may be used to characterise sorghum varieties without the need for the detailed and time consuming laboratory analyses used at present.

NIRS spectra are a comprehensive record of chemical structure and content of a biological substrate and thus are a rich source of information to highlight differences. NIRS may be used to differentiate among plant species and varieties or for screening varieties in plant breeding programmes.

5 ACKNOWLEDGEMENTS

This project was partly funded by the Overseas Development Administration (EMC X0093). Sorghum samples were imported under licence number PHF 976/37/119 issued by the Ministry for Agriculture, Fisheries and Food.

References

1. J. D. Reed, A. Tedla and Y. Kebede, *J. Sci. Food Agric.,* 1987, **39**, 113.
2. K. H. Norris, R. F. Barnes, J. E. Moore and J. S. Shenk, *J. Anim. Sci.,* 1976, **43**, 889.
3. J. S. Shenk, M. O. Westerhaus and M. R. Hoover, *J. Dairy Sci.,* 1979, **62**, 807.
4. I. A. Cowe and J. W. McNicol, *Appl. Spectrosc.,* 1985, **39**, 257.
5. R. J. Barnes, M. S. Dhanoa and S. J. Lister, *Appl. Spectrosc.,* 1989, **43**, 772.
6. GENSTAT 5 committee, GENSTAT 5 reference manual, Clarenden press, Oxford, 1987.
7. W. R. Windham, S. L. Fales and C. S. Hoveland, *Crop Sci.,* 1988, **28**, 705.
8. S.W. Coleman and I. Murray, *Anim. Feed Sci. Technol.,* 1993, **44**, 237.

The Influence of Energy Migration on Fluorescence Kinetics in Photosynthetic Systems

D. L. Andrews and A. A. Demidov[1]

SCHOOL OF CHEMICAL SCIENCES, UNIVERSITY OF EAST ANGLIA, NORWICH NR4 7TJ, UK

[1] PHYSICS DEPARTMENT, MOSCOW STATE UNIVERSITY, 119899 MOSCOW, RUSSIA

1 INTRODUCTION

The absorption of light by photosynthetic structures is accompanied by a highly efficient process of energy migration between molecules such as chlorophyll *a*, phycobiliproteins etc. in the light-harvesting antennae of photosynthetic organisms. It is this process which leads to trapping of the excitation by reaction centres where the energy of the incident photons is converted to chemical form. The whole process proceeds with a quantum efficiency of about 90%; losses due to fluorescence are nonetheless useful in that they carry information on the energy migration processes. To achieve such high efficiency it is crucial to have optimised structures and an effective mechanism for energy transfer, and these problems are currently under intensive investigation. Recently it has been established[1,2] that there is a correction term which needs to be incorporated in the well-known Förster law describing the energy migration, and which may be manifest in fluorescence signals. The corrected formula has been applied and evaluated in a computer simulation[3] of energy transfer in a model photosynthetic complex.

2 THE MECHANISMS OF ENERGY TRANSFER

The theory of multi-step energy migration within a system containing a large number of molecules requires proper modelling of the elementary single-step act of transfer. This process takes place between a molecule with excitation instantaneously localised upon it, designated the donor, and a ground-state acceptor which in some cases may be chemically identical. Each step in the overall energy transfer is crucially affected by: (a) the distance R between these molecules, (b) the mutual orientations of their transition dipole moments, and (c) their spectroscopic features - the donor radiation lifetime, the fluorescence and absorption spectra of the donor and acceptor molecules respectively. At short distances $R < 1$ nm energy transfers mainly via exchange mechanisms associated with electronic orbital overlap. This interaction has an exponential dependence on the distance R.

For distances in the range 1 nm < R < 100 nm, energy transfers mainly via the Förster mechanism, which has an R^{-6} distance-dependence: $K_{DA} = \tau^{-1}(R_0/R)^6$. Here K_{DA} is the rate of energy transfer, τ is the radiation lifetime of a 'free' donor, and R_0 the characteristic Förster radius. For distances greater than 100 nm the radiative mechanism is responsible for energy transfer: this mechanism yields an R^{-2} distance-dependence.

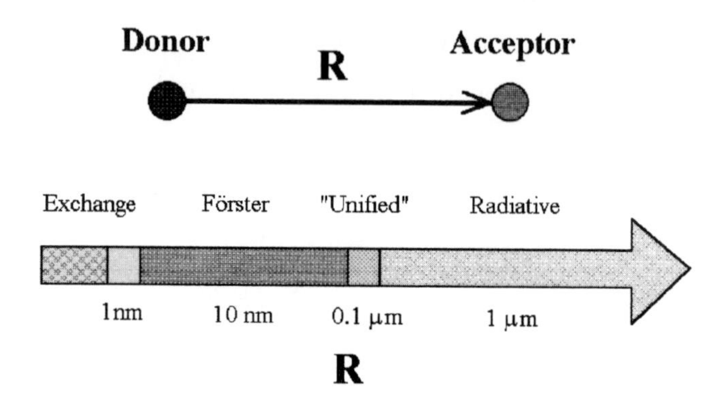

Figure 1 *Diagram of energy migration from the donor to the acceptor molecules.*

Recently it was proved[1,2] that in the range of distances around 100 nm an R^{-4} distance-dependence should operate, and a 'unified' theory of energy transfer was formulated. This theory, derived from first principles, provides a formula for the transfer of energy from the donor molecule to the acceptor which includes both Förster and radiative mechanisms, together with a term representing the intermediate range dependence. For a system of randomly oriented molecules the result can be most simply expressed in the following form:

$$K_{DA} = \alpha\left(\frac{3}{R^6} + \frac{K^2}{R^4} + \frac{K^4}{R^2}\right)$$

Here the coefficient α is essentially proportional to the overlap integral of the donor fluorescence and the acceptor absorption spectra:

$$\alpha \approx \frac{c^4}{64\pi^5 \bar{v}^4 \tau} \int f^D_{fluor.}(v)\sigma^A_{abs.}(v)dv$$

In the above equations, $\bar{v}$ is the mean value of the frequencies covered by the overlap integral, and for the photosynthetic unit the value of the parameter $K \cong 2\pi\bar{v}/c$ is about 9×10^{-3} nm^{-1}. Figure 2 shows on a log-log plot the general dependence of the energy transfer rate with the distance R. As expected, the involvement of the R^{-4} dependence is most important in the region $R \sim 100$ nm, and consequently one has to use the 'unified theory' to make reliable evaluations of energy transfer in this region.

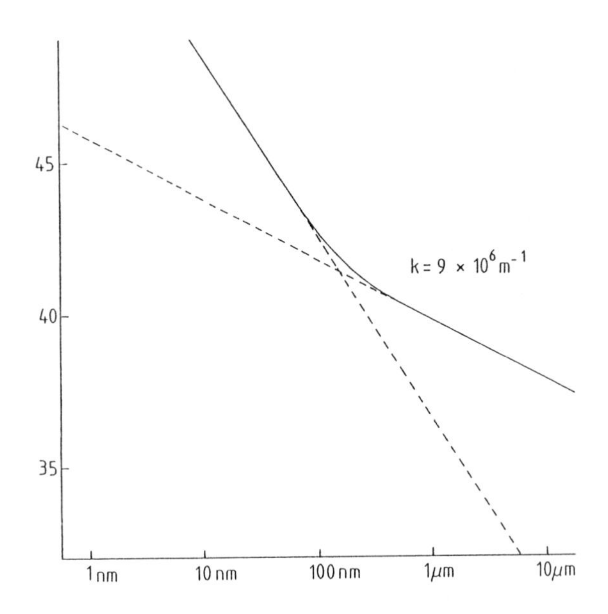

Figure 2 *Dependence of the 'one-step' energy migration rate on the distance between the donor and acceptor molecules ('unified' theory). The dashed lines show the asymptotic R^{-6} and R^{-2} behaviour.*

3 PHYSICAL MODELS OF THE PHOTOSYNTHETIC UNIT (PSU)

In natural photosynthetic organisms excitation energy is never delivered to a reaction centre (RC) from the photosynthetic 'antenna' pigment molecules in a single step. In fact excitation regularly reaches the RC after tens or hundreds of random jumps between different pigments, of which chlorophyll *a* is the most important. The photosynthetic pigments and the RC are organized in pigment-protein complexes termed photosynthetic units (PSU's). Within each PSU photosynthetic pigments serve for the light energy harvesting and for delivery of the excitation energy to the RC, where the energy is utilised in the charge separation that ultimately drives the photosynthetic chemistry.

3.1 Unicentre PSU model

The unicentre model of the PSU includes chlorophyll *a* (*Chl a*) light-harvesting antennae and one reaction centre: the number of *Chl a* molecules in the PSU is regularly about 80-300 per RC. The incident light energy (hν) is converted into *Chl a* excitation, which may be delivered to the RC after a set of random jumps (N_{jumps} = 50 - 200), or else experience non-radiative decay as well as fluorescent emission. The typical time for energy delivery to the RC is about 200 ps, and the Förster radius for the *Chl a** → *Chl a* transfer is $R_0 \approx 7$ nm.

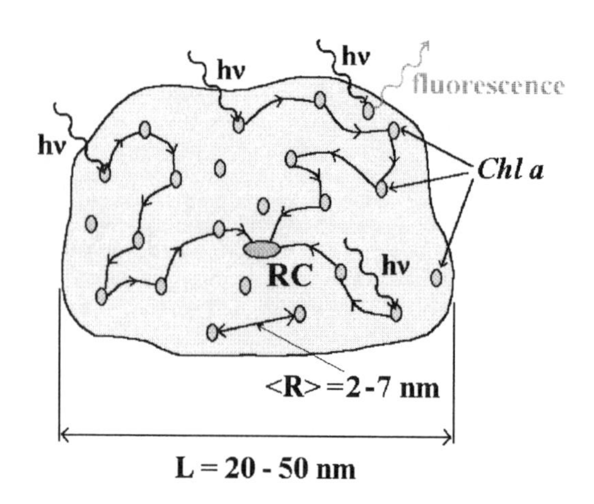

Figure 3 *Unicentre model of PSU.*

3.2 Multicentre PSU model

In the multicentre PSU model the *Chl a* light-harvesting antennae serve for a number of RC's. The average number of *Chl a* per RC is still about 80-300, the same as in the unicentre model. If the excitation reaches a 'closed' RC it is able to continue its random walk to an 'open' RC to be captured. The process of the excitation random walk to the open

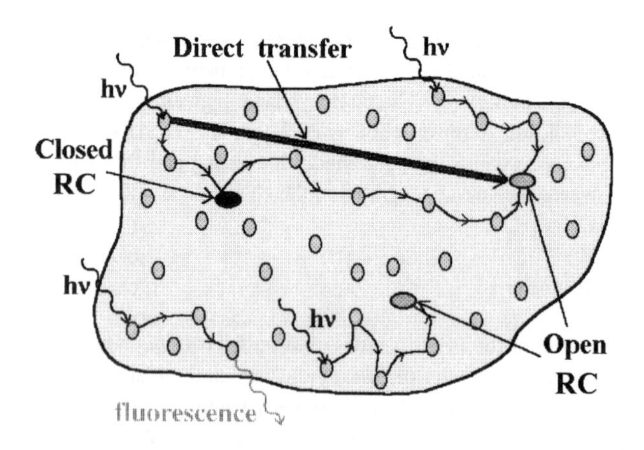

Figure 4 *Multicentre model of PSU.*

RC via the Förster mechanism could be relatively slow because of the great number of steps entailed, and the concurrent option of long-distance direct transfer to the RC by the R^{-4} mechanism might therefore be of comparable efficiency.

3.3 Complex of unicentre PSU's

Figure 5 below represents a complex of unicentre PSU units containing blocked and active reaction centres (black and grey). The major difference between this model and the previous one is that the chlorophyll light-harvesting antennae are not considered a pool for all RC's, but rather they are relatively localized around specific centres. Excitation can then be delivered to an open RC either via random walk multistep energy migration among the *Chl a* molecules, or via a direct long-range interaction ('unified theory'). These routes are concurrently available and both of them could be employed in mediating the flow of energy from the initially excited *Chl a** molecules to the RC, or to other *Chl a* molecules.

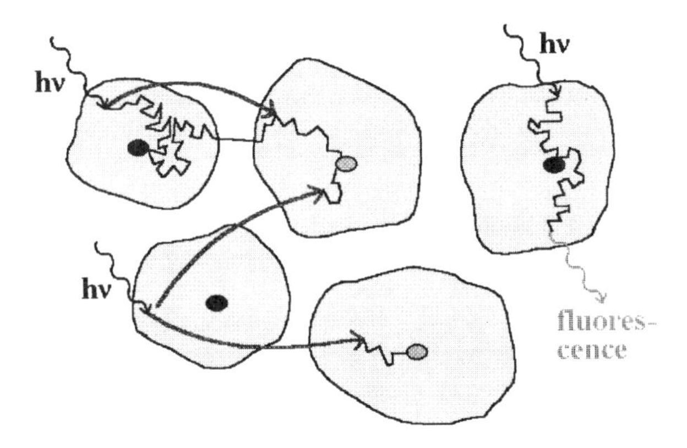

Figure 5 *Complex of unicentre PSU's.*

4 METHODS OF EXCITATION DYNAMICS ANALYSIS

There are in general two stages in analysing the excitation dynamics of molecular complexes like the PSU: (a) design of a physical model, and (b) development and treatment of a mathematical model based on the physical one. The design of a physical model for the PSU means determination of the major physical properties of the molecular complex such as its spatial structure, the spectroscopic features of individual molecules, the mechanisms of their interaction etc. The physical model should of course be as close as possible to the real physical object. Development and treatment of the mathematical model here entails appropriate mathematical representation of the physical model, with correct formulation and solution of the energy transfer dynamics. Generally the mathematical solution serves not only for explanation of experimentally observed phenomena, but also the prediction of some new ones, which could be tested by experiment.

The Unicentre and Multicentre PSU models are physical models of *in vivo* and *in situ* photosynthetic complexes. For full rigour, the 'unified theory' should be employed in rate calculations based on these models. For such systems there in general are three mathematical models for the analysis of energy migration: (a) the diffusion model, (b) the master equation approach, and (c) the Monte-Carlo method. The first implies statistical averaging of excitation random walks and is cast in terms of a 'diffusion coefficient' which represents the average speed of excitation migration. The second method is based on a system of differential balance equations, which describe the probability of finding each molecule in its excited state. The rates of energy transfer inbetween each pair of molecules are incorporated in this system and the number of equations is equal to the number of molecules. The averaging procedure commonly follows numerical solution of the master equation system.

It is the third, Monte-Carlo, method that is the focus of our attention here. This method offers the advantage of 'searching' the behaviour of individual excitations created in molecular complexes, tracking all of its jumps from one molecule to another until it decays or is captured by a RC. The calculations can only be accomplished by computer methods and the final result is obtained after statistical averaging the 'life story' of millions of such excitations.

Recent application of the Monte-Carlo method to the specific photosynthetic complex C-phycocyanin[3] has shown that this method has a very high potential and can provide a great deal of information about the excitation dynamics in molecular complexes, based only on knowledge of the spatial structure of the molecular complex and the spectroscopic properties of the individual molecules. In particular this method has allowed us to evaluate the dependence in the PSU models of the *Chl a* fluorescence kinetics. The regular fluorescence lifetime of 'free' *Chl a* molecules is about 6 ns, while in the PSU it has a value of about 200 ps or 1.4 ns when the RC's are all open or all closed respectively.

To conclude, our work in progress is the first attempt to apply the newly available unified theory of intermolecular energy transfer to molecular ensembles. The PSU is one particular example of a wide range of molecular complexes to which the theory can be applied, and is of special interest because of its biophysical significance.

References

1. D. L. Andrews, *Chem. Phys.*, 1989, **135**, 195.
2. D. L. Andrews and G. Juzeliūnas, *J. Chem. Phys.*, 1992, **96**, 6606.
3. A. A. Demidov and A. Yu. Borisov, *Biophys. J.*, 1993, **64**, 1375.

Flow Injection Procedures with Spectrophotometric Detection for the Determination of Nitrate and Nitrite in Riverine, Estuarine, and Coastal Waters

T. McCormack, A. R. J. David, and P. J. Worsfold

DEPARTMENT OF ENVIRONMENTAL SCIENCES, UNIVERSITY OF PLYMOUTH, PLYMOUTH, DEVON PL4 8AA, UK

1 INTRODUCTION

Nitrogen levels in surface waters are very closely linked with rates of algal growth. The major forms of nitrogen in these waters are nitrate, nitrite, ammonia and organic nitrogen, of which nitrate is usually the most significant. They are all interconvertible with each other and with molecular nitrogen. Elevated nitrate levels are of chemical concern because of links with methaemoglobinaemia (blue baby syndrome) and gastric and stomach cancers. This has led to legislation by the European Union that stipulates maximum admissible concentrations (MACs) and guide levels (GLs) for nitrogen species in water intended for human consumption as shown in Table 1[1].

Table 1 *Maximum admissible concentrations and guide levels for nitrogen species in water intended for human consumption*

Species	MAC ($mg\ l^{-1}$)	GL ($mg\ l^{-1}$)
Nitrate-N	11.3	5.65
Ammonia-N	0.38	0.038
Nitrite-N	0.03	-
Kjeldahl nitrogen	1	1

Nitrogen containing species in surface waters arise from natural processes and from anthropogenic sources with the major input arising from leaching of fertiliser from agricultural land (via diffuse runoff and wastewater point discharges), with secondary inputs from farm slurry and sewage. The rising levels of nitrate in freshwaters and groundwaters in recent years can therefore be directly related to increasingly intensive agricultural practices.

Therefore there is a need to monitor the levels of nitrate (and other nitrogenous species) in all types of surface waters, both to ensure compliance with legislation and to further understand the complex processes involved in the biogeochemical cycling of nitrogen in ecosystems. Due to the problems associated with sample collection and storage it is also desirable to be able to monitor these species *in situ.*

There are several possible procedures for the determination of nitrate and/or TON (i.e. nitrate plus nitrite) in surface waters as shown in Table 2. Of these, flow injection (FI) procedures offer the greatest potential for *in situ* deployment. There have been several FI procedures reported for the determination of nitrate[5,8-13] of which two[5,9] have been specifically adapted to sea water analysis.

This paper describes reliable flow injection methods for the determination of TON in freshwaters and in estuarine and coastal waters and discusses the issues of sampling, treatment and storage in the context of a potential *in situ* monitor.

2 EXPERIMENTAL

2.1 Reagents

AnalaR reagents (BDH) and Milli-Q water were used throughout. A stock nitrate solution (100 mg l^{-1}) was prepared by dissolving 0.7220 g of potassium nitrate in 1L Milli-Q water. Standards were prepared daily by serial dilution of the stock with appropriate concentrations of sodium chloride in Milli-Q water.

The concentrations of the various FI streams were as shown in Fig 1a (freshwater manifold) and Fig 1b (estuarine and coastal waters manifold). The mixed colour reagent (N-(1-naphthyl)ethylenediamine dihydrochloride (Sigma) and sulphanilamide) was stored in a brown glass bottle. The cadmium reductor column was prepared by stirring 5 g of cadmium (325 mesh, 99.5 % purity, Johnson-Matthey Metals) in 50 ml of copper sulphate pentahydrate solution (5 g l^{-1}) for a couple of minutes. The copperised cadmium was sequentially washed with hydrochloric acid (2 M) and ammonium chloride (10 g l^{-1}), packed in a glass tube (40 mm long x 2 mm i.d.) and plugged with glass wool. The column was stored in ammonium chloride (10 g l^{-1}) when not in use.

2.2 Instrumentation

Peristaltic pumps (FIAstar 5020 Analyzer) fitted with flexible, modified PVC pump tubing (Anachem) were used to propel each of the carrier streams. The flow rates were as stated in Figs 1a and 1b. PTFE tubing (0.8 mm i.d.) was used throughout the remainder of the manifold. Samples and standards were injected via a rotary PTFE valve (FIAstar 5020 Analyzer). The glass bead mixing column (20 mm long x 3 mm i.d. filled with 1.5-2.0 mm o.d. glass balls) was used for saline samples in order to overcome the refractive index effect associated with such samples.

Table 2 *Selected procedures for the determination of nitrate and TON in waters*

Species	Sample	Method	Detection	Range (mg l^{-1})	Ref.
Nitrate	Water	Nitrate selective electrode	Amperometry	1-1000	2
Nitrate	Natural water	Reduction to nitrite then reaction with 3-aminonaphthalene-1,5-disulphonic acid	Fluorimetry	10^{-4}-3	3
Nitrate	Drinking and river water	Reaction between nitrate and the uranyl ion	Polarography	10^{-2}-0.1	4
Nitrite, TON	Sea water	FIA; diazotisation using sulphanilamide and N-(1-naphthyl)ethylenediammonium chloride	Photometry (LED / phototransistor)	8×10^{-2}-8×10^{-1}	5
Nitrite, nitrate	Sea water	CFA; diazotisation using sulphanilamide and N-(1-naphthyl)ethylenediammonium chloride	Photometry	7×10^{-4}-7×10^{-2}	6
Nitrate	Sea water	CFA; diazotisation using sulphanilamide and N-(1-naphthyl)ethylenediammonium chloride	Photometry	3×10^{-5}-1×10^{-3}	7
Nitrate	River water	Automated FIA; diazotisation using sulphanilamide and N-(1-naphthyl)ethylenediammonium chloride	Photometry (LED / photodiodes)	0.03-12	8
Nitrate, nitrite	Natural waters	FIA; diazotisation using sulphanilamide and N-(1-naphthyl)ethylenediamine	Spectrophotometric	14×10^{-3}-0.35	9

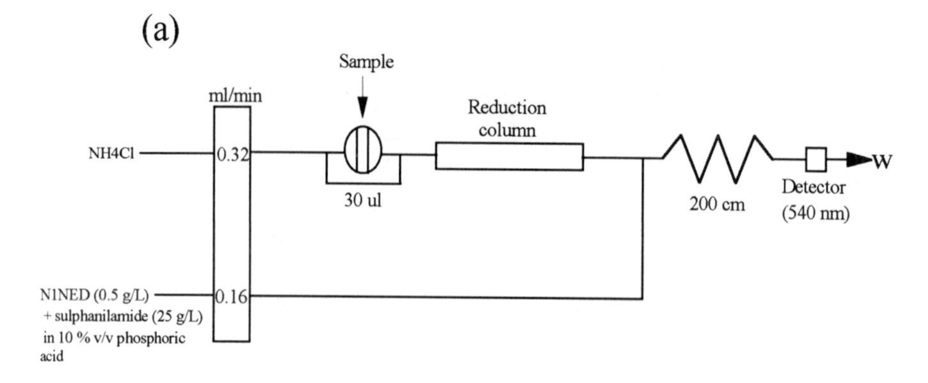

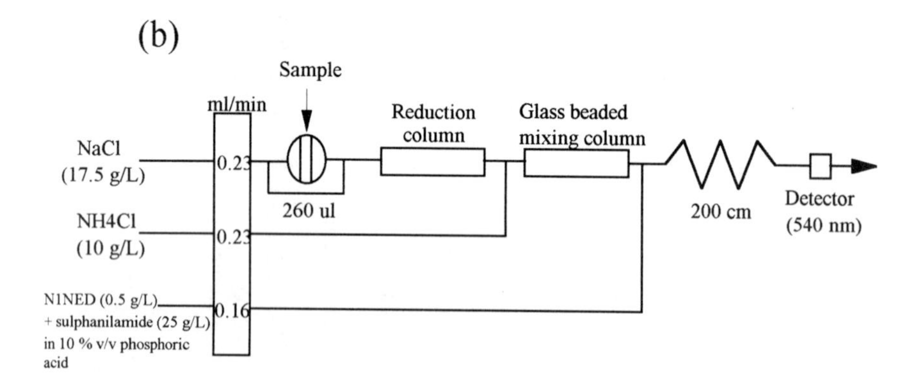

Figure 1 *Flow injection manifolds for the determination of TON in (a) freshwaters and (b) estuarine and coastal waters*

3 RESULTS AND DISCUSSION

3.1 Sample Collection and Storage

There are various protocols for the collection and storage of surface waters[14-16] but there is general agreement that glass or hard polyethylene bottles should be used to minimise biological activity. Samples should be analysed within 24 h but the addition of a preservative (e.g. 0.1 % v/v chloroform) and storage at 4 °C can stabilise them for at least 28 days[15]. It has also been reported that freezing to -20 °C for up to 28 days followed by slow thawing (12 h) shows no systematic difference as compared with immediate analysis[15]. To obtain operationally defined "dissolved" and "particulate" fractions, filtration through a 0.45 μm filter is used. To determine total nitrogen the sample must be digested e.g. by autoclaving in the presence of an oxidising agent such as potassium persulphate. Autoclaving has also been reported as a stabilisation process for sea water nutrients (TON and phosphate) and has been used in the preparation of a reference material for the analysis of sea water nutrients[16].

To minimise contamination and microbial growth during sample treatment and reagent preparation in this work the glassware was cleaned by soaking in Decon 90 for two days, washed with hydrochloric acid (10 % v/v), washed with Milli-Q water and oven dried.

3.2 Freshwater Analysis

Figures of merit for this manifold are shown in Table 3. The linear range adequately covers the concentrations required for monitoring compliance with legislation and the reproducibility over this range is well within the ± 5 % tolerance permitted[1]. The accuracy of this approach was assessed by participation in an intercomparison exercise involving fifteen European laboratories. The results obtained for two samples were 210 ± 5 and 879 ± 11 μmol kg^{-1} NO_3 as compared with the overall means of five replicates for all fifteen laboratories of 214 ± 4 and 886 ± 13 μmol kg^{-1} NO_3 respectively. For the FI method, the two samples were analysed on five consecutive days using fresh reagents on each day and the errors represent the standard deviations of the five results. The samples were bracketed between the two closest standards on each occasion and were analysed sequentially from low to high concentrations.

Table. 3 *Analytical figures of merit for the freshwater (Fig 1a) and estuarine and coastal (Fig 1b) manifolds*

	Freshwater	Estuarine and Coastal
Linear range (mg l^{-1} N)	0-5	0-1
Linear range (μM N)	0-360	0-72
Limit of detection* (mg l^{-1} N)	1×10^{-2}	1×10^{-3}
Limit of determination** (mg l^{-1} N)	5×10^{-2}	7×10^{-3}
Reproducibility (n = 5)	0.5 % (at 1 mg l^{-1}) 2 % (at 5 mg l^{-1})	0.1 % (at 1×10^{-3} mg l^{-1}) 1 % (at 0.5 mg l^{-1})
Throughput (h^{-1})	40	30

* based on 3σ
** based on 10σ

3.3 Estuarine and Coastal Water Analysis

The major significance of this manifold is that the response is completely uniform for samples with salinities ranging from 0-35 ‰, i.e. it is appropriate for monitoring TON concentrations in riverine, estuarine and sea water samples without the need for refractive index corrections. Figures of merit for this manifold are shown in Table 3. The limit of detection (0.1 μM NO_3-N) is sufficient for the analysis of TON in coastal waters. The accuracy of the approach was assessed by participation in a nutrients in sea water

intercomparison exercise involving over 100 international laboratories organised by the International Council for the Exploration of the Sea[17].
The mean results obtained for three sea water samples using the FI procedure were 2.1, 10.5 and 27.8 μM N and the modal concentration ranges on a histogram of results from all the participating laboratories were 1.2-1.6, 10.4-10.8 and 27.2-27.6 μM N respectively.

3.4 *In Situ* Analysis

The ultimate objective of this work is to remotely deploy the FI manifolds discussed above as part of a self contained submersible nutrient sensor. The results show that the manifolds can achieve the required analytical performance (linear range, limit of detection, accuracy and precision) in a laboratory environment. It has also been shown that all of the reagents and standards used for the above work are stable for at least 30 days in an environment with a fluctuating temperature (14-25 °C), and a sampling frequency of 8 h^{-1} can be attained (duplicate determinations of the sample and one standard) which is adequate for monitoring environmental changes in TON concentration. The manifold and associated reagents have been configured in a remote, battery powered housing which will be field tested in the near future.

4 CONCLUSIONS

The flow injection manifolds presented are capable of analysing TON in surface waters at the required concentrations with acceptable accuracy and precision. Particular care must be taken in the collection and storage of surface water samples and in the cleaning of glassware used for samples and reagents.

Acknowledgements

The authors would like to thank NERC for the award of a Research Grant (GST/02/669) and two CASE studentships under the SIDAL special topic initiative to support this work.

References

1. J. Gardiner and G. Mance, United Kingdom Water Quality Standards Arising from European Community Directives, Technical Report TR 204, Water Research Centre, Marlow, 1984.
2. L. Ebdon, J. Braven and N. C. Frampton, Analyst, 1991, **116**, 1005.
3. S. Motomizu, H. Mikasa and K. Toei, Anal. Chim. Acta, 1987, **193**, 343.
4. M. Noufi, C. Yarnitzky and M. Ariel, Anal. Chim. Acta, 1990, **234**, 475.
5. K. S. Johnson and R. L. Petty, Limnol. Oceanogr., 1983, **28**, 1260.
6. C. Oudot and Y. Montal, Mar. Chem., 1988, **24**, 239.
7. P. Raimbault, G. Slawyk, B. Coste and J. Fry, Mar. Biol., 1990, **104**, 347.
8. J. R. Clinch, P. J. Worsfold and H. Casey, Anal. Chim. Acta, 1987, **200**,523.
9. L. Anderson, Anal. Chim. Acta, 1979, **110**, 123.
10. Gine, M. F., Bergamin, H., Zagatto, E. A. G. and Reis, B. F.,Anal. Chim. Acta, 1980, **114**, 191.

11. Madsen, B. C., Anal. Chim. Acta, 1981, **124**, 437.
12. Van Staden, J. F., Anal. Chim. Acta, 1982, **138**, 403.
13. Van Staden, J. F., Joubert, A. E. and Van Vliet, H. R., Fresenius' Z Anal. Chem., 1986, **325**, 150.
14. Oxidised Nitrogen in Waters, Methods for the Examination of Waters and Associated Materials (HMSO, London, 1981) 61pp.
15. Results of the Comparative Studies of Preservation Techniques for Nutrient Analysis on Water Samples, Report PB88-142898 (US Department of Commerce, Springfield, VA, 1986) 89pp.
16. Aminot, A. and Kerouel, R., Anal. Chim. Acta, 1991, **248**, 277.
17. ICES Report on the Results of the Fourth Intercomparison Exercise for Nutrients in Sea Water (ICES, Copenhagen, 1991) 83pp.

Quantitative Determination of Chlorophyll A

Keyhandokht Kavianpour, Pedro W. Araujo, and Richard G. Brereton

SCHOOL OF CHEMISTRY, UNIVERSITY OF BRISTOL, CANTOCK'S CLOSE, BRISTOL BS8 1TS, UK

1 INTRODUCTION

Quantitation of chlorophylls is a major problem in many areas of science[1], especially when studying degradation reactions. Naturally occurring mixtures of chlorophylls and their degradation products are important in environmental monitoring[2-4] where the amount of chlorophyll reflects productivity. The relative amount of degradation products in oceans and lakes can signify the recent history of the water column. Numerous approaches have been applied in oceanography, including High Performance Liquid Chromatography[5] (HPLC), Electronic Absorption Spectroscopy (EAS) and Laser Fluorometry[6].

Because chlorophylls have vacant co-ordination sites, pure chlorophyll in the absence of external ligands cannot be obtained. A difficulty with chlorophyll *a* is that pure chlorophyll is hard to estimate by weighing, as it is highly hygroscopic and often hard to get free of grease. Conventional methods using electronic absorption spectroscopy suffer because of interferents in mixtures, and published equations provide widely different estimates of chlorophyll concentration in identical samples[7-12]. Diode Array Detector (DAD) High Performance Liquid Chromatography[1,5,13-15] is a promising approach for chlorophyll estimation, but internal standards have yet to be established[16]. This paper reports a new method for quantitative determination of chlorophyll *a*.

In this new approach the amount of chlorophyll is estimated using atomic spectroscopy to determine Mg in a sample, and this is calibrated to electronic absorption of pure standards. In turn, EAS is used to determine concentrations of pure standards which are calibrated to peak areas in DAD-HPLC, used to analyse mixtures of different chlorophylls.

2 EXPERIMENTAL METHODS

2.1 Apparatus

A Pye Unicam Model SP9 atomic absorption spectrometer with air-acetylene flame and a Pye Unicam Mg hollow-cathode lamp resonating at 285.2 nm with a 0.5 nm spectral width and 4 mA lamp current, was used for Atomic Absorption Spectroscopy (AAS) measurement. EAS were recorded using a Pharmacia LKB BioChrom Ultrospec III UV-visible spectrophotometer Model 80-2097-62 equipped with deuterium and tungsten halogen lamps. The spectra were obtained from a Wavescan software package with a resolution of 1 nm.

The chromatograms were recorded using a Waters 990 DAD High Performance Liquid Chromatograph equipped with a 600 E multi-solvent delivery system and a C_{18}RP column (300 × 39 mm id). Acetone, methanol and water were used for the mobile phase with a flow rate of 1 ml / min. Chromatograms were recorded between 350 and 800 nm and digitised every 2 s and 2 nm.

2.2 Reagents

General Purpose Reagent (GPR) methanol, distilled dioxane and GPR diethyl ether were used for extraction of chlorophyll *a*. Petroleum ether (bp 30-40°C) and *n*-propanol were used for open column chromatography. Stock solutions of $Mg(NO_3)_2$ containing 1 mg/ml Mg and HNO_3 (69%) were from BDH Chemicals. HPLC grade acetone and methanol (Aldrich) were used for chromatography, electronic and atomic spectroscopy. Ultrapure water was obtained from a Milli-Q water system (Millipore Corporation). All test solutions were prepared immediately prior to use.

2.3 Preparation and purification of chlorophyll *a*

Chlorophyll *a* was extracted from spinach using methods based on those of Svec[17], Iriyama *et al.*[8]and Rahmani *et al.*[19] 500 g of deribbed spinach leaves were blended together with ice cold methanol and the extract filtered. Pre-distilled 1,4-dioxane (methanol: dioxane; ratio 7:1 v/v) was added. Complete precipitation was obtained by adding dropwise ice cold water, then the solution was centrifuged for 30 min. A green solid was collected, dissolved in diethyl ether and rotatory evaporated to complete dryness. The chlorophyll was purified by column chromatography using icing sugar with anti-caking agents E341 (Tate and Lyle) as described by Strain and Svec[20].

3.5 kg of sugar was compressed into a large glass column (7.5 cm long, 10 cm diameter) by pounding the top with a metal plunger. A space of about 10 cm was left at the top of the column and a filter paper was placed on the surface. The column was washed with one bed volume of petroleum ether. The solid extract was dissolved in 10 ml diethyl ether and the column was developed with petroleum ether and *n*-propanol, increasing from 0 to 2 % v/v. The column was eluted with diethyl ether and the solution was dried . After repeating this procedure 5 times the pure chlorophyll *a* was obtained. The purity of chlorophyll *a* was monitored by HPLC.

3 ANALYTICAL PROCEDURE

Figure 1 shows all the analytical procedures used in this work.

3.1 Atomic Absorption Spectroscopy

The pure dry chlorophyll *a* was dissolved in 30 ml acetone. An aliquot of 1 ml of this solution was taken and the acetone evaporated. The dry solution was digested in 3 ml

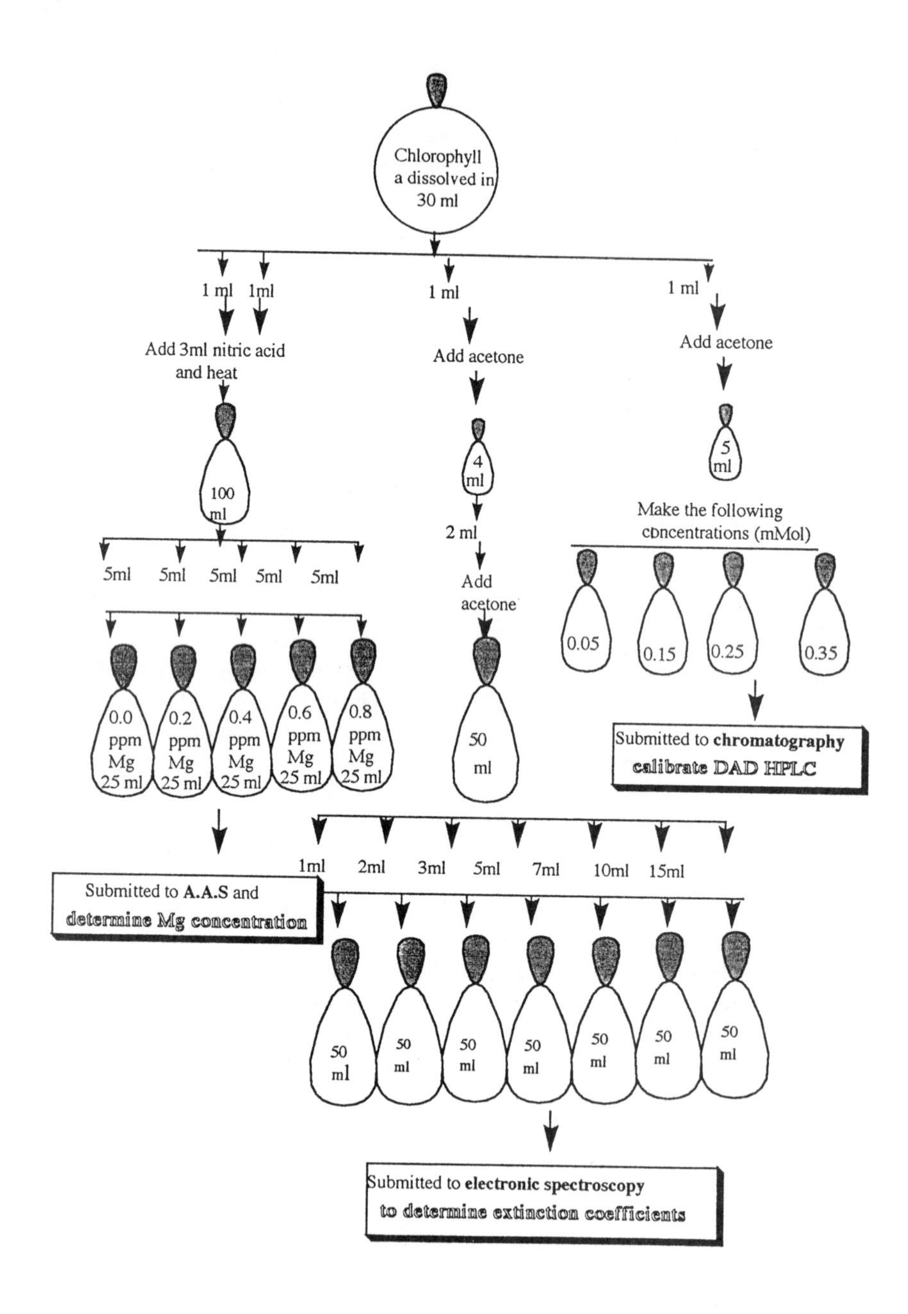

Figure 1 Analytical procedure

of nitric acid by gently heating until there was expulsion of nitrous oxide and total dissolution of the chlorophyll *a*. The sample solution was diluted with deionized water in a 100 ml volumetric flask and with this solution a standard addition curve was prepared, varying the concentration of Mg standard between 0 and 0.8 mg / ml and adding a constant volume (5 ml) of the sample solution. The final volume was 25 ml. The reproducibility was checked by taking duplicate aliquots of 1 ml.

3.2 Electronic Absorption Spectroscopy

1 ml of the chlorophyll *a* dissolved in 30 ml of pure acetone was dissolved in 4 ml of acetone. 2 ml of this solution placed in a volumetric flask (50 ml), diluted with pure acetone, and different volumes (1, 2, 3, 5, 7, 10 and 15 ml) of this solution diluted with pure acetone to give 50 ml, and EAS were recorded.

3.3 Diode Array High Performance Liquid Chromatography

A calibration curve was constructed by injecting in the DAD-HPLC concentrations of chlorophyll *a* in the range of 0.05 - 0.45 mM. The concentration was estimated by AAS and EAS. The solvent gradient used in this experiment is shown in Table 1.

Table 1 *Chromatographic system gradient*

Time(min)	*% Acetone*	*% Water*	*% Methanol*
0	0	7	93
2	0	7	93
17	40	0	60
25	0	7	93
30	0	7	93

4 RESULTS AND DISCUSSION

4.1 Atomic Spectroscopy

Two replicate samples were analysed by AAS. The analytical characteristics of the standard addition curves, amount of Mg and chlorophyll *a* estimated, are given in Table 2.

4.2 Electronic Absorption Spectroscopy

Knowing the amount of chlorophyll *a* estimated by AAS it is possible to determine extinction coefficients in pure acetone by electronic absorption spectroscopy. The calibration curves at three maxima are shown in Fig. 2.

Table 2 *The analytical characteristics of the standard addition curves*

Standard addition curve	*Slope AU × L × mg^{-1}*	*Intercept (AU)*	*Mg estimated(mg)*	*Chlorophyll estimated (mg)*
Curve 1	0.759	0.109	2.68	80.4
Curve2	0.759	0.108	2.67	80.0
				Average: 80.2

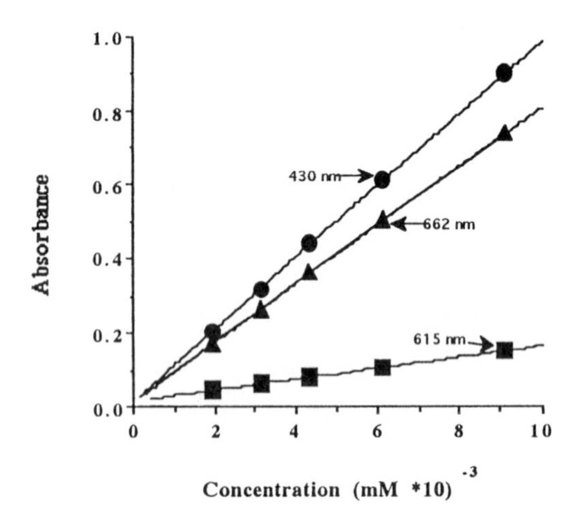

Figure 2 *Calibration curves for EAS.*

The absorbances at maxima close to 430, 615 and 662 nm were recorded and extinction coefficients calculated using linear calibration. The values of extinction coefficients which correspond to the slope of the curves are given in Table 3.

Table 3 *Extinction coefficients: values in brackets have been reported in the literature.*[8]

Extinction Coefficient units	*λmax 430 (nm)*	*λmax 615 (nm)*	*λmax 662 (nm)*
L × g^{-1} × cm^{-1}	108.610 (112.360)	16.553 (17.250)	88.770 (92.450)
L × mmol^{-1} × cm^{-1}	97.040 (100.391)	14.790 (15.413)	79.310 (82.602)

There are small solvent and spectrometer calibration shifts, and in this work we use the actual maxima even if the position changes shortly according to spectra.

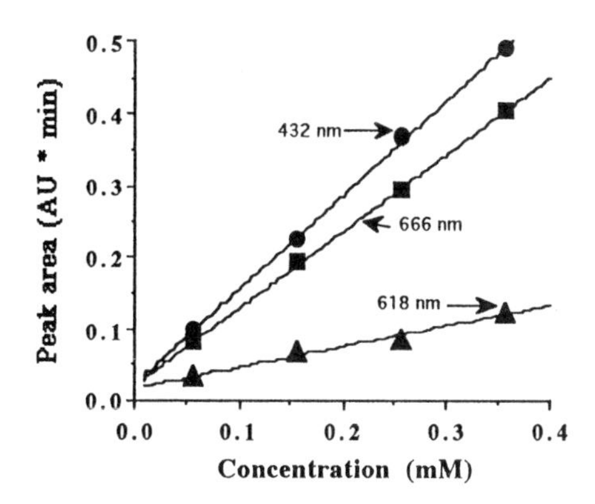

Figure 3 *Calibration curves for DAD HPLC.*

4.3 Diode Array High Performançe Liquid Chromatography

Peak areas at maxima close to 432, 618, and 666 nm estimated by EAS were recorded in Figure 3. Analytical characteristics of these curves are:

$A_{432} = 0.022 + 1.313[\text{chl } a]$ (1)
$A_{618} = 0.007 + 0.293[\text{chl } a]$ (2)
$A_{666} = 0.017 + 1.074[\text{chl } a]$ (3)

where A represents the peak area in Absorbance units $\times$ min and [chl *a*] shows the concentration of the chlorophyll *a* in mM, as estimated from EAS.

The discrepancies between the maxima and ratio of extinction coefficients obtained by EAS and by DAD-HPLC chromatography are due to the fact that in the former, extinction coefficients at different wavelengths are measured in pure acetone whereas in the latter, these values are measured in an acetone methanol mixture. Because of time lag between changing solvent conditions and detection, it is not easy to determine the exact solvent mixture in which the DAD-HPLC spectra are recorded, this being dependent on flow rate and column length; moreover both techniques can be used to estimate the amount of chloropigments with relatively good precision. This last observation is consistent with the following fact: the ratio of the slopes in Eqs (1) and (3) (1.223) is identical to the ratio to the extinction coefficients at 430 and 662 nm. We do not show comparison with the slope of Eq.2 because of noise and baseline problems.

Whereas the extinction coefficients as measured by EAS are independent of instrumental conditions, it is important to realise that the HPLC calibration curves are dependent on instrument (e.g. detector geometry), so the HPLC calibration has to be performed separately for each machine.

5 CONCLUSION AND FURTHER WORK

The approach developed in this study allows quantitation of chlorophyll by HPLC in the absence of internal standards. The method can be extended to the study of mixtures of chlorophylls such as in degradation reactions. The approach involves preparing pure degradation products (e.g. allomers) and using the method proposed above to calculate extinction coefficients and so calibrate the HPLC. Finally, chemometric factor analysis can be used quantitatively to resolve out individual elution profiles[1,5,21]. Chromatograms of a mixture typical of (allomers) are shown in Figure 4. Using the methods proposed above, it should be possible to isolate each pure allomer, calibrate HPLC areas to true concentrations and using factor analysis detemine concentrations of each compound.

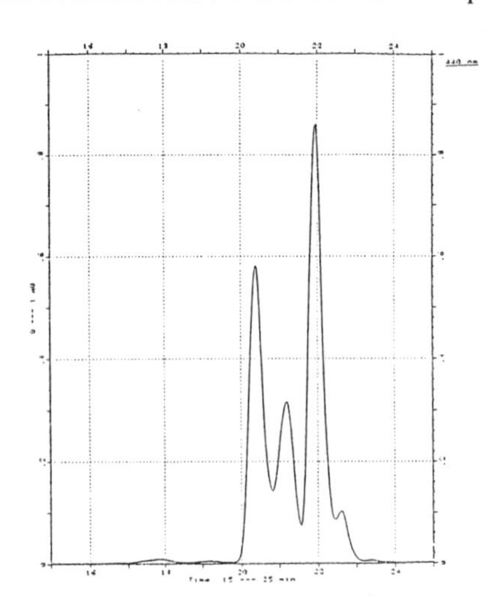

Figure 4 *Chromatograms of a typical mixture of allomers at 440 nm*

6. ACKNOWLEDGEMENTS

The Consejo Nacional de Investigaciones Cientificas y Tecnologicas de Venezuela (CONICIT) is thanked for financial support (PA), and Dr. D. J. Roberts for his valuable assistance in the AAS measurements.

References

1. Y.-Z. Liang, R.G. Brereton, O.M. Kvalheim, A. Rahmani, *Analyst,* 1993, **118**, 779.
2. J. Newton Downs, C.J. Lorenzen, *Limnol. Oceanogr.*, 1985, **30**,1024.
3. S.B. Brown, J.D. Houghton, G.A.F. Hendry, in *Chlorophylls* (Edited by H. Scheer), CRC press, Boca Raton, 1991, 456-489.
4. G.A.F. Hendry, J.D. Houghton, S.B. Brown, *New Phytol.*, 1987, **107**, 255.
5. Y.Shioi, in *Chlorophylls,* (Edited by H. Scheer), CRC Press, Boca Raton, 1991, 59-88.
6. A.F. Pasternak, A.A. Demidov, A.V Drits, V.V Fadeev, *Oceanology,* 1987, **27**, 5.
7. R.J. Porra, W.A. Thompson, P.E. Kriedemann, *Biochim. Biophys. Acta,* 1989, **975,** 384.
8. H.K. Lichtenhaler, *Methods Enzymol.*, 1987, **148**, 350.

9. R.J. Porra, L.H. Gimme, *Anal. Biochem.*, 1974, **57**, 255.
10. J.H.C. Smith, A.Benitez, *Modern Methods of Plant Analysis,* (Edited by K.Paech, M.V. Tracey), Springer-Verlag, Berlin, 1955, **4,** 42.
11. R. Zieler, K. Egler, *Beitr. Biol. Pflanz,* 1965, **41,** 11.
12. R.G. Brereton, A. Rahmani, Y.-Z. Liang, O.M. Kvalheim, *Photochem. Photobiol.*, 1994, **59**, 99.
13. R.F.C. Mantoura, C.A. Llewellyn, *Anal. Chim. Acta.*, 1983, **151,** 297.
14. R.B. Van Breeman, F.L. Canjura, S.J. Schwartz, *J. Chromatogr.* , 1991, **542**, 373.
15. P.M. Schaber, J.E. Hunt, R. Fries, J.J. Katz, *J. Chromatogr.*, 1984, **316,** 25.
16. J. De Figuereido, MSc thesis, *University of Bristol,* 1991.
17. W. A. Svec, in *Porphyrins* (Edited by D. Dolphin), Academic Press, New York, 1978, **5,** 341.
18. K. Iriyama, N. Ogura, A. Takamiya, *J. Biochem.*, 1974, **76,** 901.
19. A. Rahmani, C.B. Eckardt, R.G. Brereton, J.R. Maxwell, *Photochem. Photobiol.*, 1993, **57**, 1048.
20. H.H. Strain, W. A. Svec, in *Chlorophylls* (Edited by L. P. Vernon, G.R. Seely), Academic Press, New York, 1966.
21. E. Malinowski, *Factor Analysis in Chemistry,* Wiley, Chichester, 1991.

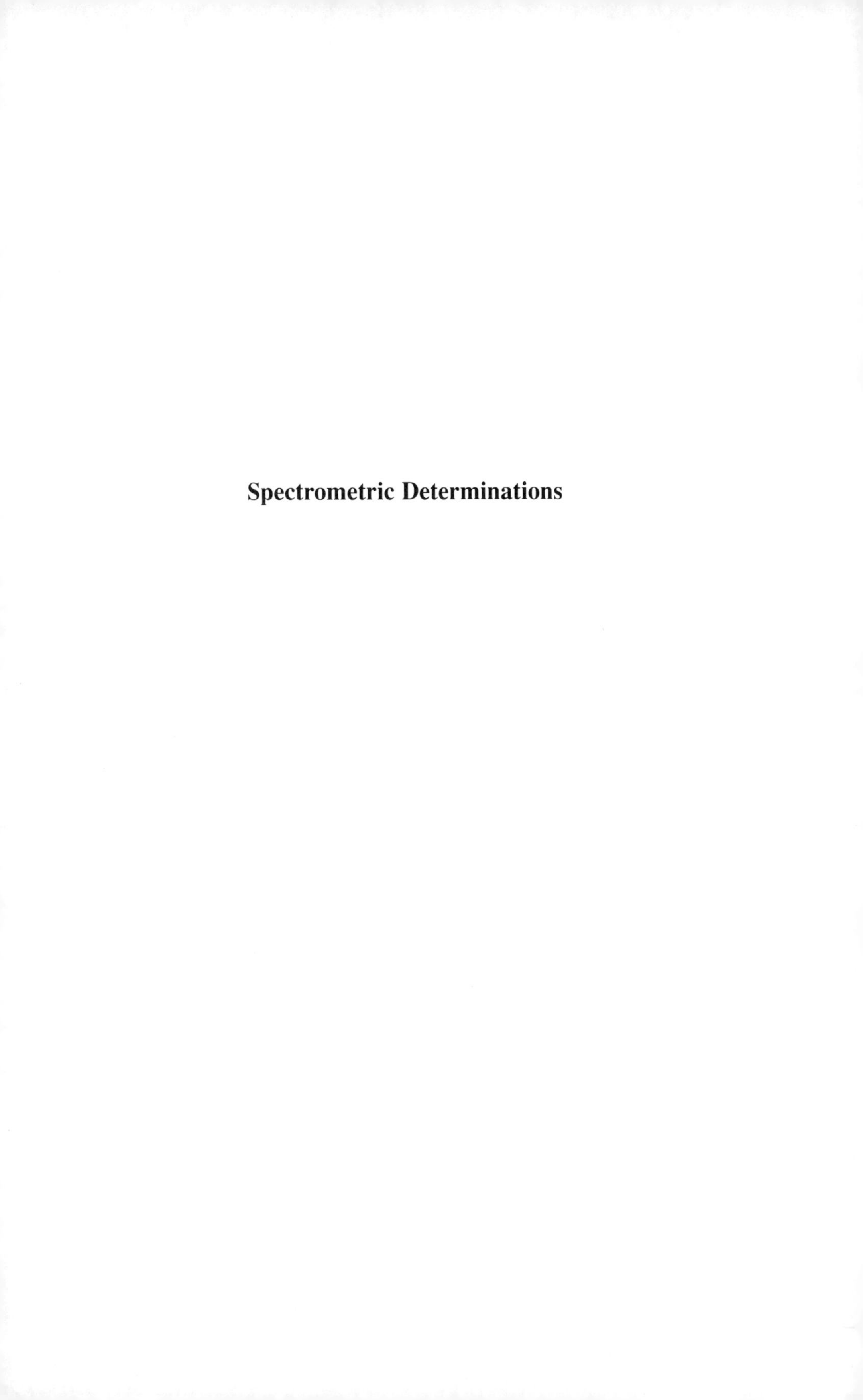

Spectrometric Determinations

Use of Mass Spectrometry Aboard United States Spacecraft

T. F. Limero

KRUG LIFE SCIENCES, 1290 HERCULES DRIVE, SUITE 120, HOUSTON, TEXAS, USA, 77058

1 INTRODUCTION

The first mass spectrometers (MS) built by the National Aeronautics and Space Administration (NASA) were flown in 1962 (Explorer 17) on unmanned probes with the goal of characterizing the chemical composition of the atmosphere of planets, including Earth. Later, more ambitious unmanned probes used mass spectrometry as one tool to seek evidence of the existence of life on other planets. As the space program was expanded to include human space flight, concerns for crew health cast the mass spectrometer in a new role as the key sensor for a metabolic analyzer. Concerns about the quality of spacecraft atmospheres led to intense efforts to develop trace-contaminant monitors. The complexity of volatile organic compound mixtures in spacecraft atmospheres required separation of those compounds by means of a gas chromatograph (GC) before the MS analysis. As NASA prepares for future missions of three months or more, in addition to continuous monitoring of major atmospheric components (i.e., O_2, N_2, CO_2, and H_2O), the potential use of regenerative life support systems requires periodic, onboard monitoring of some life support processes. Advanced mass spectral techniques are being explored to address the technical challenges associated with monitoring life support processes and assessing the air quality aboard spacecraft.

The challenges associated with developing spacecraft mass spectrometers tend to be specific to mission types; however, many design constraints are common for unmanned probes, manned spacecraft, and advanced space flight applications. The design driver for spacecraft mass spectrometers is to meet technical objectives in a reliable, small package that requires minimal resources. First and foremost, all components must be compatible with microgravity; this constraint severely limits pump options for maintaining MS vacuum. Also, instruments must be designed to survive the shock and vibration of lift-off. Auxiliary resources (i.e., carrier gas) that are not absolutely necessary are to be minimized or eliminated. Finally, any mass spectrometer to be used on spacecraft must be small, lightweight, and consume little power. The very stringent restrictions on weight, volume, and power for equipment aboard space station has driven NASA to consider gas chromatography/ion mobility spectrometry (GC/IMS) instead of gas chromatography/mass spectrometry (GC/MS) to monitor trace contaminants on spacecraft. For still longer missions, e.g., to Mars, these constraints will be more severe. However, for other potential missions involving humans, e.g., establishing a lunar base, resource constraints could be slightly relaxed.

The following material describes some of the instruments and mission goals for spacecraft mass spectrometers, and will conclude by describing investigations underway at NASA to address the technical challenges of advanced space missions.

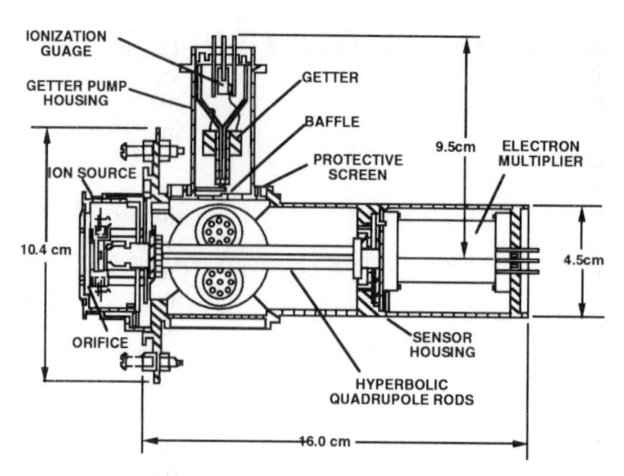

Figure 1 *Schematic of the upper-atmosphere quadrupole mass spectrometer used on the Pioneer-Venus Probe. Adapted from ref. 1*

1.1 Unmanned Missions

Mass spectrometers were used in unmanned missions to identify compounds associated with the atmospheres or surfaces of the planets being explored. The goal of these efforts was to better understand the solar system, planetary evolution, and the origins of life. Mass spectrometers flown on unmanned probes were subject to several unique design constraints above and beyond those described in the introduction. These instruments had to travel undamaged through the extremes of the space environment for months or years, and then, in response to remote commands, start up and perform perfectly. Instrument components had to be able to withstand radiation, temperature extremes, and G forces associated with entering planetary atmospheres. Finally, communications equipment had to transmit data accurately to Earth, where investigators analyzed and interpreted the information received.

1.1.1 Pioneer-Venus. Five mass spectrometers, customized for specific data collection tasks, were flown on the Pioneer-Venus probe. Two of these mass spectrometers will be described as examples of this instrument genre. Each mass spectrometer was designed and built to function in a specific strata of the Venusian atmosphere. A quadrupole-based mass spectrometer system was selected to measure constituents in the upper atmosphere of Venus, and a single focus, scanning magnetic-sector mass spectrometer was used to determine the gas composition of the lower atmosphere.[1-4]

Beyond the substantial engineering constraints on instruments, the major technical challenges were to obtain representative samples and to complete the analyses without significantly modifying the sample composition. These themes continue to drive the design of spacecraft analytical instrumentation even today. The approach selected to obtain a representative sample, to distinguish spacecraft and instrument generated species from sample species, and to perform all analytical processes without modifying the sample serves to illustrate the ingenuity and science incorporated into each component of the mass spectrometer system.

The lower-atmosphere mass spectrometer used a continuous sampling scheme based on a fixed ceramic micro-leak (CML) made of passivated stainless steel.[5] The CML was linked by a control valve to a pump, which was adjusted continually to maintain a constant pressure in the ionization source. A controlling feedback mechanism maintained the necessary 10^7 dynamic range during spacecraft descent. Velocity of the molecules relative to the probe was used to discriminate spacecraft-generated contaminants from the samples themselves. The sample inlet was designed to prevent the sample from reacting with the walls by constructing the walls of inert materials and by minimizing the contact between the sample gas molecules and the walls. Despite this clever design, early ground-based testing revealed a problem that required two years and a change of materials from stainless steel to tantalum before the CML could permit sulfuric acid, certain to be in the Venus atmosphere, to reach the ionization region of the MS.

The upper atmosphere was assessed with a quadrupole mass spectrometer connected to the ionization source by molecular leak (Figure 1). The ionization and analyzer regions were equipped with pumps to eliminate build-up of background contaminants during travel and to flush the sample quickly. Ion and getter pumps were selected because they were compatible with microgravity and consumed minimal power. In the open source mode, the velocity differential was used to discriminate sample from spacecraft contaminants. Electron ionization energy could be selected from 70 eV or 27 eV to discriminate constituents of equal mass. The entire assembly weighed 3.8 kg and required only 12 watts of power.

The lower atmosphere composition was investigated from a slightly different perspective, i.e., the use of scanning magnetic sector mass spectrometry to achieve wide dynamic and mass ranges. The instrument scanned from 1 to 208 amu in 64 seconds, and achieved a sensitivity of 1 ppm with dual collectors for the mass range 1-16 and 15-208. Figure 2 illustrates how modifications of the electron energy can be used to aid in identifying unknown substances. This instrument contained dual filaments, weighed 10.9 kg, and used 14 watts of power.

The results of the Pioneer-Venus experiments, shown in Table 1, provided a clearer understanding of the atmospheric composition of Venus.

1.1.2 Viking Lander (Mars). Objectives for unmanned probes sent to Mars presented a substantially new challenge by requiring the identification of organic compounds in soil samples. The detection of organic compounds in the Martian soil would have been strong evidence for the existence of life on Mars. This radical departure from previous planetary atmospheric analysis demanded a significant increase in instrument sophistication. In addition to more complex analyses, the Viking instrument had to survive the rigorous journey to Mars as well as the rough descent and landing on the planet surface.

On the Martian surface, organic compounds were extracted from soil samples by thermal volatilization and transferred to a GC column by a stream of radioactively labeled carbon dioxide.[6,7] The GC, with its hydrogen carrier gas, was needed to achieve the qualitative and quantitative goals of the mission. In order to preserve the mass spectrometer's ion pumps, a hydrogen separator was incorporated in the interface between the GC and MS. Additionally, the ion-pump current was used in a feedback mechanism to a series of restrictors that controlled the amount of GC effluent that entered the MS. This feedback arrangement protected the MS pumps from widely varying component concentrations. The MS was a double-focusing magnet sector instrument. A block diagram of this instrument is shown in Figure 3. The instrument weighed 20 kg.

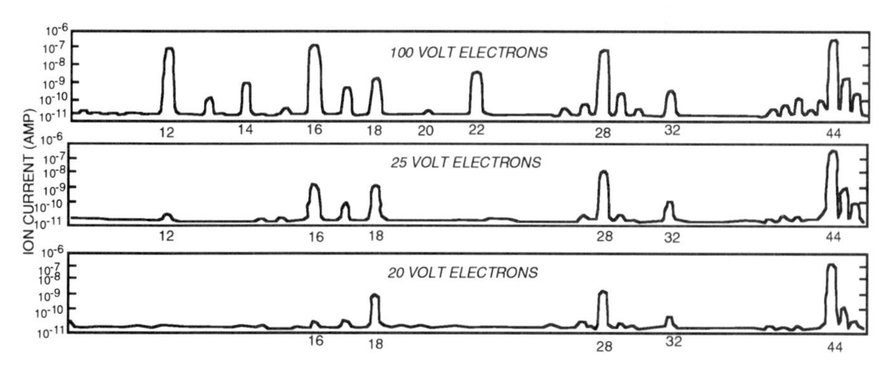

Figure 2 *Effect of electron beam energy on mass spectra of CO_2. Adapted from ref 2*

Table 1 *Mixing ratios in the lower atmosphere of Venus determined by Pioneer-Venus probe. Adapted from ref. 8*

Gas	Amount, ppm
Argon	40 -120
40/36	1.03 - 1.19
38/36	0.18
Carbon dioxide	96%
Carbon monoxide	20 - 28
Krypton	0.05 - 0.5
Neon	4.3 - 15
Nitrogen	3.41% (at 24km);4%
(percentages)	3.54% (at 44km)
	4.60% (at 54km)
Oxygen	16 (at 44 km); <30
	43 (at 55km)
Sulfur dioxide	185 (at 24km)
	<10 (at 55km)
Water	20 (at surface)
	60 - 1350 (at 24km)
	150 - 5200 (at 44km)
	200 - <600 (at 54km)

The packed GC column was filled with Tenax™ coated with a polymetaphenoxylene to minimize the interference of water and carbon dioxide with the organic analysis. The GC was temperature-programmed from 50°C to 200°C over 18-54 minutes. The ion source of the MS was operated at 225°C, with a filament energy of 70 eV. The masses were scanned from 11.5 to 215 in 10.24 seconds.

A typical total ion chromatogram (all masses above m/e 47) is shown in Figure 4. The large peak in spectrum 10015 was identified as methyl chloride, at a concentration of approximately 15 ppb. The only other organic contaminant detected in the Viking Lander samples was freon-E. These compounds were detected in prior testing on Earth; therefore their origin was assuredly terrestrial. The failure to detect organic compounds in the soil at two different locations on Mars strongly suggested that life was not flourishing on Mars.

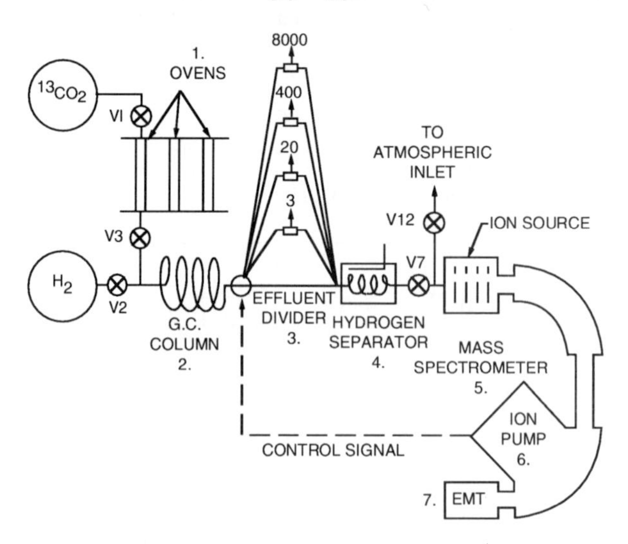

Figure 3 *Schematic of the gas chromatograph/mass spectrometer flown on Viking Landers. Adapted from ref. 5.*

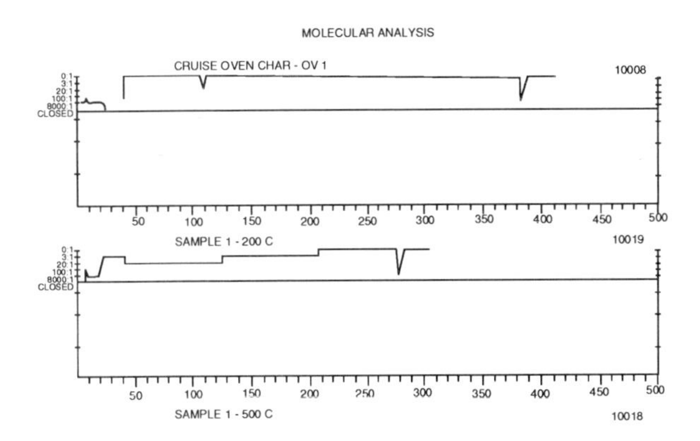

Figure 4 *Total ion currents (TIC) obtained from Viking GC/MS analysis during the cruise to Mars and from a sample of Martian Soil. Trace at top of each TIC is the split ratio fluctuation during the run. Adapted from ref. 5.*

1.2 Manned Missions

Once the technology had been established to propel humans safely into space, questions began to arise regarding how microgravity affected human physiology, particularly during long missions. Skylab provided the first opportunity to investigate putative changes in human physiology during extended stays in microgravity, and MS was an important part of these early experiments. Although complete automation of equipment was not strictly necessary for missions involving human crews, simple operation was and is preferable during space flight.

1.2.1 Skylab. The MS played a central role in a medical experiment (M-171) to investigate physiological responses during exercise in space.[8,9] A part of a device to analyze samples during exercise, the MS instrument evolved from the mass spectrometers developed for the Explorer and Viking programs. The MS was a magnetic-sector

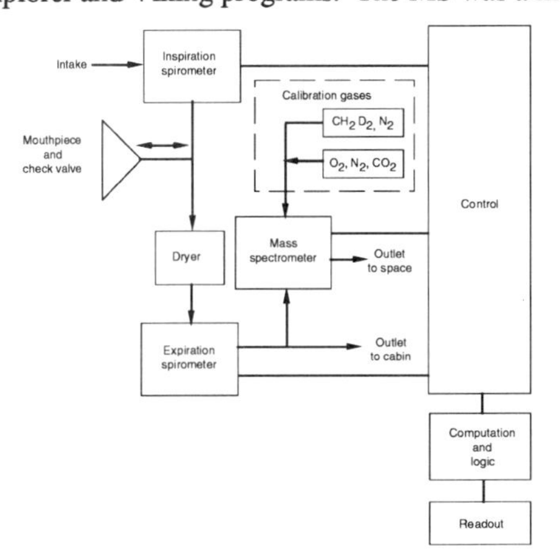

Figure 5 *Block diagram of the Skylab metabolic analyzer. Adapted from ref. 6.*

instrument that monitored preselected masses passing through exit slits in the range from 2 to 60 amu. The target substances were nitrogen, water vapor, oxygen, and carbon dioxide. A calibrant mixture was provided in the instrument to ensure accurate results. A block diagram of this instrument is shown in Figure 5.

The results from the Skylab experiment indicated that physical performance in flight could be maintained by adequate periods of strenuous physical exercise. Crews spent up to 1.5 hours/day performing both aerobic and isokinetic exercises on the last and longest (84 days) Skylab mission. A slight decrement after landing suggested that the Skylab crews had adapted to some aspect of the space flight environment as evidenced by decreased total blood volumes by the time they returned to Earth.

1.2.2 Space Shuttle. Mass spectrometers continue to be used in the U.S. Space Shuttle program for medical experiments, but in the design phase of the Shuttle program, it was recommended that a trace gas analyzer (TGA) be used to monitor atmospheric contaminants in the Spacelab module. Once again, the instrument design was based upon Viking Lander technology.[10,11] The Shuttle device was to be a GC/MS system that could detect up to 40 compounds at ppm concentrations (with preconcentration) in a package weighing 61.4 kg and using 150 watts of power. The addition of a preconcentrator, GC, valves, and heaters for the entire assembly is responsible for the large increase in weight and power consumption over previous MS flight units. Unfortunately, funding shortfalls and the management perception that such an instrument was not essential on Spacelab resulted in cancellation at the breadboard stage.

A gas analyzer mass spectrometer (GAMS), a magnetic sector instrument similar to that used on Skylab missions, was flown on STS-40 in 1991 and STS-58 in 1993 for medical experiments to document cardiopulmonary function in weightlessness.[12] In addition to the substances that could be detected by the Skylab version, the GAMS had additional collectors and electrometers to detect $C^{18}O$, N_2O, Ar, He, C_2H_2, and total hydrocarbon. Results from the STS-40/STS-58 investigations suggested that cardiac stroke volume and pulmonary capillary blood volume are elevated on orbit. Furthermore, the large gradients of lung function on Earth have only a modest influence on single-breath tests such as single breath nitrogen (SBN) washout. Therefore, these studies did not demonstrate completely the expected ventilation/perfusion relationships in space.

A new device, the gas analyzer for metabolic analysis physiology (GASMAP), is a quadrupole-based instrument meant to replace the GAMS units for metabolic analyses.[13] The GASMAP is being designed specifically for metabolic experiments, with improvements over the GAMS that include better dynamic range (i.e., 0-10 and 0-800 torr for N_2), quicker response time (approximately 100 msec), and extension of the mass range that permits detection of metabolic experiment compounds such as SF_6. The digital programming capability of GASMAP permits implementation of mass scans unique for a given study, a distinct advantage over the GAMS.

1.3 Space Station

Two separate requirements for MS instruments evolved during early design phases of the International Space Station. The first addressed the critical requirement to monitor major atmospheric constituents and provide feedback to the atmosphere revitalization subsystem of the environmental control and life support (ECLS) system. NASA selected a fixed-collector, magnetic-sector MS derived from the Skylab analyzer. All electronics were upgraded with state-of-the-art high reliability components.

The second requirement, to monitor trace volatile organic contaminants that might accumulate in the station, was addressed by two groups, each of which had their own perspective on mission goals and priorities. The ECLS group focused on trace contaminants as a marker of system functioning; the other group, toxicologists and chemists, focused on crew health and safety with regard to the contaminants themselves. Unexpected incidents of atmospheric contamination aboard Shuttle[14] and the Russian space station Mir[15] have confirmed the need to monitor trace contaminants on a space

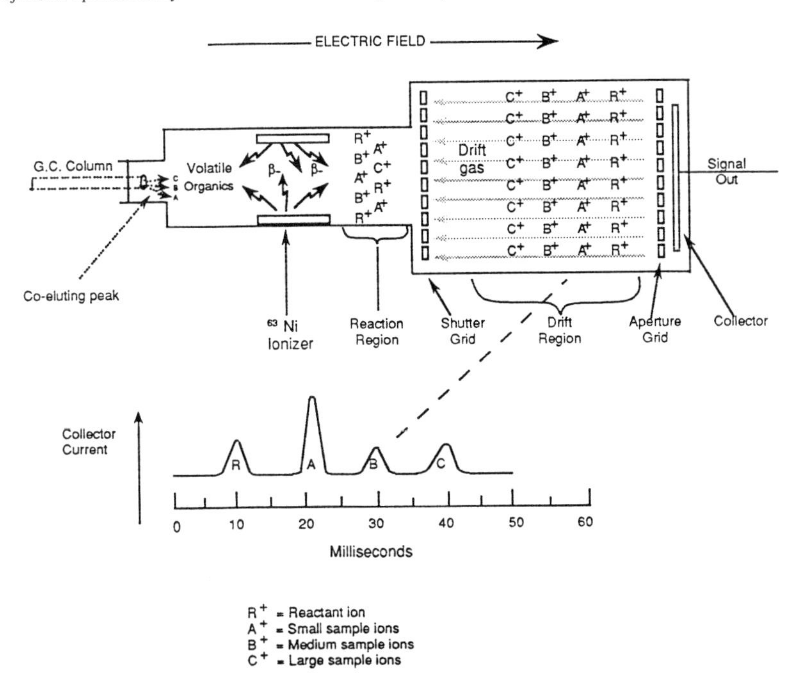

Figure 6 *Operating principle of ion mobility spectrometer*

station for the following purposes: to monitor contaminant accumulation, to warn of ECLS system malfunction, to detect leaks, and to assist in decontamination procedures after accidental chemical releases.

The ECLS group favors a trace contaminant monitor (TCM) based on Viking Lander technology. The manufacturer has stated that the modified version will be able to identify and quantify up to 200 compounds using a preconcentrator and a gas-chromatographic column for separation. The mass spectrometer will be improved to increase the scan speed to 2 seconds/scan, and the GC column will be a 30 meter hybrid column (polar and nonpolar phases) to achieve separation of a broad range of compounds.[16]

The toxicology group at Johnson Space Center investigated an ion trap detector and Hewlett-Packard 5971 MS coupled to a GC for use on a space station. As plans for an international station have been scaled back, severe limitations on resources forced exploration of other technologies. The toxicology group quickly focused on ion mobility spectrometry, the technology used in a hydrazine monitor flown on several Shuttle missions.[17] A gas chromatography/ ion mobility spectrometry (GC/IMS) instrument has been investigated to analyze volatile organic compounds in spacecraft because resources are conserved, since the IMS operates at atmospheric pressure and the nitrogen carrier gas for the GC simplifies the overall system (Figure 6).[18] The GC/IMS instrument can easily meet the performance requirements for a space station which include periodically quantifying 28 target compounds (alcohols, aromatics, aldehydes, etc.) at concentrations below their specified limits.

The major atmospheric constituents (O_2, N_2, CH_4, etc.) will be monitored aboard space station by an upgraded version of the type of magnetic-sector instrument flown on Skylab. Data from this instrument will be used to control some environmental conditions, e.g., oxygen concentration.

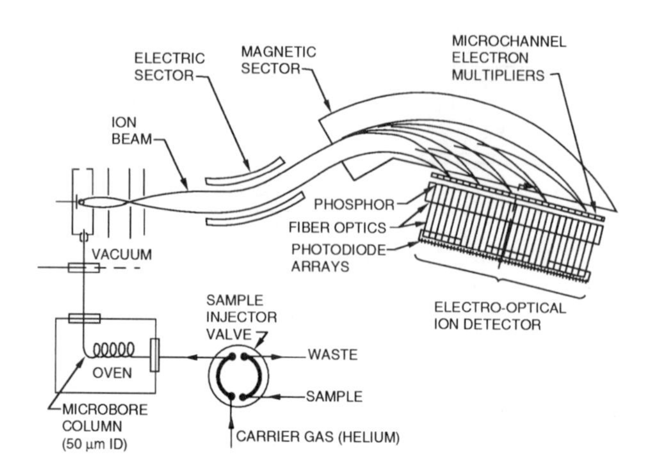

Figure 7 *Schematic of the microbore capillary column gas chromatograph and the focal plane mass spectrograph assembly with electro-optical ion detector. Adapted from ref. 18*

1.4 Advanced Missions

Two programs to develop contaminant monitors that use advanced mass spectral techniques are currently underway at the NASA Ames Research Center and the Jet Propulsion Laboratory (JPL). The group at Ames is applying MS/MS techniques to the problem of detecting a range of compounds quickly at low concentrations.[19] An important task of this effort is to develop expert systems combined with deconvolution algorithms to accurately analyze the data. This approach may lead to eliminating the GC, which would decrease instrument complexity and increase rate of analysis.

At JPL, investigators are coupling a small focal plane mass spectrometer (Mattauch-Herzog) to an array detector (electro-optical ion detector) to achieve very quick analysis.[20] Since the magnet need not be scanned, this instrument is compatible with capillary and even microbore GC columns, and can be extremely sensitive. A block diagram of this instrument is shown in Figure 7.

2 CONCLUSIONS

The unique nature of the space environment, coupled with the extreme constraints on resources owing to the expense and harshness of space flight, encourage the development of creative engineering solutions to monitoring air quality. Mass spectrometers developed by NASA have helped characterize the atmosphere of many planets, have searched for life on Mars, and have provided a tool for medical science to investigate the effects of microgravity on human physiology. In the future, the portable, analytically powerful instruments under development at NASA will find applications in monitoring Earth-based hazardous sites and in meeting specific monitoring requirements in the industrial-commercial community.

3 ACKNOWLEDGEMENTS

I would like to thank Dr. Charles Sawin, Dr. John James, and Ms. Chris Wogan for contributions.

References

1. R. Fimmel, L. Colin, and E. Burgess, 'Pioneer Venus (NASA SP-461)'. Scientific and Technical Information Branch, Washington, DC, 1983, pp.63, 73-75.
2. H. Niemann and W. Kasprzak, *Adv. Space Res.,* 1983, **2**, No.12, 261.
3. N.W. Spencer, 'Proceedings, Outer Planet Probe Technology Workshop', Sect. VIII, NASA, Goddard Space Flight Center, Greenbelt, MD, 1975, p 16.
4. J. H. Hoffman, 'Proceedings, Outer Planet Probe Technology Workshop, Sect. VIII, NASA, Goddard Space Flight Center, Greenbelt, MD, 1975, p 44.
5. R. Fimmel, L. Colin, and E. Burgess, 'Pioneer Venus, NASA SP-461, Scientific Technical Information Branch, Washington, D C, 1983, p 23.
6. K. Biemann, J. Oro, P. Toulmin III, L. Orgel, A. Nier, D. Anderson, P. Simmonds, D. Flory, A. Diaz, D. Rushneck, J. Biller, and A. LaFleur, *J. Geophysical Res.* 1977, **82**, No. 28, p 4641.
7. K. Biemann, *Origins of Life,* 1974, **5**, p. 417.
8. E. Michel, J. Rummel, C. Sawin, M. Buderer, and J. Lem, Results of Skylab Medical Experiment MI7I- Metabolic Activity. In: 'Biomedical Results of Skylab (NASA SP 377) ', R. Johnston and L. Dietlein, Scientific and Technical Information Branch, Washington, DC, 1977, pp. 372, 445.
9. E. Michel, J. Rummel, and C. Sawin, *Acta Asironautica,* 1975, **2**, 351.
10. F. White and G. Wood, 'Mass Spectrometry: Applications in Science and Engineering', John Wiley & Sons, New York, 1986, p 530.
11. Perkin-Elmer Aerospace Division, 'TGA Final Configuration Review'. TGA NAS9-15432, 1979.
12. J. West, H. Guy, G. Prisk, and P. Wagner, Pulmonary function in weightlessness. In: 'Spacelab Life Sciences-1: 180-Day Preliminary Results', unpublished report. NASA-Johnson Space Center, Houston, Tx, p 3.1.3-1.
13. Medical Sciences Division, 'Gas Analyzer System for Metabolic Analysis Physiology: Statement of Work', NASA- Johnson Space Center, Houston, Texas.
14. J. T. James, T.F. Limero, H.J. Leano, J.F. Boyd, and P.A. Covington, *Aviat. Environ. Med.,* 1994.
15. Unpublished data, from Dr. V. Savina, Institute of Medical and Biological Problems Moscow, Russia, 1994.
16. W. Niu, The development of an atmosphere composition monitor for the environmental control and life support system. (Society of Automotive Engineers Technical Paper Series No. 921149). 22nd International Conference on Environmental Systems, Warrendale, PA, 1992.
17. T. Limero, J. Cross, S. Beck, J. James, H. Johnson, N. Martin, D. Davis, G. Eiceman, and J. Brokenshire, 'Preliminary studies of IMS with 5-nonanone chemistry to perform hydrazine monitoring functions aboard spacecraft at reduced

pressures. In: 'Proceedings of the 1992 Workshop on Ion Mobility Spectrometry', 1993.

18. T. Limero, J. Brokenshire, C. Cuming, E. Overton, K. Carney, J. Cross, G. Eiceman, and J. James, A volatile organic analyzer for space station: description and evaluation of a gas chromatograph/ion mobility spectrometer. (Society of Automotive Engineers Technical Paper Series No. 921385). 22nd International Conference on Environmental Systems, Warrendale, PA, 1992.
19. P. Palmer and C. Wong, Development of ion trap mass spectrometric methods to monitor air quality for life support applications. (Society of Automotive Engineers Technical Paper Series No. 932206). 23rd International Conference on Environmental Systems, Warrendale, PA, 1993.
20. M.P. Sinha, Development of a microbore capillary column GC-focal plane mass spectrometograph with an array detector for field measurements. In: 'Proceedings of the Second International Symposium on Field Screening Methods for Hazardous Wastes and Toxic Chemicals', Las Vegas, NV, 1991.

Determination of Rare Earth Elements in Their Mixtures Using Inductively Coupled Plasma Atomic Fluorescence Spectroscopy

A. A. Galkin, G. N. Maso, and G. G. Glavin

DEPARTMENT OF CHEMISTRY, MOSCOW STATE UNIVERSITY, 119899 MOSCOW, RUSSIA

1 INTRODUCTION

In recent years the demands for ultra-pure rare earth elements (REE) have grown [semiconductors, lasers etc.]. REE are found naturally in mixture because their chemical properties are closely related, resulting in difficulties in separation of these elements. It has been suggested to separate compounds of REE by means of fractional sublimation[1]. In this connection, to control completeness of separation, the problem arises of how to determine trace amounts of one REE in the presence of the REE matrix. We suggest Inductively Coupled Plasma Atomic Fluorescence Spectroscopy (ICP-AFS) for this purpose. This technique combines high sensitivity, low detection limits, large dynamic ranges, small interelement interferences and a simultaneous multielement capability. Besides, ICP-AFS prevents most of Atomic Absorption Spectroscopy (AAS) and Inductively Coupled Plasma Atomic Emission Spectroscopy (ICP-AES) drawbacks[2].

The aim of this work is to develop a procedure for analysis of REE mixtures using ICP-AFS.

2 EXPERIMENTAL

Three REE were selected for investigations (Dy, Eu and Er). Eu and Er are elements of Cerium and Holmium subgroups respectively, and Dy is a borderline element. Standard solutions of these elements were prepared by dissolving $EuCl_3 \cdot 6H_2O$, $ErCl_3 \cdot 6H_2O$ and $DyCl_3 \cdot 6H_2O$ (Aldrich Chemical Co., 99.99 % pure) in 5 % spectroscopic grade nitric acid. The composition and concentration of all solutions are presented in table 1.

Eu and Dy in acid solutions were determined by ICP-AFS using a commercially available spectrometer AFS-2000 manufactured by Baird, USA. The fluorescence was excited by pulsed hollow cathode lamps (HCL) at wavelengths 459.4, 462.7, 466.2 nm for determination of Eu and 416.8, 4118.7, 419.5, 421.2 nm for determination of Dy. The desired band of the spectrum was cut out with an optical interference filter.

The optimization of operating parameters was done for achieving the lowest limits of detection (DL) for Eu and Dy. The influence of the following operating parameters on the DL was studied: the RF power; carrier gas (Ar) and reductant gas (methane) flows; the observation height above the tip of the plasma torch. We found the following operating parameters optimum: RF power - 950 W for Dy and 700 W for Eu; Ar inlet pressure 35 psi, which provides carrier gas flow of 1.7 l/min (Meinhard type nebulizer), methane flow - 150-180 ml/min (visually it corresponds to a tall and wide green plasma tailplume); observation height 95 mm for Eu and 90 mm for Dy (above the inner tube of the ICP torch); sample uptake rate - 1.6 ml/min. Each measurement included 10 HCL pulses of 0.5 sec each. A standard mode of instrument two point calibration was used for calibration of the detector.

It is known that the coefficient of fractional sublimation of REE mixtures is not over 3[3]. Thus a standard solution of Eu and Dy dipivaloylmethanates ($Dy(Dpm)_3$, $Eu(Dpm)_3$) was prepared with molar ratio of Dy and Eu 1 : 5. Test - portion of 30 mg was dissolved in 50 ml of 5 % nitric acid. The theoretical and measured concentrations of Dy in obtained solution were 65.4 and 63.8 ppm respectively. Reversible standard deviation (RSD) for Dy in this solution was 2.71 %.

3 RESULTS AND DISCUSSION

The atomization efficiency in ICP and, consequently, the precision of analyses are temperature and plasma composition dependent. In an effort to estimate the degree of atomization that depends on the temperature, calculations of an equilibrium plasma composition have been accomplished for Yttrium, since thermodynamic data for Eu, Dy and Er are not on hand. Chemical properties of Y and REE are closely related to each other and the temperature dependence of the atomization degree for Y is likely analogous for REE. An equilibrium composition was calculated for temperatures between 1000 and 14000 K using the principle of maximum entropy. For temperatures above 6000 K thermodynamic properties of some substances were obtained by extrapolation.

Table 1 *The composition and concentration of standard REE solutions*

Number of solution	*Concentration of REE, ppm*		
	Dy	*Eu*	*Er*
1	84.81	-----	-----
2	84.81	79.20	-----
3	84.81	192.00	-----
4	84.81	300.01	-----
5	84.81	408.01	-----
6	84.81	504.01	-----
7	84.81	-----	79.98
8	84.81	-----	194.54
9	84.81	-----	302.62
10	84.81	-----	399.90
11	84.81	-----	497.17
12	-----	79.2	-----

All components which were taken into account are presented in table 2. The logarithm of the quantity of Yttrium particles depending on the temperature is shown on figure 1. The maximum of relative quantity of Y atoms is exhibited at 5000 K but at this temperature only 1 % of Y is found in atomic form. This is explained by formation of strongly dissociative Yttrium oxide compounds; therefore, to increase the atomization efficiency, a reductant gas should be introduced into the plasma, e.g. methane.

In the presence of methane the RF power - dependence of the photodetector current has been obtained for standard solutions of Eu and Dy (see figures 2 and 3). It is evident that determination of Dy and Eu needs to be accomplished at RF powers of 950 and 700 Watts, respectively, because in each case the atomic fluorescence signals are maximal.

The influence of REE presence on determination of Eu and Dy was investigated by the example of systems Dy-Eu and Dy-Er. For these purposes, measurements of solutions 2-6 and 7-11 were carried out (see table 1). The influence of the Eu and Er content in solutions on the measured Dy concentration is shown on figures 4 and 5, respectively. The enhancement of Eu and Er concentration in solutions tends to decrease the Dy signal. These results indicate that the matrix of solutions involving REE has to be made in accordance with the matrix of standard REE solutions.

The reproducibility and DL for Eu, Dy, Dy in the presence of Eu, Dy in the presence of Er, and Eu in the presence of Dy have been estimated. For this purpose, measurements of solutions 12, 1, 2, 7 and 2 have been carried out (see table 1). Average values, RSD and DL for Dy and Eu in all selected solutions are presented in table 3. Obtained DL are comparable with analogous values in AAS and ICP-AES[4]. Based on the results obtained we can conclude that the presence of a second REE in solution does not affect the precision of analysis. Low detection limits permit the use of small quantities of substance in analysis. This is very important because REE compounds are costly.

The ICP-AFS method was used for investigations of the separation completeness of Dy and Eu volatile compounds. In the case of $Dy(Dpm)_3$ and $Eu(Dpm)_3$ the separation by means of sublimation in vacuum was not observed, and after an addition of pivalic acid into this mixture the separation coefficient became equal to 3.

Table 2 *List of components used for thermodynamic calculations*

A set of elements	*A set of components*
Ar	$O, O^+, O^-, O_2, O_2^+, H, H^+, H^-, H_2, H_2^+, OH,$
H	$OH^+, OH^-, HO_2, H_2O, H_2O^+, Ar, Ar^+, N, N^+,$
O	$N_2, NO, NO^+, NH, NH^+, N^-, Ar_2^+$(**), O_2^-(*),
Y	O_3(*), H_2O_2(*), H_3O^+(*), H_3^+(*), NO_2(*),
N	NH_2(*), NH_3(*), HNO(*), HNO_2(*), Y, Y^+, YO, YO^+, YO_2(*), electrons.

(*) - these components were taken into account in the range from 1000 to 7000 K.
(**) - these components were taken into account in the range from 8000 to 14000 K.

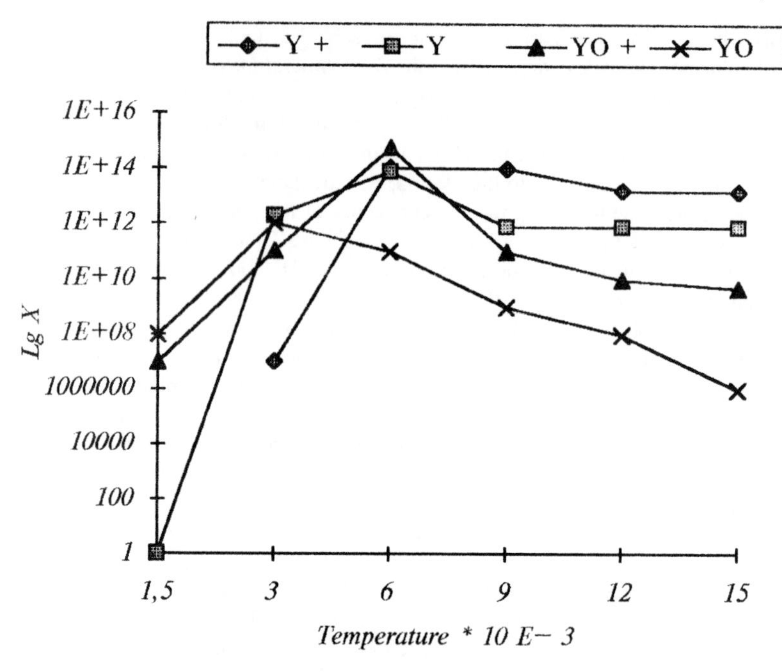

Figure 1 *Logarithm of quantity of Yttrium particles depending on the temperature*

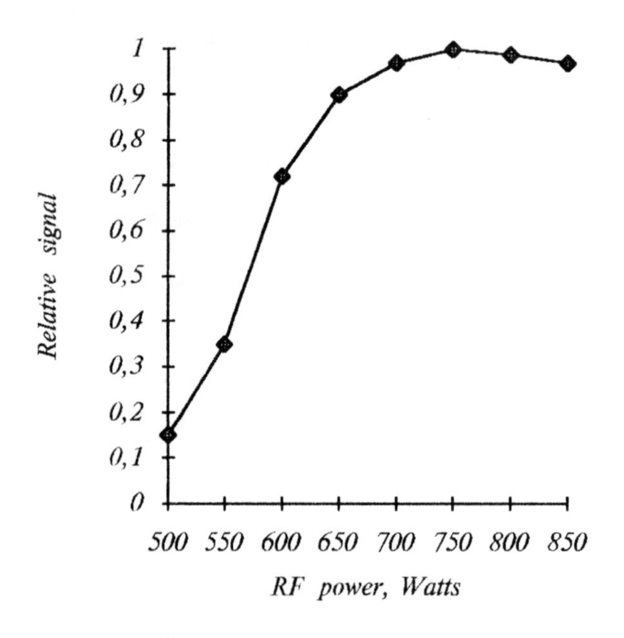

Figure 2 *Relative signal of Eu depending on the RF power*

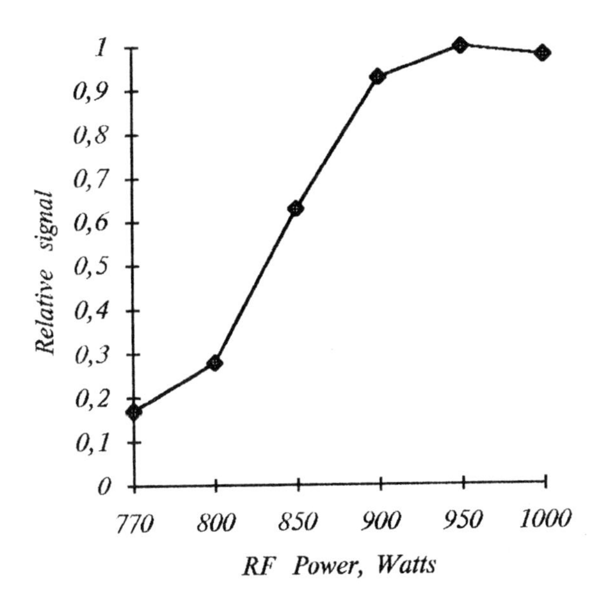

Figure 3 *Relative signal of Dy depending on the RF power*

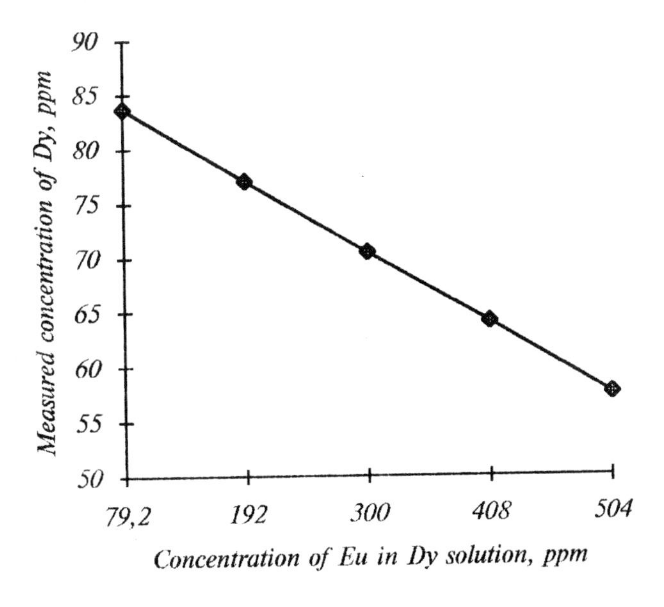

Figure 4 *Measured concentration of Dy depending on the concentration of Eu in solution*

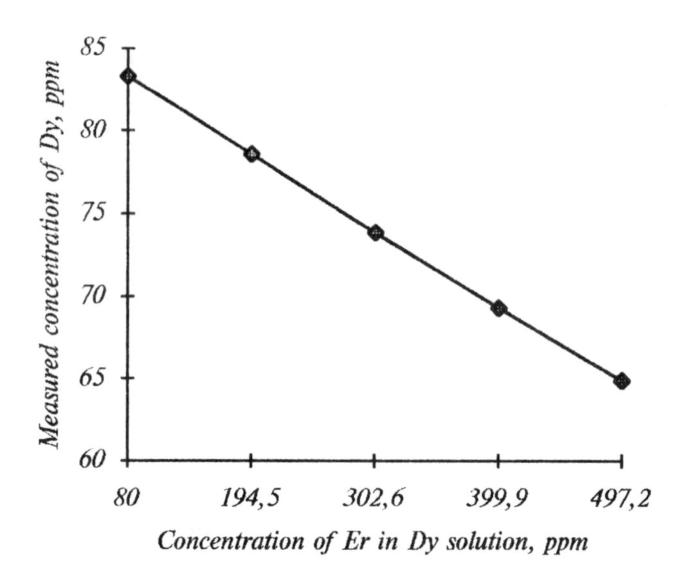

Figure 5 *Measured concentration of Dy depending on the concentration of Er in solution*

4 CONCLUSION

In the present work the technique for determination of REE in their mixtures using ICP-AFS was developed. It can be applied to the control for separation completeness of REE compounds.

Table 3 *Average values, RSD and DL for Eu and Dy in acid solutions*

Number of solution	*Element*	*Average value, ppm*	*RSD, %*	*DL, ppm*
12	Eu	78.92	2.18	0.065
1	Dy	84.54	2.03	0.072
7	Dy in the presence of Er	83.33	2.26	0.078
2	Dy in the presence of Eu	82.78	2.56	0.075
2	Eu in the presence of Dy	78.96	2.44	0.069

References

1. A.T. Zoan, N.P. Kuzmina, L.I. Martinenko, G.N. Maso, *Coordination Chem. (Russia)*, 1992, **18**, 649.
2. D.R. Demers, C.D. Allemand, 'The spectral line and background interference maze in ICP-AES : a way out via a new approach', Baird Corporation, Bedford, 1981.
3. L.M. Aistring, 'Chemistry and technology of REE', Mir, Moscow, 1970.
4. M. Thompson, J.N. Walsh, 'A Handbook of Inductively Coupled Plasma Spectrometry', Blackie, London, 1983.

Chemiluminescence Based Detection of Aldehydes and Carboxylic Acids in Engine Oils

A. N. Gachanja and P. J. Worsfold

DEPARTMENT OF ENVIRONMENTAL SCIENCES, UNIVERSITY OF PLYMOUTH, PLYMOUTH, DEVON PL4 8AA, UK

1 INTRODUCTION

Oxidation is one of the ageing processes that degrades the performance of an engine oil. One of the major pathways for oxidation is the formation of peroxide intermediates by a radical chain mechanism, which then react to yield a variety of polar species such as alcohols, aldehydes, ketones, and carboxylic acids. These compounds interact with foreign particles and water producing suspended solids and sludges which block filters, oil lines and lubrication grooves[1].

It is therefore desirable to have a mechanistic understanding of the oxidation pathways, which necessitates analytical procedures for monitoring chemical parameters in oils. Carboxylic acids are major end products of oil oxidation and aldehydes are intermediates in the oxidation process. Both are therefore important indicators of the degree of oil oxidation.

The analysis of used engine oils for carboxylic acids and aldehydes is complicated by the highly coloured and viscous sample matrix with high suspended solid content. Quantification of the total polar fraction can be achieved by thin layer chromatography coupled with flame ionisation detection while acid content of the oils may be determined by potentiometric titration. These methods however do not differentiate between individual acidic components in the oil, such as aliphatic and aromatic carboxylic acids, sulphonic acids and phenols.

2 EXPERIMENTAL

A series of oxidised oils sampled from a test engine after different periods of operation (0, 16, 24 and 32 h) were analysed for aliphatic aldehydes and carboxylic acids.

2.1 Pre-column Derivatisation

2.1.1 Aldehydes. Aldehyde standards in toluene, fresh oil dialysate spiked with aldehyde standards and used oil dialysate samples (100 μl) were added to a mixture of 3-aminofluoranthene (1.0 x 10^{-3} mol l^{-1})in glacial acetic acid-toluene (2:7 v/v; 2.0 ml) and borane-pyridine complex (5 μl). Toluene was added to bring the final volume to 2.25 ml. The reaction mixture was shaken for 3 min.

2.1.2 Carboxylic acids. Carboxylic acid standards in heptane, fresh oil dialysate spiked with carboxylic acid standards and used oil dialysate samples (250 μl) were separately mixed with 1.0 ml of 9-anthracenemethanol in dichloromethane (2.1 x 10^{-2} M), 1.0 ml 4-pyrrolidinopyridine in dichloromethane (1.0 x 10^{-2} M) and dicyclohexylcarbodiimide (0.25 g: DCC), dissolved in 15 ml heptane and refluxed for 15 min. After allowing the reaction vessel to cool for 5 min, the solvent was removed by rotary evaporation and the residue redissolved in 5 ml hexane.

2.2 Solid Phase Extraction Methodologies

2.2.1 Aldehydes. Neutral alumina Sep-Pak cartridges (Waters) were used to remove the excess 3-aminofluoranthene before liquid chromatography (LC) analysis of the aldehyde derivatives. The cartridge was precleaned with 10 ml hexane. 100 μl sample, dissolved in the reaction solvent mixture, was loaded on the alumina cartridge and the aldehyde derivatives eluted with 15 ml of 20% (v/v) ethyl acetate in hexane. The solvent was removed by rotary evaporation and the residue redissolved in 10 ml acetonitrile:tetrahydrofuran (4:1) for LC analysis.

2.2.2 Carboxylic acids. Silica Sep-Pak cartridges (Waters) were used to remove the excess label before LC analysis of the carboxylic acid derivatives. The cartridge was precleaned with 10 ml hexane. 200 μl sample was loaded on the cartridge and the carboxylic acid derivatives eluted with 20 ml of 20 % dichloromethane in hexane. The solvent was removed by rotary evaporation and the residue redissolved in 5 ml acetonitrile:tetrahydrofuran (40:60) for analysis.

2.3 Flow Injection and Liquid Chromatographic Analysis

Flow injection (FI) was used to optimise the chemiluminescence (CL) detection conditions for carboxylic acid and aldehyde derivatives prior to their detection after LC separation.

The manifold shown in Fig. 1 was used for analysis of the carboxylic acid derivatives. A single reagent stream of mixed bis (2,4,6-trichlorophenyl)oxalate (TCPO) and hydrogen peroxide in acetonitrile was used for analysis of the aldehyde derivatives. The detectors used were a charge coupled device (CCD; 270M imaging spectrograph, 1200 grooves mm^{-1} grating and 578 x 385 pixels, liquid nitrogen cooled CCD chip (Instruments SA)) and a photodiode (PD) based CL detector (Camspec Ltd, Cambridge).

After simplex and univariate optimisation of the CL signal-to-noise ratio with the FI system, the carrier stream was replaced with the eluent from an ODS2-5 LC column (250 x

3.2 mm, Phenomenex). Chromatograms were recorded on a chart recorder or a PC based integration system.

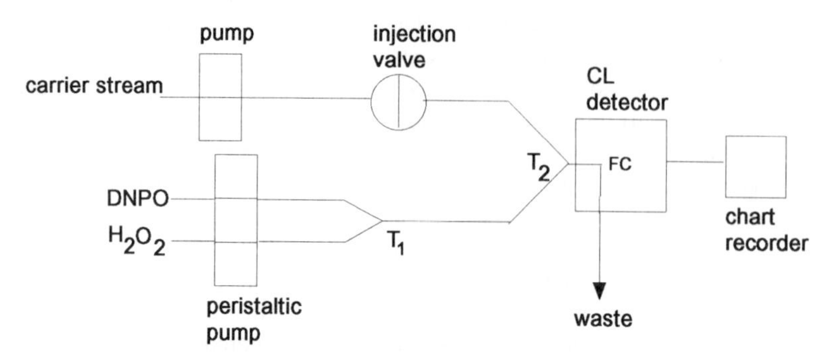

Figure 1 *FI manifold for the determination of the carboxylic acid derivatives with chemiluminescence detection. Injection volume = 20 µl, FC = Flow cell, T_1 & T_2 = mixing T-pieces, DNPO- bis (2,4-dinitrophenyl)oxalate.*

3 RESULTS AND DISCUSSION

3.1 Pre-column Derivatisation Reactions

The oil samples were initially fractionated using a continuous dialysis system[2]. A solution of oil dissolved in light petroleum spirit (60/80 °C) and contained in a semipermeable membrane (dry, hypo-allergenic incontinence sheath rubber; nominal molecular mass cut-off 1000 daltons) was continually extracted with warm petroleum spirit (60/80 °C) for 24 h. This was done to remove polymeric additives, organometallic oxidation products and solid debris which could interfere with the pre-column derivatisation.

Aliphatic aldehydes and carboxylic acids do not possess chromophoric substituents but are amenable to precolumn derivatisation, using fluorescence labels followed by LC separation and fluorescence or chemiluminescence detection. A procedure for labelling of aldehydes and ketones, based on reaction with 3-aminofluoranthene in non-aqueous media followed by fluorescence and chemiluminescence detection has been reported[3]. The reaction scheme involved reductive amination of the aldehyde with 3-aminofluoranthene in the presence of a reducing agent, borane-pyridine complex and is shown in Figure 2.

Esterification of carboxylic acids in non-aqueous media with 9-anthracenemethanol utilising a coupling agent dicyclohexylcarbodiimide (DCC), has been shown to proceed rapidly[4] and is complete in 15 min. The carboxylic acid reacts with DCC to form an intermediate with a good leaving group. The carboxylic acid-DCC derivative then reacts with the labelling alcohol to form dicyclohexylurea and the ester. The reaction is enhanced in the presence of a base catalyst, pyridine or 4-pyrrolidinopyridine. The equation for the reaction is given in Figure 3.

In derivatisation reactions, the label should be present in excess to ensure complete labelling of the analytes but the unreacted label often degrades the chromatographic separation of early eluting compounds. Low efficiency solid phase extraction cartridges are often used to separate the excess label from the derivatives, minimising the possibility of further reactions during storage. In addition, the clean-up step removes many other compounds in the oil matrix, e.g. phthalate esters, which would otherwise accumulate in the LC column resulting in rapid deterioration. The choice of solid phase material and solvent mixture is determined by the nature and polarity of the label and derivatives. The solid phase extraction methodologies applied to the separation of 3-aminofluoranthene label from the aldehyde-3-aminofluoranthene derivatives, and 9-anthracenemethanol label from 9-anthracenemethyl esters, gave >97 % recoveries of the derivatives.

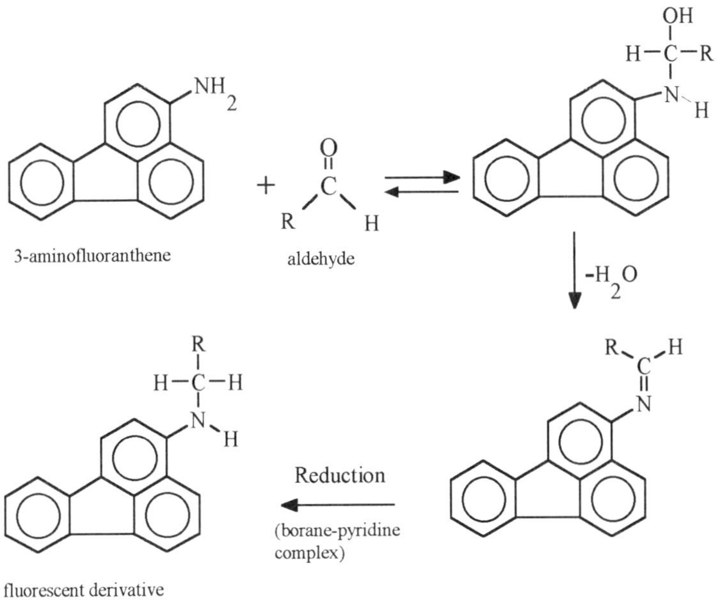

Figure 2 *Reductive amination of aldehydes with 3-aminofluoranthene in the presence of borane-pyridine complex as the reducing agent.*

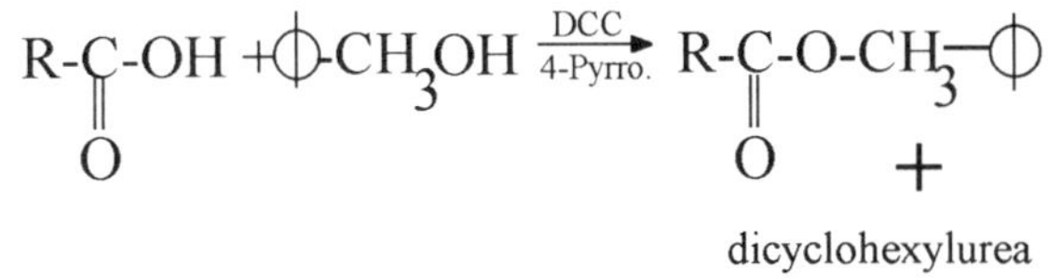

$$R\text{-}C(=O)\text{-}OH + \phi\text{-}CH_3OH \xrightarrow[\text{4-Pyrro.}]{\text{DCC}} R\text{-}C(=O)\text{-}O\text{-}CH_3\text{-}\phi + \text{dicyclohexylurea}$$

Figure 3 *Esterification reaction of carboxylic acids catalysed by 4-pyrrolidinopyridine. 4-pyrro. = 4-pyrrolidinopyridine, DCC = dicyclohexylcarbodiimide & ϕ = 9-anthracenemethyl.*

3.2 Instrumentation for CL Detection

Photomultiplier tubes (PMTs) are the most commonly used light sensing detector and typically cover the spectral range from 200 - 700 nm with a radiant sensitivity >30 mA/W. A PMT requires a stable high voltage power supply (ca 1.0 kV) and gives a low background signal at room temperature. PMTs with bi-alkali photocathodes are usually selected for CL detection since they give the lowest background signal and highest sensitivity at the most common CL emission wavelengths (350 - 500 nm).

Recent advances in solid state technology have provided cheap, robust and reliable photodiodes which can be used as alternative light detection devices in analytical instrumentation. The photosensitive area of photodiodes ranges from 1 mm^2 to 100 mm^2 with a radiant sensitivity > 0.1 A/W in the spectral range 200 - 900 nm. Photodiodes are powered by low voltages (ca. 15 V) and give maximum sensitivity at 700 nm. Diode arrays and CCDs for molecular spectroscopy exploit this solid state technology.

3.3 Optimised FI Conditions

3.2.1 Aldehyde analysis. The blank gave a peak due to the reaction between the acetic acid and 3-aminofluoranthene. The optimised conditions using TCPO/H_2O_2 mixed reagent with 3-aminofluoranthene as the analyte are shown in Table 1.

Table 1 *Optimal CL conditions using a TCPO/H_2O_2 mixed reagent in an FI system.*

Factor	PD CL detector
Carrier flow rate	0.5 ml min^{-1}
Mixed CL reagent flow rate	1.0 ml min^{-1}
[TCPO]	1.3 x 10^{-3} M
[H_2O_2]	8.8 x 10^{-2} M
[Imidazole]	2.5 x 10^{-2} M
pH	7.5

3.3.2 Carboxylic acid analysis. The optimised conditions obtained using DNPO in an FI system are given in Table 2.

Table 2 *Optimal conditions for CL detection using a PD and CCD based CL detection.*

		CL Detector	
Factor	Unit	PD	CCD
Carrier flow rate	ml min^{-1}	1.4	0.37
[DNPO]	mM	3.42	2.27
DNPO flow rate	ml min^{-1}	0.5	0.85
[H_2O_2]	M	0.88	2.2
H_2O_2 flow rate	ml min^{-1}	0.5	0.85

Calibration data for CL signal (mV) versus concentration (M) of the ester derivatives of C_6 - C_{18} aliphatic acids and benzoic acid over the range 0 - 36.3 x 10^{-7} M were linear with RSDs (%) ≤ 2.7 (n=10) using the PD CL detector (see Table 3). An overlay of the CL spectra of the 9-anthracenemethyl ester derivatives obtained using the CCD is shown in Figure 4:

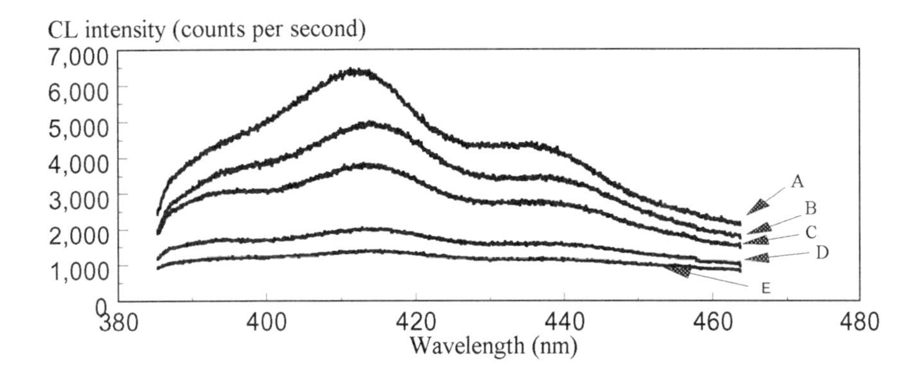

Figure 4 *CL spectra of 9-anthracenemethyl ester derivatives of carboxylic acids. A - 9-anthracenemethanol (2.04×10^{-4} M); B - hexanoate ester (7.43×10^{-4} M); C - octanoate ester (1.77×10^{-4} M); D - hexadecanoate ester (4.54×10^{-4} M); E -benzoate ester (2.86×10^{-4} M).*

3.4 LC Calibration Data

Calibration data for straight chain aldehydes (C_6, C_8, C_{10}, C_{12} and C_{14}) spiked on the 0 h oil matrix were all linear over the range 0 - 5.0×10^{-7} mol ml $^{-1}$ ($0.9980<r^2<0.9997$) and the limits of detection (S/N = 3) were in the range 0.7 - 75 fmol on-column (5 μl injection). Similarly, calibration data for aliphatic carboxylic acids (C_6, C_7 ,C_8, C_9, C_{10}, C_{12}, C_{16}, C_{18}, C_{20}) spiked on the 0 h oil matrix were all linear over the range 0 - 1.85×10^{-7} mol ml^{-1} ($0.9933<r^2<0.9992$) and the limits of detection (S/N = 3) were in the range 1.8 - 4.5 pmol on-column (5 μl injection).

Table 3 *Calibration data for carboxylic acids in an FI system*

Ester	Gradient	Intercept	r^2	DL*(pg)
benzoate	26911990	0.48	0.9998	208
C_6	30226881	0.44	0.9994	183
C_8	30064063	0.56	0.9995	200
C_9	32456619	0.92	0.9999	191
C_{16}	27402166	1.73	0.9996	293
C_{18}	20505266	1.68	0.9977	417

*DL = detection limit calculated at 3 x baseline noise.

3.5 Oil Analysis

Analysis of the oil dialysate samples for aldehydes showed that C_6 - C_{17} aliphatic aldehydes appeared in the 16 h used oil dialysate sample and concentrations of the aldehydes increased with test duration[5]. Analysis of the oil dialysate samples for carboxylic acids showed that aliphatic carboxylic acids (C_6, C_7 ,C_8, C_9, C_{10}, C_{12}, C_{16}, and C_{18}) appeared in the 24 h used oil dialysate and also increased with test duration[4]. An LC-CL chromatogram showing the analysis of carboxylic acids in the 24 h used oil dialysate with peak identities is shown in Figure 5. In the LC analysis, the peaks were identified using capacity factors and confirmed by spiking standards on the samples before derivatisation.

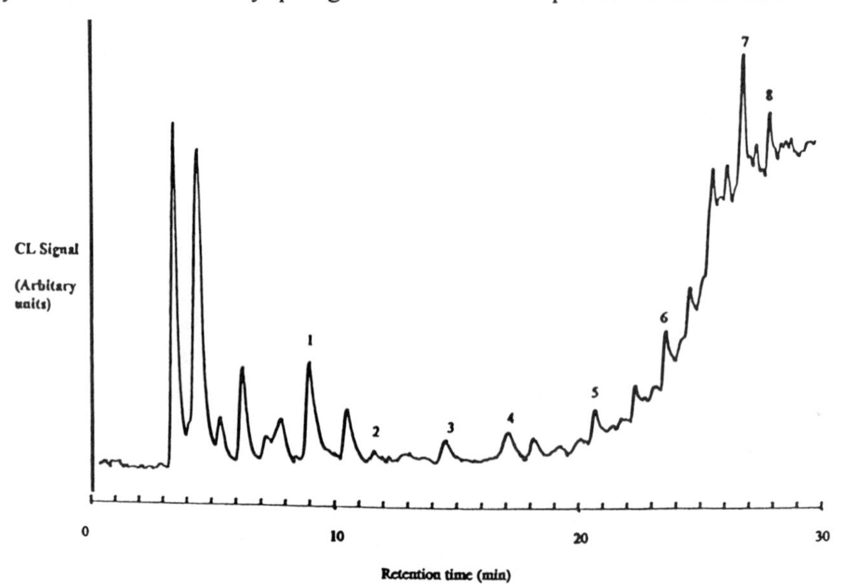

Figure 5 *LC-CL of a 24 h used engine oil dialysate. Peak identities: 1 = C_6, 2 = C_7, 3 = C_8, 4 = C_9, 5 = C_{10}, 6 = C_{12}, 7 = C_{16}, 8 = C_{18}. (From Ref. 4, with permission).*

Acknowledgements

This study was funded by SERC grant GR/H49528 as part of a DTI TAPM LINK award. The authors would like to thank Tony Moss (Camspec Ltd, Cambridge) for providing the PD based CL detector.

References

1. D. Klamen, 'Lubricants and related products', Verlag Chemie, Basel, 1984.
2. S. W. Lewis, P. J. Worsfold and E. H. McKerrell, J. Chromatogr. 1994, **667**, 91.
3. B. Mann and M. L. Grayeski, J. Chromatogr. 1987, **386**, 149.
4. A. N. Gachanja and P. J. Worsfold, Anal. Chim. Acta, 1994, **290**, 226.
5. A. N. Gachanja, S. W. Lewis, E. H. McKerrell and P. J. Worsfold, submitted to J. Chromatogr., 1994.

Chemiluminescence Based Detection of Hydrogen Peroxide in Seawater

D. Price,[1] P. J. Worsfold,[1] and R. F. C. Mantoura[2]

[1] DEPARTMENT OF ENVIRONMENTAL SCIENCES, UNIVERSITY OF PLYMOUTH, PLYMOUTH, DEVON PL4 8AA, UK

[2] PLYMOUTH MARINE LABORATORY, WEST HOE, PLYMOUTH, DEVON PL1 3DH, UK

1 INTRODUCTION

Hydrogen peroxide is present at elevated concentrations at the sea surface (up to 500 nM) where it can affect the redox state of many chemical species. The major source is via photochemical reactions involving certain dissolved organic chromophores (Org) and molecular oxygen[1].

$$Org + h\upsilon = Org^*$$
$$Org^* + O_2 = Org^+ + O_2^{\cdot-}$$
$$2O_2^{\cdot-} + 2H^+ = H_2O_2 + O_2$$

The production of H_2O_2 therefore results in the conversion or destruction of a portion of the organic carbon pool in the ocean which has far reaching implications on the "Greenhouse Effect". Quantifying H_2O_2 in seawater can also be used to trace photochemically active areas of water[2]. Other minor sources (< 5%) to the ocean include wet/dry deposition of photochemically produced H_2O_2 in the atmosphere[3] and extracellular production from certain species of phytoplankton[4].

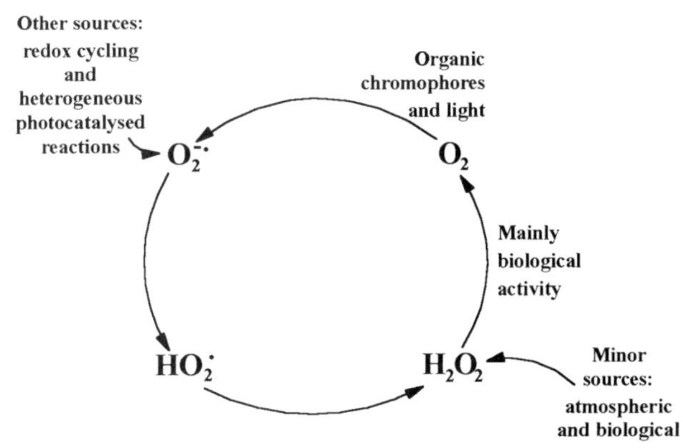

Figure 1 *Diagrammatic Representation of the Cycling of Hydrogen Peroxide in Seawater*

1.1 Why Chemiluminescence ?

The most important analytical requirement for monitoring H_2O_2 in seawater is rapid analysis due to the reactivity of the analyte. The consequences of this are that the monitor must be deployed *in situ*, i.e. at sea, and allow a high sample throughput. Most liquid phase chemiluminescence (CL) reactions are very fast (< 30 s) and are ideal for use in flow sytems[5]. The CL reaction used in this method is the H_2O_2 induced oxidation of luminol (5-amino-2,3-dihydro-1,4-phthalazinedione). The reaction is optimal at pH 10.8 and cobalt(II) is the most efficient inorganic catalyst. The emitted light (λ_{max} = 440 nm) is detected at a coiled glass flow cell by an end-window photomultiplier tube (PMT) with internal amplifier.

2 METHOD DEVELOPMENT

The method was optimised to allow fast and reproducible determination of H_2O_2 within an expected ambient concentration range of <5 - 500 nM. Full details of the development work are published elsewhere[6].

2.1 Flow Injection Manifold

A simple flow injection (FI) manifold (Figure 2) was used which gave a high sample throughput and the most efficient CL response. A small volume of sample (100 µl) was introduced into the luminol stream, transported to a T-piece for mixing with the cobalt(II) catalyst and the emitted light detected at a coiled glass flow cell positioned close to the end-window of a photomultiplier tube (PMT). A simplex optimisation was used to achieve the most efficient operational settings (e.g. reagent concentrations, flow rate, pH).

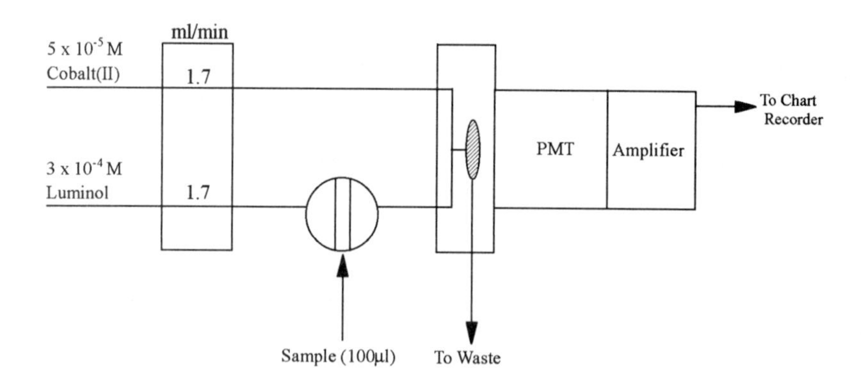

Figure 2 *FI Manifold for the Determination of Hydrogen Peroxide in Seawater*

2.1.1 Automation. The FI manifold was automated using a custom-made in-house designed 8052 micro-controller enabling the control of pumps and injection valve.

2.2 Hydrogen Peroxide Free Water

Hydrogen peroxide was found to be present in source waters such as Milli-Q at concentrations of up to 50 nM. If untreated Milli-Q was used for preparing the reagents a high CL background would cause a serious limitation to the monitor's sensitivity. MnO_2 chemically bonded to Amberlite XAD-7 polymeric beads was therefore used to produce highly efficient and stable scavenging columns to remove all H_2O_2 from the Milli-Q water.

2.3 Figures Of Merit

Table 1 *Figures of Merit for the Determination of Hydrogen Peroxide in Seawater*

Limit of Detection ($S/N = 3$)	5 nM
Linear Range	5 - 500 nM
Correlation Coefficient (R^2)	0.9983
Reproducibility* (n = 4)	RSD = 1.3 %
Repeatability (100 nM H_2O_2; n =5)	RSD = 0-2 %
Sample Injection Rate	120 h^{-1}

* Four separate samples collected at sea at precisely the same location and time. The mean concentration was 42.3 nM.

3 SHIPBOARD DEPLOYMENT

A berth of opportunity was obtained on R.R.S. Discovery in the western Mediterranean for July 1993. The entire analytical system including all supporting equipment was transported in purpose made boxes by road and ship to the port of Gibraltar. The cruise was composed of two legs; the first began at the Straits of Gibraltar, headed for the Straits of Sicily before docking at Monaco. The second leg left Monaco for a more intensive study of the Gulf of Lions and terminated at Nice.

3.1 Shipboard Procedures

Reagents and standards were freshly prepared each day. Samples were collected from "Go-Flo" bottles surrounding the CTD (a frame containing sensors and other equipment which is winched over the side of the ship on an oceanographic wire to km depths) and transferred to 50 ml amber bottles for immediate analysis. The analysis of 12 samples including standard additions took just 45 minutes to complete.

3.2 Results

Depth profiles of H_2O_2 were collected from a wide variety of oceanographic environments in the western Mediterranean. Figure 3 shows a typical profile collected at 0045 hours (local time). The profiles were found to be affected by various parameters including; time of day, latitude, cloud cover, light attenuation in the water column and supply of suitable organic chromophores. Highest surface H_2O_2 concentrations were found just before sunset at low latitude in mesotrophic waters.

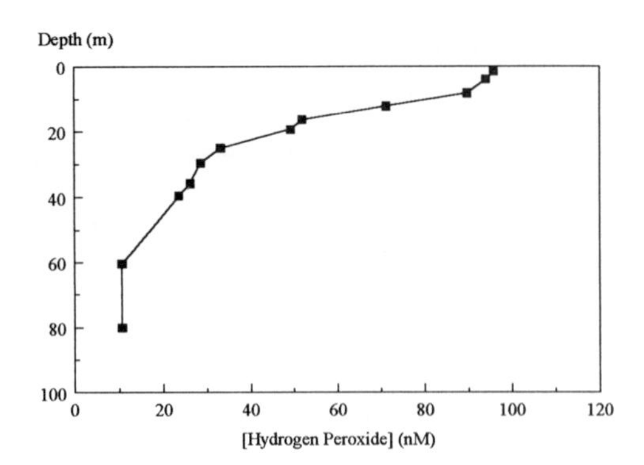

Figure 3 *Depth Profile of Hydrogen Peroxide at Station MD1 at 42° 52' N and 3° 47' E*

Other experiments onboard ship included diurnal Lagrangian surveys and deck based photochemical generation studies using seawater samples contained in sealed silica tubes.

4 CONCLUSIONS

The FI-CL monitor has suitable figures of merit to allow shipboard measurements of H_2O_2. The analytical performance at sea was excellent throughout with no problems over the thirty days in the Mediterranean Sea. Environmental results were in agreement with spectrophotometric[7] and fluorimetric[8] techniques but have the additional benefits of being faster, more sensitive and automated.

The application of simple CL based detection systems coupled with FI has been found to be ideal for demanding environments such as working onboard a research vessel. The robustness of the system was also validated by arduous transportation to and from the Mediterranean.

Acknowledgements

Funding for this project was made available by a NERC CASE award in collaboration with Plymouth Marine Laboratory.

References

1. W. J. Cooper, R. G. Zika, R. G. Petasne and J. M. C. Plane, *Environ. Sci. Technol.*, 1988, **22**, 1156.
2. R. G. Zika, P. J. Milne and O. C. Zafiriou, *J. Geophys. Res.*, 1993, **98-C2**, 2223.
3. W. J. Cooper, E. S. Saltzman and R. G. Zika, *J. Geophys. Res.*, 1987, **92-C3**, 2970.
4. S. E. Stevens, Jr., C. O. P. Patterson and J. Myers, *J. Phycol.*, 1973, **9**, 427.
5. K. Robards and P. J. Worsfold, *Anal. Chim. Acta*, 1992, **266**, 147.
6. D. Price, P. J. Worsfold and R. F. C. Mantoura, *Anal. Chim. Acta* (in press).
7. K. S. Johnson, C. M. Sakamoto-Arnold, S. W. Willason and C. L. Beehler, *Anal. Chim. Acta*, 1987, **201**, 83.
8. C. A. Moore, C. T. Farmer and R. G. Zika, *J. Geophys. Res.*, 1993, **98-C2**, 2289.

The 'Co-master' Concept in Developing Robust Near Infrared Calibrations

I. A. Cowe, C. G. Eddison, N. Hewitt, and A. M. C. Davies[1]

MULTISPEC LTD, WHELDRAKE, YORK Y04 6NA, UK

[1] NORWICH NEAR INFRARED CONSULTANCY, 75 INTWOOD ROAD, CRINGLEFORD, NORWICH NR4 6AA, UK

1. INTRODUCTION

Near Infrared Analysis (NIR) is almost unique among analytical techniques for measuring chemical composition in that it relies entirely upon multivariate mathematical modelling. Unlike more conventional methods, an instrument cannot be calibrated by measuring a series of absolute physical or pure chemical standards. Instead, by taking a number of empirical measurements, we develop a calibration model which is valid only in the context of the matrix of constituents which are present in a series of representative samples.

A crucial element in the success of NIR analysis has been the development of transferable calibrations allowing models developed at one site to be installed in other instruments at diverse locations. Producing calibrations which function to commercially acceptable standards on a large number of instruments adds a further level of complexity to calibration development. With a single instrument "Accuracy", defined as the Standard Error of Prediction (SEP) comparing predicted values by NIR with reference chemical data, is the only criterion. However, because commerce demands that all instruments in a market predict the same value for a sample, "Transferability", defined as the SEP between predicted NIR values for the same set of samples on two instruments, is of equal if not greater importance than Accuracy.

Even with instruments standardised to be as optically similar as possible, there are always small residual differences which lead to variations in performance under field conditions. The need to minimize these differences has led Multispec to adopt a co-master calibration strategy where data from several instruments are used to develop a model. This produces a calibration which is robust to minor instrument differences and minimises the SEP for transferability.

2. MATERIALS AND METHODS

2.1 Instrumentation

The instrumentation used in this work was the Foss Electric Grainspec (developed by Multispec Ltd., York, England). This is an eleven filter transmission instrument covering the Herschel region of the spectrum (800 to 1075 nm). The patented design is unusual in that the angle at which the light beam meets any interference filter can be varied using a device called a "tuning wheel". As well as a "centre wavelength" this provides two additional observations from each filter. The "first offset" shifts the output approximately 8% towards the visible while the "second offset" shifts the beam approximately 15% towards the visible. The spectrum obtained consists therefore of thirty three points.

This geometry means that there are three different optical paths and, as a consequence, three different energy levels reaching the detector. These differences are minimised by adopting a proprietary data pre-treatment "Three Dimensional Normalisation" (3DN) which converts each input spectrum, typically covering the range 1.5 to 3.5 OD, to a normalised series of values with an intrinsic range of 0.8 to 1.2 arbitrary units.

Although the filters used in Grainspec are manufactured to an extremely high specification there are still small, but significant, differences which must be taken into account to make all Grainspecs appear spectrally similar and therefore capable of accepting calibrations derived on other Grainspecs. A series of grain samples (wheat and barley) are measured on each new Grainspec and also on a "Standardisation Master" Grainspec which is held permanently at Multispec. Linear regression is used to produce for each Grainspec a standardisation map of 33 slopes and 33 intercepts relating instrument response to that of the Master. These coefficients become part of the data pretreatment modifying each sample spectrum prior to 3D-normalisation and application of the regression model.

Partial Least Squares[1] is used to relate spectra to reference chemical data. A Grainspec calibration consists therefore of a 33 point mean spectrum, a series of 33 PLS coefficients and two coefficients (a slope and bias term) to allow adjustment of a calibration for factors such as residual instrument differences, differing seasons or climatic conditions, and the appearance of new varieties in the market.

2.2 Samples

To illustrate the principles behind the co-master system, calibrations were derived for moisture in rice and moisture in barley. Rice data was obtained from Ricegrowers Co-operative Ltd, Leeton, N.S.W., Australia, and barley data from Dalgety Seamans, Norwich, England.

The moisture contents of the rice samples were created artificially by tempering common varieties to known moisture levels and then by oven drying a subsample (AACC method 44-18, two stage method with a final drying stage of 2 hours at 135°C) to confirm the value. This work was carried out to produce an interim calibration which would allow data to be collected until sufficient fresh harvest samples were available to produce a calibration which would represent truly the current rice crop in Australia. A description of the samples and experimental design can be found in Blakeney *et al.*[2]

Barley samples were collected by Dalgety Seamans in 1993 as part of their commercial trading operation for malting and feed barley and are representative of the varieties traded in the UK.

As part of their intake assessment, moisture content of the samples was obtained by oven drying (130°C for 2 hours, Method ISO712) and their reference values are used in this paper.

2.3 The co-master concept

The co-master strategy involves selecting for calibration development several instruments, typically three to five, which span the variation in manufacturing tolerances of Grainspec. The sequence of operations needed to produce a calibration and install it in

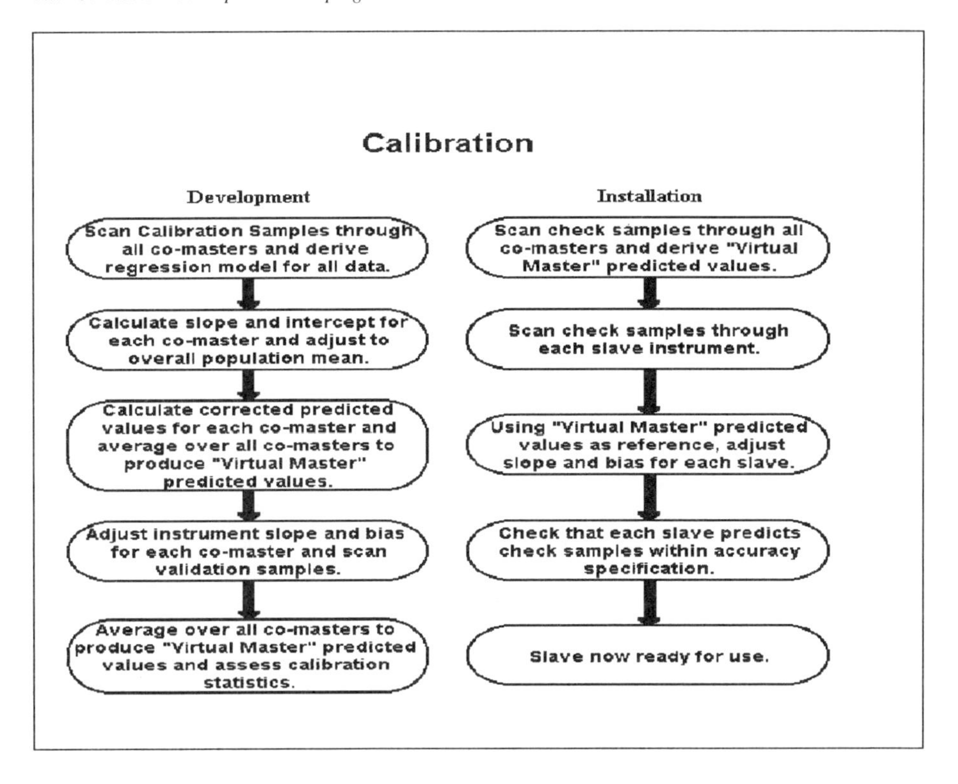

Figure 1. *Schematic diagram for co-master calibration development and installation.*

slave instruments is shown schematically in Figure 1. Initially all samples are scanned through each of the co-masters and all of the data are used to derive a PLS model. At this stage, a plot of reference against NIR predicted values hides the fact that there are several sub-populations, one for each co-master, which can be characterised by their slopes, intercepts and SEC.

We now adjust the slope and intercept for each instrument to be the same as the overall response and adjust each fitted value accordingly. This "shrinks" the distribution towards the overall or "Virtual Master" regression line and allows the calculation of an average value across all instruments (the "Virtual Master" predicted value) for each sample.

Having adjusted slope and bias terms for each co-master, a series of validation samples can be scanned and predictive statistics for each instrument (SEP) can be calculated. As with calibration data, we obtain a virtual master response consisting of the average for all co-masters for each sample in the validation data.

Up to this point we have considered only the co-masters. However in a large market we may have several hundred slave instruments. In order to optimize installation of the calibration in these instruments we must scan a subset of validation samples through each slave to establish the instrument response and then, using the virtual master predicted values as reference, calculate the slope and bias required to match the slave instrument to the virtual master.

Table 1. *Rice calibration statistics. SEC and SECa are Standard Errors of Calibration (% concentration).*

Instr.	Slope	Int.	SEC	SECa
All	0.999	0.01	0.550	0.493
Co-Mst1	0.962	0.49	0.562	0.482
Co-Mst2	1.019	-0.04	0.601	0.517
Co-Mst3	1.005	-0.14	0.500	0.496
V-Mstr.	1.001	0.01	-	0.495

By using the virtual master fitted values as reference rather than the chemical data we maximise Transferability rather than Accuracy. At this stage, our aim is to make all instruments perform identically. If a particular sample cannot be predicted well by the model it should be predicted badly by every instrument. The ability to cope with any sample should be a model response rather than an instrument response.

If a slave instrument cannot meet the accuracy specification then we begin by having the instrument checked to determine whether it requires servicing. If this resolves the difficulty then the co-master calibration model is accepted for commercial use. If problems cannot be resolved then we must derive a new model which can meet the specifications.

For periodic calibration maintenance, a series of check samples representing the latest harvest are scanned through the co-masters and virtual master predicted values are generated. These check samples can then be sent to slaves to check if slope or bias changes are necessary to cope with current conditions.

3. RESULTS AND DISCUSSION

3.1 Moisture in rice

The aim of this work was to produce a starting calibration which could be used at harvest to gather data for a more complete calibration covering all the varieties which are current in New South Wales. Without a starting calibration, which can be used to set up several instruments, it is difficult to survey a large number of samples during the restricted harvest period.

Figures 2 and 3 show calibration data for rice. In Figure 2, regression lines for each of the co-masters are shown. For each sample, the predicted value has not been corrected for instrument slope and bias. In Figure 3, the values shown are corrected for slope and bias and only the virtual master regression line is shown. After adjustment, the distribution of values about the virtual master regression line is tighter than before. Table 1 shows the effects on the Standard Error of Calibration before and after adjustment for slope and bias. With a single regression equation, and no adjustment for slope and bias, the SEC for each instrument is raised artificially because of small slope and intercept differences between instruments which are still present despite standardisation.

In Table 2 we see that when virtual master predicted values for a set of check samples (a subset of the validation data) were used to set the slope and bias terms for each slave instrument, the resulting Standard Errors of Prediction for both Transferability (TSEPa) and Accuracy (ASEPa) were reduced considerably when compared to uncorrected values, with

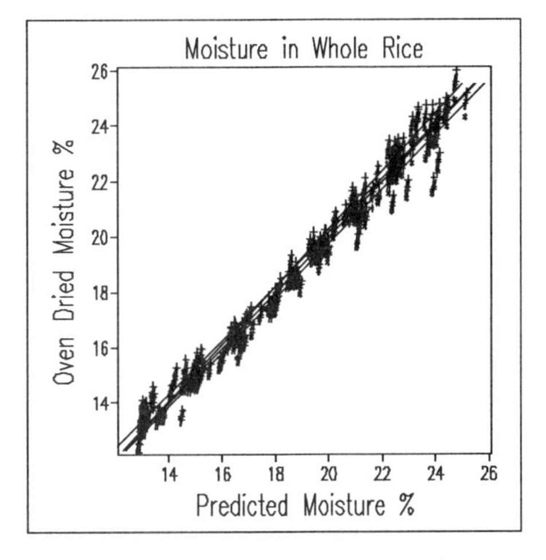

Figure 2. *Calibration data for rice before adjustment for slope and bias.*

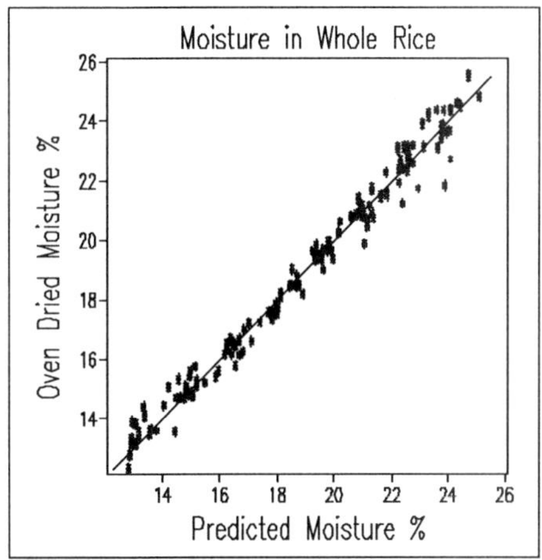

Figure 3. *Calibration data for rice after adjustment for slope and bias.*

bias primarily the mechanism by which they were reduced.

Table 2 illustrates also the difference between Transferability and Accuracy. After optimisation for slope and bias, Transferability values (TSEPa) are consistently much lower than Accuracy values (ASEPa). As stated previously, Accuracy is a function of model and instrumental error while Transferability is mainly the residual manufacturing differences between instruments. Comparing the two values for each slave indicates the potential for improvement either in instrument performance or in the PLS model.

Figures 4 and 5 show, for a typical slave instrument, scatter plots of predicted values versus virtual master predicted values (Transferability) before and after correction for slope and bias effects. There are relatively few observations, but the aim of this exercise is not primarily to validate the regression equation but to adjust the slope and bias of the slave instrument to obtain the optimal agreement between instruments. Figures 4 and 5 highlight the fact that, in both cases, the points are consistently tight to a regression line close to 45° and that the higher SEP before adjustment was due mainly to bias effects.

Table 2. *Rice validation statistics. TSEP and TSEPa are Standard errors (% concentration) for Transferability before and after slope and bias correction; ASEP and ASEPa are equivalent values for Accuracy.*

Instr.	Slope	Int.	TSEP	TSEPa	ASEP	ASEPa
Slave 1	0.981	1.15	0.814	0.041	0.841	0.217
Slave 2	1.005	0.98	1.103	0.055	1.117	0.188
Slave 3	1.029	0.39	0.953	0.045	0.974	0.208
Slave 4	0.991	0.23	0.086	0.051	0.241	0.231
Slave 5	1.010	0.31	0.502	0.027	0.544	0.211
Slave 6	0.980	0.89	0.536	0.032	0.572	0.208

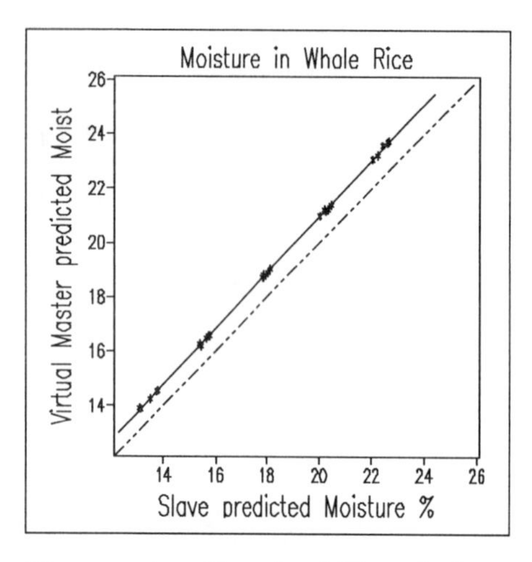

Figure 4. *Transferability before adjustment to slope and bias.*

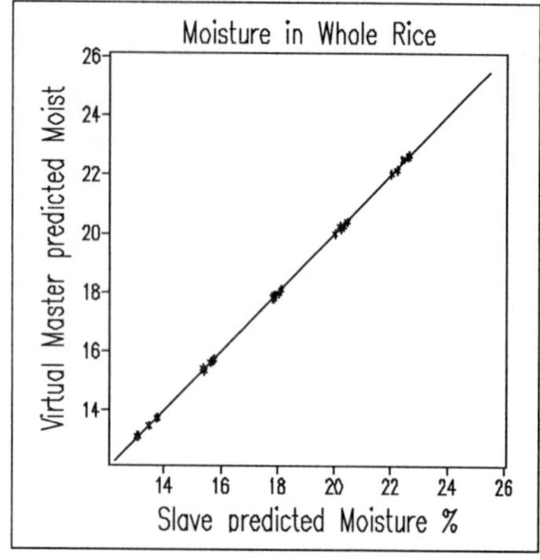

Figure 5. *Transferability after adjustment for slope and bias.*

3.2 Moisture in barley

The samples used to generate moisture data were collected in 1993 as part of an evaluation on barley of a single Grainspec by Dalgety Seamans. Subsequently, the samples were scanned through nine other Grainspec units to evaluate whether a commercially viable transferable calibration for moisture could be obtained.

Tables 3 and 4 summarise the results for moisture in UK barley. In Table 3 we see that all three co-masters had similar slopes and intercepts and, after adjustments for instrument differences, improvements to SEC for any of the co-masters was relatively small.

With the slave instruments we see a different picture. There was a much greater variation both in slope and intercept (Table 4).

For Slave 4, the slope of 0.958 was largely responsible for the high values for TSEP and ASEP. For Slave 5, the slope is close to 1.00 and it is a bias effect which is largely responsible for the ASEP of 0.421 and TSEP of 0.399. In both cases, adjustment for instrumental slope and bias effects produces acceptable Standard Errors of Prediction.

Table 3. *Barley calibration statistics. SEC and SECa are Standard Errors of Calibration (% concentration).*

Instr.	Slope	Int.	SEC	SECa
All	1.000	0.00	0.312	0.309
Co-Mst 1	1.002	0.01	0.288	0.284
Co-Mst 2	0.983	0.24	0.280	0.278
Co-Mst 3	1.013	-0.22	0.290	0.287
V-Mstr.	1.004	0.01	-	0.278

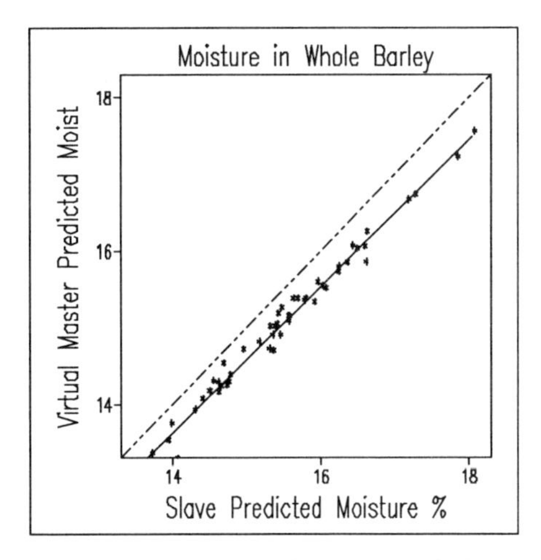

Figure 6. *Transferability plot before correction for slope and bias.*

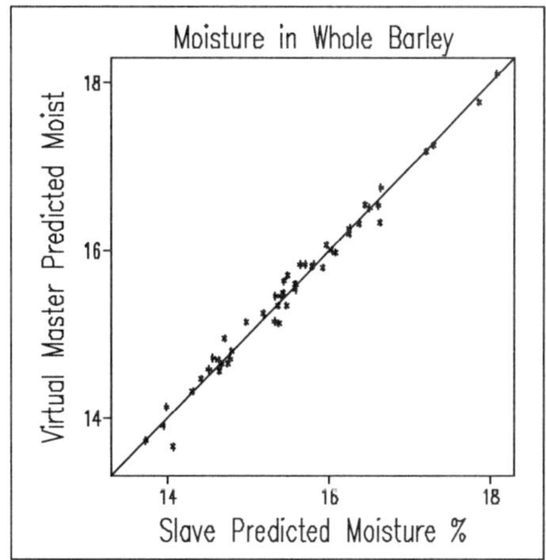

Figure 7. *Transferability plot after correction for slope and bias.*

Figures 6 and 7 show for Slave 4, the effects on Transferability of slope and bias correction. Figures 8 and 9, which show equivalent plots for Accuracy, suggest that we have not changed the intrinsic relationship between the reference and NIR predicted values but have simply skewed and biased the instrument in such a way that the calibration equation functions optimally.

As with rice, the scatter plot for Transferability (Figure 7) is tighter to the optimal regression line than its equivalent Accuracy plot (Figure 9) indicating that to improve Accuracy we must concentrate on improving the model rather than the instrument. With barley we have also used all the validation samples rather than pick a subset of check samples.

4. CONCLUSIONS

Transferable calibrations are an essential feature of all current near infrared instrumentation. Regulatory authorities often insist that the same calibration be installed in every instrument of a particular type from which trading information is derived.

Table 4. *Barley validation statistics. TSEP and TSEPa are Standard errors for Transferability (% concentration) before and after slope and bias correction; ASEP and ASEPa are equivalent values for Accuracy.*

Instr.	Slope	Int.	TSEP	TSEPa	ASEP	ASEPa
Slave 1	0.984	0.17	0.098	0.087	0.229	0.215
Slave 2	0.967	0.33	0.230	0.145	0.345	0.279
Slave 3	1.010	0.05	0.293	0.168	0.342	0.265
Slave 4	0.958	0.22	0.438	0.131	0.504	0.240
Slave 5	0.980	0.66	0.399	0.159	0.421	0.245

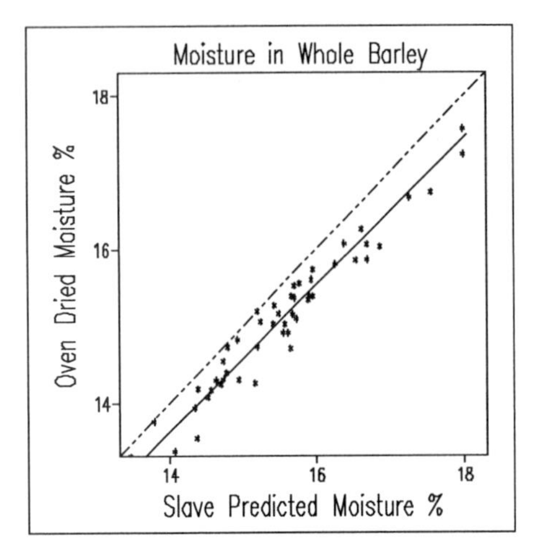

Figure 8. *Accuracy plot before correction for slope and bias.*

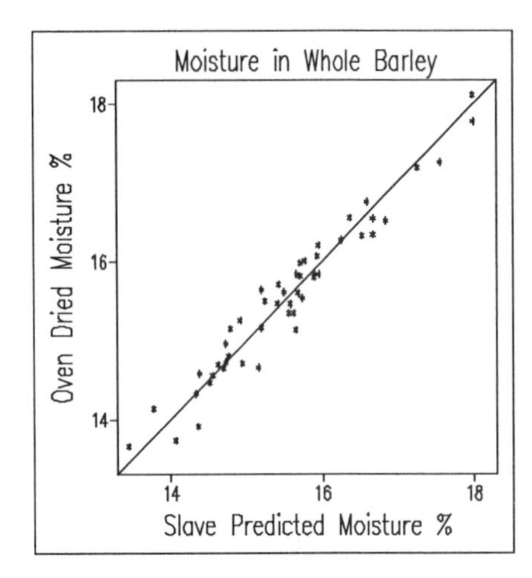

Figure 9. *Accuracy plot after correction for slope and bias.*

When a universal calibration is installed in a number of instruments, the regulator may allow adjustment for bias but insist that there will be no adjustment for slope differences between instruments.[3] This situation is caused by the fear that indiscriminate slope adjustments potentially can cause large changes in predicted values upon which commercial decisions are made. However, if an instrument has a response characterised by a slope which is considerably different from 1.00 it will always have a high Standard Error of Prediction, even although a test set of data can be shown to form a tight distribution about a regression line.

Optimisation of instrument slope and bias on calibration installation overcomes this problem. Thereafter, the slope can be locked for the lifetime of the calibration to prevent the problems mentioned previously.

The use of virtual master predicted values, rather than reference chemical data, to adjust slope and bias in slave instruments is an attempt to separate the performance of the instrumentation from that of the regression model. Measuring Transferability confirms whether an instrument functions correctly, which is the overriding concern when managing a large pool of instruments.

By confirming that all instruments respond similarly to a set of check samples, we can then concentrate on increasing the Accuracy of the system by improving the regression model perhaps by using more accurate chemical reference values or improved data pretreatments.

References

1. H. Martens and T. Naes, 'Multivariate Calibration', John Wiley & Sons Ltd., Chichister, England, 1989

2. A. B. Blakeney, *et al*, In 'Proceedings of the 6th International Conference on Near Infrared Spectroscopy.', Lorne, Victoria, Australia, 18-22 April, 1994, In Press.

3. 'NIST Handbook 44', H. V. Oppermann and T. G. Butcher, Eds., United States Department of Commerce Technology Administration, Technology, National Technical Information Service, Springfield, Va 22161, USA, 1994

Acknowledgements.

Multispec would like to thank Janelle Reece and Tony Blakeney of NSW Agriculture, John Sharman from Ricegrowers Co-operative Ltd, Leeton, N.S.W., Australia, for permission to use spectra and reference chemical data for rice and Frank Connor of Radiometer Pacific Pty. Ltd., Terrey Hills, N.S.W., Australia for co-ordinating data collection in Australia. Thanks are due also to John Franks of Dalgety Seamans, Norwich, England for the use of reference chemical data and samples collected as part of their commercial transactions.

The Application of Inelastic Neutron Scattering to Advanced Composites

Stewart F. Parker[1] and John N. Hay[2]

[1] ISIS DIVISION, RUTHERFORD APPLETON LABORATORY, CHILTON, DIDCOT, OXON OX11 0QX, UK

[2] DEPARTMENT OF CHEMISTRY, UNIVERSITY OF SURREY, GUILDFORD, SURREY GU2 5XH, UK

1 INTRODUCTION

Advanced composites are engineering materials that offer similar mechanical properties to metal alloys but are lighter than them. The materials consist of fibres embedded in a polymer matrix and there is a need for a spectroscopic technique that can examine the cured resins in the presence of the fibres to aid the understanding of the cure chemistry. Inelastic neutron scattering (INS) has considerable potential in this regard since two common fibre types, glass and carbon are, effectively, invisible to neutrons.

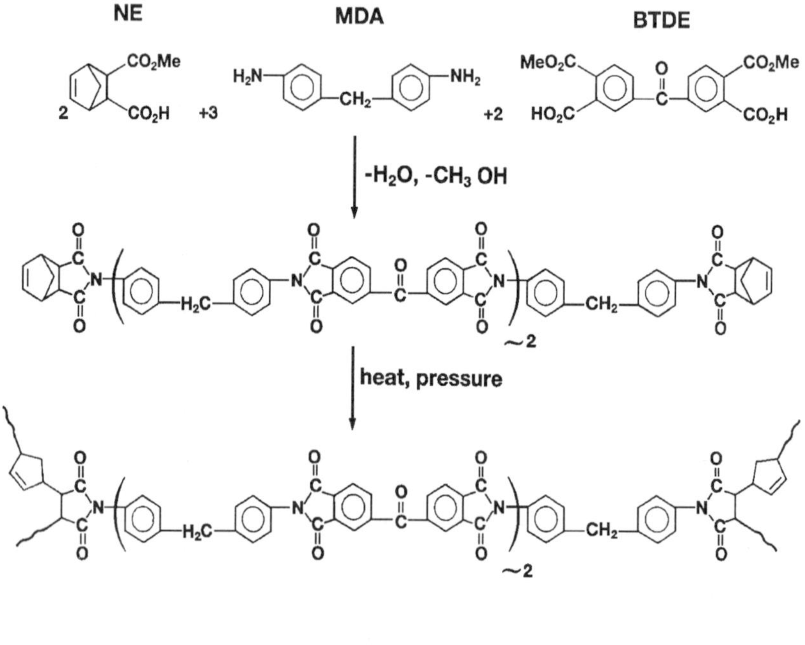

Figure 1 *The chemistry of the cure process of PMR-15*

The materials are commonly made by coating the fibres with the monomers to give a precursor fabric ("prepreg") that can be moulded to the desired shape. This is then heated to cure the resin and give the finished product. The chemistry of the cure processes is complex and frequently involves a number of stages. The ability to study the reaction(s) is greatly hampered by the nature of the products: they are often highly cross-linked and thus insoluble, and the presence of the fibre matrix makes them difficult to study spectroscopically.

A wide variety of polymers have been used including epoxies, bismaleimides and polyimides. One of the most common polyimides is PMR-15[1]. The chemistry is complex, see Figure 1, but consists essentially of two stages; polymerization to give a norbornene end-capped oligomer followed by reaction of the norbornene group to give the cross-linked polymer. The temperature at which the cross-linking reaction is carried out has a major effect on the mechanical properties of the finished product, particularly its susceptibility to microcracking[2].

2 EXPERIMENTAL

The polyimide samples were produced by compression moulding of PMR-15 prepreg with a final cure temperature of 270 and 330°C and were the same as used in a previous study[2]. *endo*-N-phenyl-5-norbornene-2,3-dicarboximide (N-phenylnadimide, see Figure 2 for the structure) was prepared by the method of Delvigs[3], *endo*-N-perdeuterophenyl-5-norbornene -2,3-dicarboximide (N-phenylnadimide-D5) was prepared by the same method but starting from perdeuteroaniline (MSD, 99% D).

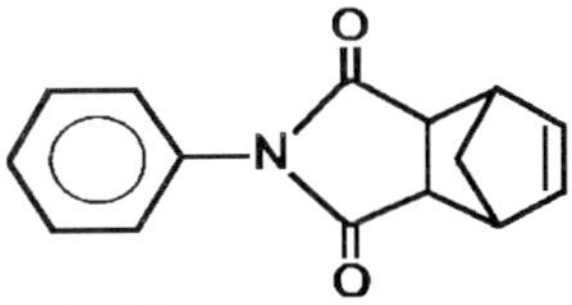

Figure 2 *Structure of N-phenylnadimide*

The inelastic neutron scattering (INS) experiments were performed using the high resolution broadband spectrometer (TFXA) at the ISIS pulsed spallation neutron source at the Rutherford Appleton Laboratory, Chilton, UK. A schematic of the instrument is shown in Figure 3. This is an inverted geometry time-of-flight spectrometer where a pulsed, polychromatic beam of neutrons illuminates the sample at 12 m from the source. The backscattered neutrons are Bragg reflected by a pyrolytic graphite analyser and those with a final energy of ~32 cm^{-1} are passed to the He^3 detector bank. Energy transfer and spectral intensity is then calculated using standard programs to convert to the conventional $S(Q,\omega)$. TFXA offers high resolution, ~2% $\Delta E/E$ between 16 and 4000 cm^{-1}. The samples (5 cm x 2 cm) were clamped to a centre stick and cooled to ~30K. The measurement times varied from a few hours for the model compounds to 24 hours for the polymers.

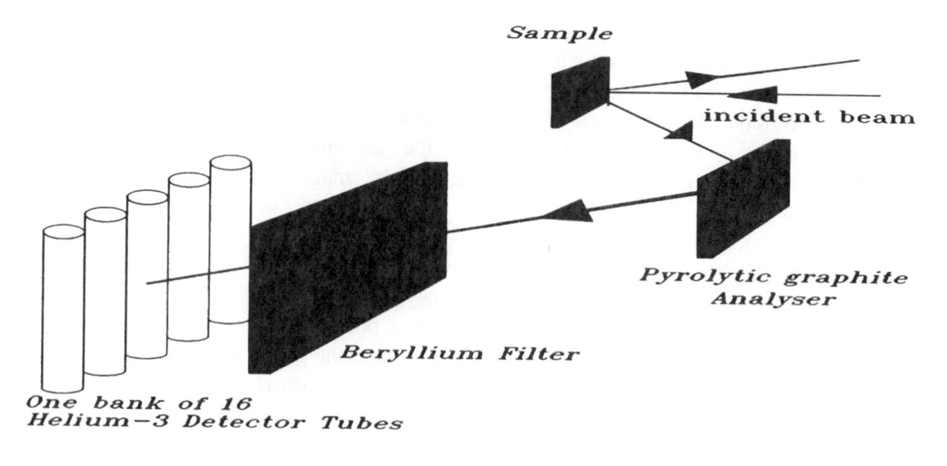

Figure 3 *Schematic diagram of the inelastic neutron scattering spectrometer TFXA*

3 RESULTS

INS spectra of the composites cured at 270 and 330°C are shown in Figure 4a and 4b respectively. There are clearly differences between the two spectra; bands at 1031 and 1114 cm^{-1} have diminished in intensity and there are indications of changes in the region 200 - 400 cm^{-1} and at 638, 720 and 1273 cm^{-1}. To aid the assignment, spectra were also recorded from N-phenylnadimide and N-phenylnadimide-D5 as shown in Figure 5a and 5b respectively.

4 DISCUSSION

While vibrational spectroscopy, particularly infrared spectroscopy, has been widely used for the characterisation of composites[4], this report is the first application of INS spectroscopy to advanced materials. The examples studied here are particularly difficult samples because by the cure temperatures employed here, a significant proportion of the end-groups have already reacted. In addition the composites are only ~30% by weight resin, of which only 2/7 of the molecules are the end-cap. Nonetheless, as Figure 4 shows, differences between the two samples are apparent. Furthermore, since neutrons are highly penetrating, the spectra are characteristic of the bulk of the polymer. This is in contrast to the infrared methods that only sample the top few microns of the cured composite.

N-phenylnadimide has been used very successfully[5] as a model compound for the norbornene endcap and the cross-linking reaction. The intensities of the bands in INS spectroscopy depend on the amplitude of vibration and the scattering cross-section of the atoms. Since the cross-section for hydrogen is ~80 and that of C, N and O are all less than 5, the spectrum is dominated by the hydrogen atom motions. The cross-section is isotope dependent and that of deuterium is also small. Thus by deuterating the phenyl ring it is possible to eliminate the spectral features of the phenyl ring and just leave the norbornene endcap. This results in a significant simplification of the spectrum as shown in Figure 5b

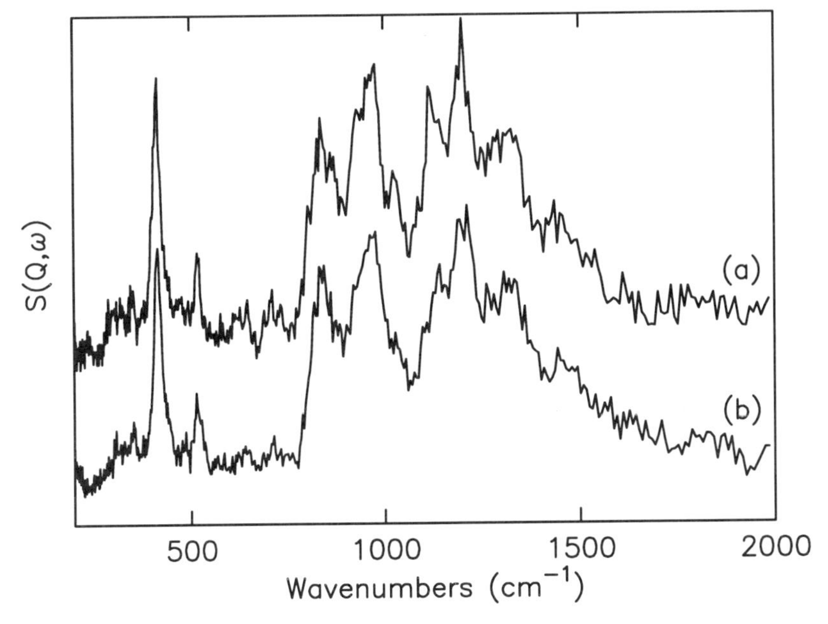

Figure 4 *INS spectra of PMR-15 cured at (a) 270 and (b) 330°C*

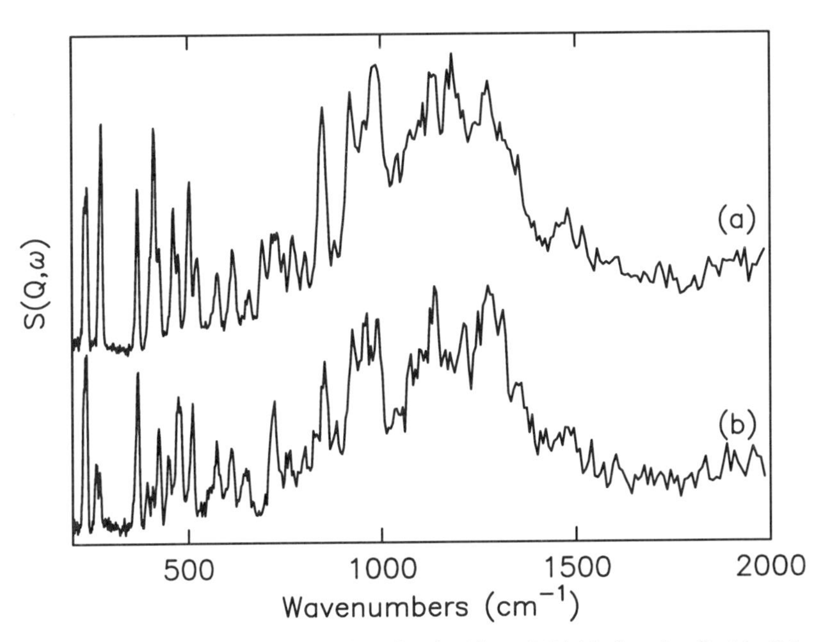

Figure 5 *INS spectra of (a) N-phenylnadimide and (b) N-phenylnadimide-D5*

and provides an excellent model compound for the endcap. Comparison of Figures 4 and 5 suggests that the decrease in the 1114 cm^{-1} and the changes in the 200 - 400 and 600 - 800 cm^{-1} regions can reasonably be assigned to loss of the endcap. The 1031 cm^{-1} band does not fit this pattern and may be assignable to one of the species that are responsible for the cross-linking.

References

1. D. Wilson, *Brit. Polymer J.*, 1988, **20**, 405.
2. J.K. Wells, J.N. Hay, D. Lind, G.A. Owens and F. Johnson, *SAMPE J.*, 1987 (May/June), 35.
3. P. Delvigs, 'High Temperature Polymer Matrix Composites', Cleveland, 1983, p23.
4. S.F. Parker, *Vib. Spec.*, 1992, **3**, 87.
5. J.N. Hay, J.D. Boyle, S.F. Parker and D. Wilson, *Polymer*, 1989, **30**, 1032.

A Fibre Optic Detecting System for the Simultaneous Determination of Oxygen and Carbon Dioxide Based on Absorption Spectroscopy*

M. F. Choi and P. Hawkins

FACULTY OF APPLIED SCIENCES, UNIVERSITY OF THE WEST OF ENGLAND, BRISTOL, COLDHARBOUR LANE, FRENCHAY, BRISTOL BS16 1QY, UK

INTRODUCTION

There are many instances in biomedical and environmental monitoring and in analytical chemistry where it is necessary to determine the concentrations of oxygen (O_2) and, or, carbon dioxide (CO_2) in a gaseous or liquid sample. Examples of this include: blood gas and respiratory gas analysis in medicine; investigations of landfill and combustion gas analysis in environmental monitoring; atmosphere control in horticulture; and in biosensors where oxygen consumption or carbon dioxide evolution are often used as indicators of enzyme and microbial activity. The requirements for the sensors in these different applications are frequently different with respect to sensitivity, response time, sensor lifetime, etc. and no single method for determining oxygen ($[O_2]$) and carbon dioxide concentrations ($[CO_2]$) is entirely satisfactory in all instances. For this reason, there have been many investigations into new oxygen and carbon dioxide sensor designs over the years and the research is continuing. Fibre-optic based sensors are a relatively new approach which offer several advantages over other methods such as intrinsic safety, so that the sensors can be used with explosive gases, freedom from electrical noise and interference, ease of miniaturisation, etc. Although there have been many new designs of oxygen and carbon dioxide sensors published, only a few detecting systems have been reported for the simultaneous determination of both oxygen and carbon dioxide. Albery and Barron[1] used the electrochemical reduction technique for the determination of O_2 and CO_2 in which a high applied potential was required for the electro-reduction of the carbon dioxide. Arquint *et al.*[2] combined pO_2, pCO_2 and pH sensors in a micro-chip. Gehrich *et al.*[3] and Miller *et al.*[4] placed three optical fibres together in a catheter to which individual analyte-sensitive dyes were attached for sensing pH, pCO_2 and pO_2. Wolfbeis *et al.*[5] and Yim *et al.*[6] also used a similar technique of positioning separate oxygen-sensitive and carbon dioxide-sensitive dyes together for the fibre optic sensing of the two gases. In all of these cases, separate detecting media were used for each gas and, to date, no sensor using a single detecting medium for both gases has been reported. In this paper, we describe an investigation into a single solvent-dye solution which can be used for simultaneously and independently detecting both gaseous oxygen and carbon dioxide based on absorption spectroscopy.

* For this work M.F. Choi was awarded the 1994 Harry Willis Prize for Analytical Spectroscopy.

MATERIALS AND EXPERIMENTAL

Different oxygen concentrations (in the range 0 to 100%) in a gas stream were produced by controlling the flow rates of the oxygen and the diluent nitrogen (N_2) gases entering a mixing chamber. The gas mixture was passed through a portable oxygen meter where the oxygen concentration in the gas mixture was determined before passing through 3.0cm^3 of solvent-dye solution (10.6μM fluorescein (Fl) and 168μM tetrabutylammonium hydroxide (TBuAOH) in 50:50 (v/v) solvent mixture of N,N'-dimethyl-p-toluidine (DMT) and N,N-dimethylformamide (DMF)) contained in a 10mm path-length cuvette. Similarly, carbon dioxide standards in nitrogen (0 to 10% [CO_2]) were obtained by replacing the oxygen gas with carbon dioxide and their concentrations were determined by a carbon dioxide detector. The absorption spectra of the gas saturated (either oxygenated or carbonated) solvent-dye solutions were then measured on a spectrophotometer against an air-saturated solvent solution.

In the fibre optic detecting system of oxygen and carbon dioxide, a laboratory-made optical arrangement was used. A modulated beam of light of a known wavelength was produced by a Bentham IL1A 100W quartz halogen lamp (the light intensity was stabilised by a Bentham 505 current stabilised filament lamp power supply), a Bentham 218 optical chopper at a frequency of 141Hz and a Bentham M300 monochromator controlled by a Bentham SMD3B stepping motor drive unit before being launched into a plastic optical fibre (1m long and 2mm diameter). The incident light from the fibre irradiated the front surface of a 10mm quartz cuvette containing 0.75cm^3 of the solvent-dye solution with either gaseous oxygen standards or gaseous carbon dioxide standards at a flow-rate of 45cm^3/min continuously passing through it. Another plastic fibre (0.5m long, 2mm diameter) was placed on the back surface of the cuvette to collect the transmitted light. The other end of this fibre led directly onto the front surface of a 100mm^2 silicon photo-voltaic detector and the output signal connected to a current pre-amplifier and a lock-in amplifier which was synchronised to the modulated light beam by a reference signal provided by the light chopper. The signal from the amplifier was recorded on a Servoscribe 1S chart recorder or displayed on a Bentham 217 digital unit.

To avoid a slight signal drift caused by a small evaporation of the solvent, the solvent-dye solution was replaced with a fresh one after about two hours.

RESULTS AND DISCUSSION

It is well known that some organic solvent molecules exhibit contact charge transfer absorption (CCTA) when in contact with molecular oxygen[7] and the change of absorbance is proportional to the concentration of the applied gaseous oxygen[8]. DMT has a strong CCTA spectrum lying in the ultraviolet/visible regions and the change of absorbance at 400nm was found to be directly proportional to the concentration of applied oxygen and can serve as a sensing medium for oxygen. It has also been known that some dyes change colour when exposed to gaseous carbon dioxide because the change in pH of the environment changes a deprotonated dye molecule into a protonated form. The dye is usually incorporated in an aqueous buffer solution. This presents a problem in this case as DMT is an almost completely non-aqueous solvent, probably accounting for why we observed that a solution of Fl in DMT changes colour on exposure to carbon dioxide but that the change is not entirely reversible. However, we found that a solution of Fl in DMF does respond reversibly to gaseous carbon dioxide as DMF contains a small amount of dissolved water (about 0.03%). Unfortunately, DMF has a CCTA spectrum which is too far into the ultraviolet region to be of practical use in a fibre-optic based sensor. After further experimentation, we devised a way around this problem by using a solution of Fl dye and TBuAOH in a DMT/DMF solvent mixture for the simultaneous, independent and reversible measurements of carbon dioxide using colour changes in the Fl and oxygen using the CCTA in the DMT.

The observed changes in absorption spectrum of the solvent-dye solution as a function of $[O_2]$ and $[CO_2]$ are shown in Figure 1a which shows that there was no cross interference between the two gases. Note that the reference solutions are saturated with air ($[O_2]=21\%$ approx.) which explains why the test solutions have negative absorbances when their oxygen concentrations are less than the air-saturated value. Figure 1b shows the increase in absorbance of the test solution as the concentration of oxygen increases. A graph of absorbance (abs) at 400nm against applied oxygen ($[O_2]\%$) is plotted in Figure 1b. The results agree with Beer-Lambert's law and the absorbance is linearly proportional to the applied oxygen concentration. The change of absorbance is caused by the CCTA of the DMT and O_2 complex.

$$\mathrm{DMT} + \mathrm{O_2} \rightleftharpoons \mathrm{DMT}\cdots\mathrm{O_2} \overset{h\upsilon}{\rightleftharpoons} (\mathrm{DMT}\cdots\mathrm{O_2})^*$$

The change of absorbance at 520nm is plotted against applied $[CO_2\%]$ (Figure 1c). It is observed that the absorbance decreases non-linearly with increasing $[CO_2]$. The change of absorbance arises from the conversion of the orange dianion form of fluorescein to the colourless, neutral lactonic form. These changes were brought about from the change of the pH of the solution with the dissolution of CO_2 in the solvent mixture and can be envisaged in the following reactions.

$$\mathrm{OH^-} + \mathrm{CO_2} \rightleftharpoons \mathrm{HCO_3^-} \quad (1)$$
$$\mathrm{H^+} + \mathrm{OH^-} \rightleftharpoons \mathrm{H_2O} \quad (2)$$
$$\mathrm{Fl^{2-}} + 2\mathrm{H^+} \rightleftharpoons \mathrm{H_2Fl} \quad (3)$$

Dissolved CO_2 removes the OH^- ions in the solvent mixture: as a result, the H^+ ion concentration is increased which shifts the equilibrium of equation (3) from left to right. In the fibre optic system for the determinations of oxygen and carbon dioxide, the response, reproducibility and total signal change of the sensing system monitoring at 400nm when subjected to step changes from 100% $[N_2]$ to 100% $[O_2]$ and from 100% $[O_2]$ to 10% $[CO_2]$ in N_2 and from 10% $[CO_2]$ in N_2 to 100% $[O_2]$ were investigated (Figure 2a). The reversibility of the system is good and, again, there is no cross interference from CO_2. The response time is 1.0min and 1.3min for 90% signal change from N_2 to O_2 and from O_2 to N_2 respectively. The response of the system when subjected to different levels of oxygen was investigated (Figure 2b) and was shown to be reversible. The decrease in signal level with increasing oxygen concentrations is non-linear (Figure 2c.i). The sensitivity (indicated by the slope of the graph) decreases as the concentration of oxygen increases and the most sensitive region is between 0 to 20% $[O_2]$ with about a 30% change of signal level. The system responds over the range 0 to 100% $[O_2]$ with a response which is again in agreement with Beer-Lambert's Law so that a linear graph is obtained when

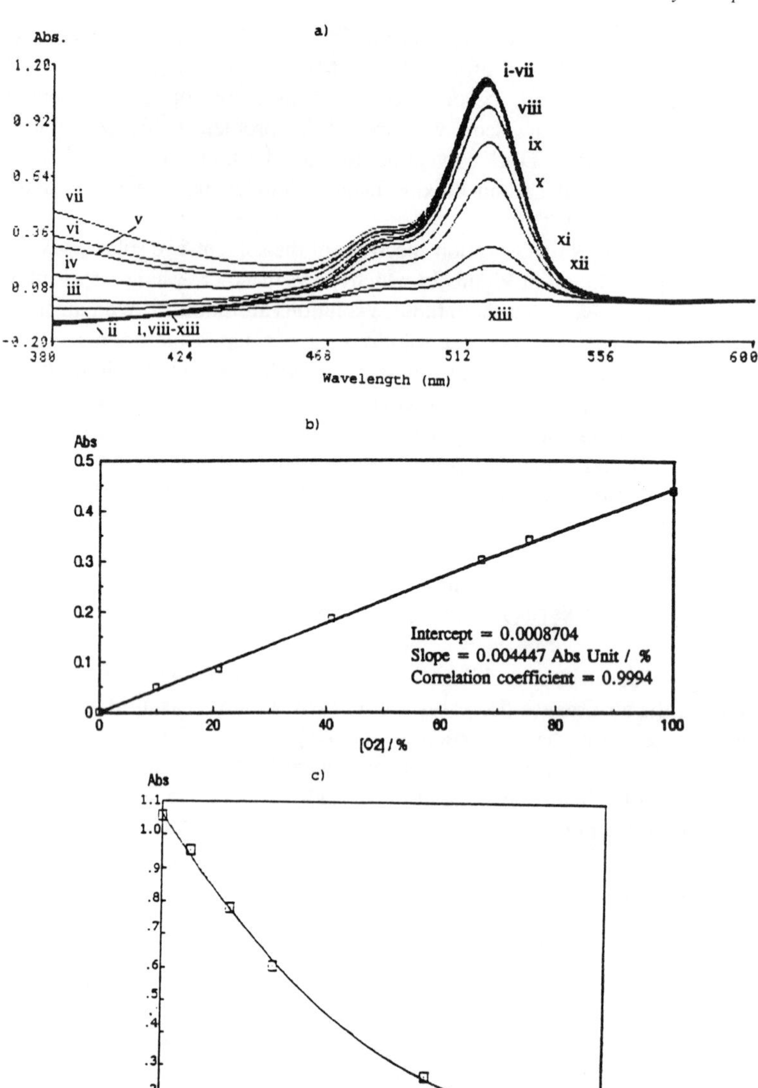

Figure 1

a) Effect of oxygen and carbon dioxide on the absorption spectrum of the solvent-dye solution (10.6 μM Fl, 168 μM TBuAOH in 50:50 (v/v) of DMT and DMF solvent mixture). Optical path-length : 10 mm; Reference : air-saturated solvent solution. i: N_2; ii: 10.0% $[O_2]$ in N_2; iii: air-saturated; iv: 40.9% $[O_2]$ in N_2; v: 67.3% $[O_2]$ in N_2; vi: 75.4% $[O_2]$ in N_2; vii: 100%$[O_2]$; viii: 0.64% $[CO_2]$ in N_2; ix: 1.55% $[CO_2]$ in N_2; x: 2.55% $[CO_2]$ in N_2; xi: 6.05% $[CO_2]$ in N_2; xii: 10% $[CO_2]$ in N_2; xiii: 100% $[CO_2]$.

b) Plot of Abs against $[O_2]$% at 400 nm using the data in Figure 1a.

c) Plot of Abs against $[CO_2]$% at 520 nm using the data in Figure 1a.

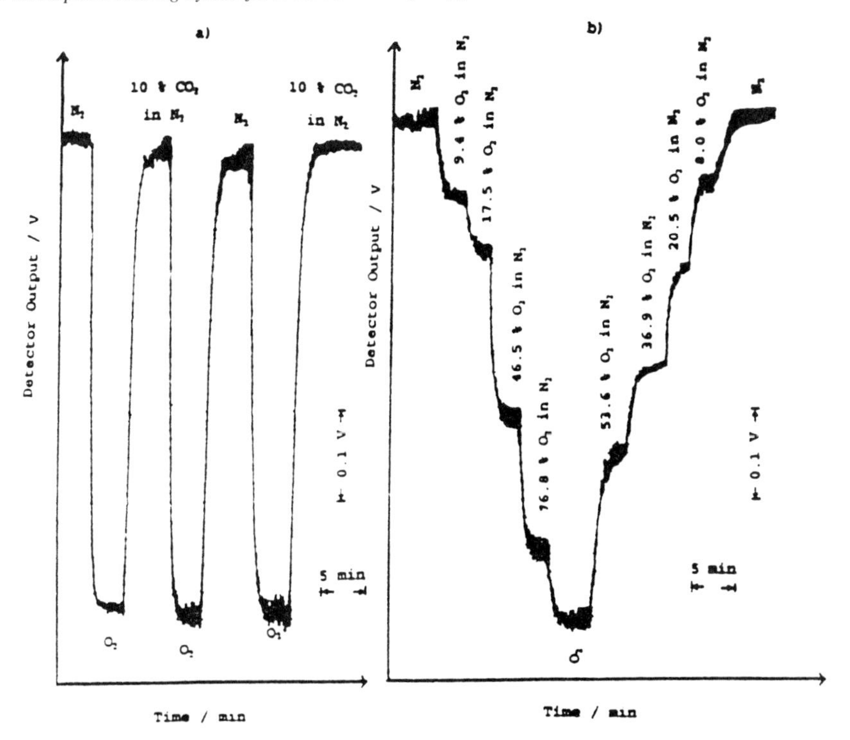

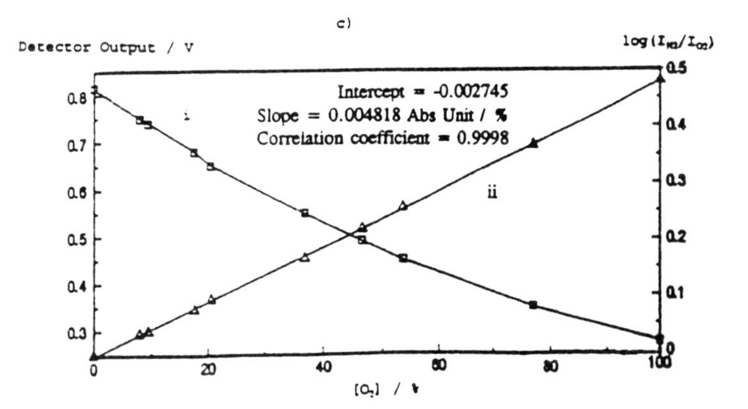

Figure 2

a) *Response time, reproducibility and total signal change of the fibre optic system when changing between: 100% $[N_2]$ to 100% $[O_2]$; 100% $[O_2]$ to 10% $[CO_2]$ in N_2; and 10% $[CO_2]$ in N_2 to 100% $[O_2]$. 0.75 cm^3 of the solvent-dye solution in a 10 mm cuvette was used. Wavelength: 400 nm; Gas flow rate: 45 cm^3/min.*

b) *Response of the fibre optic system when subjected to different oxygen concentrations. Wavelength: 400 nm; Gas flow rate: 45 cm^3/min; the amplification is 10 times that used in Figure 3.*

c) *Response of the fibre optic system with oxygen.*
   i) Plot of Detector Output against $[O_2]$.
   ii) Plot of $\log(I_{N2}/I_{O2})$ against $[O_2]$ using the data in Figure 2b.

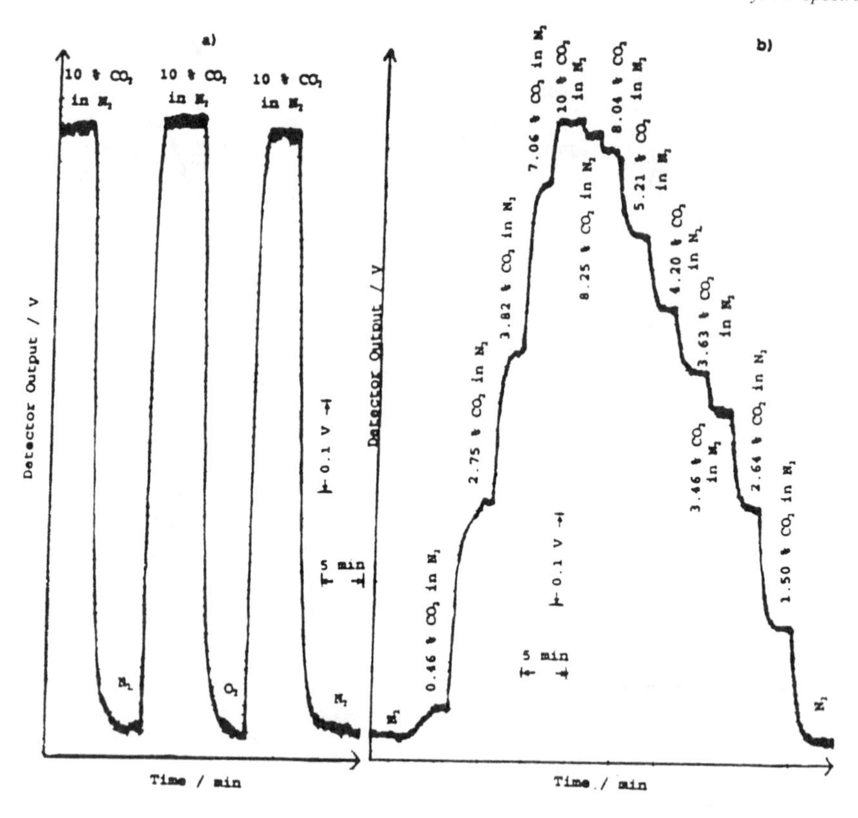

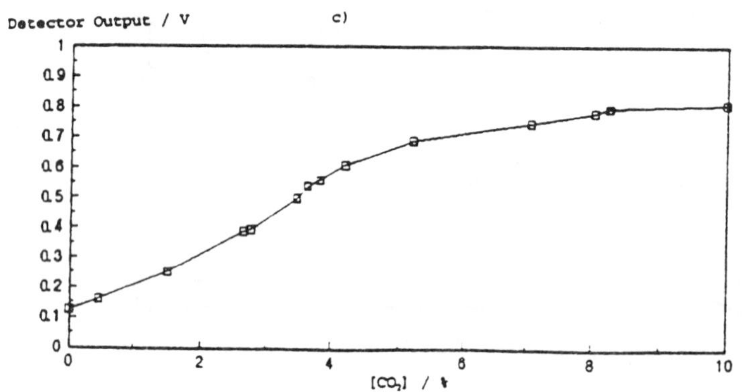

Figure 3

a) *Response time, reproducibility and total signal change of the fibre optic system when changing between: 10% $[CO_2]$ in N_2 to 100% $[N_2]$; 100% $[N_2]$ to 10% $[CO_2]$ in N_2; and 10% $[CO_2]$ in N_2 to 100% $[O_2]$. 0.75 cm^3 of solvent-dye solution in 10 mm cuvette was used. Wavelength: 520 nm; Gas flow rate: 45 cm^3/min.*

b) *Response of the fibre optic system when subjected to different carbon dioxide concentrations. Wavelength: 520 nm; Gas flow rate: 45 cm^3/min.*

c) *Response of the fibre optic system with carbon dioxide. Plot of Detector Output against $[CO_2]$ using the data in Figure 3b.*

log(I_{N2}/I_{O2}) is plotted against [O_2]% where I_{N2} is the signal level recorded with the sensing system in nitrogen only (i.e. [O_2] = 0%) and I_{O2} is the level recorded in a gas mixture having an oxygen concentration of [O_2] (Figure 2c.ii).

The response, reproducibility and total signal change of the sensing system monitoring at 520nm when subjected to step changes from 10% [CO_2] in N_2 to 100% [N_2], from 100% [N_2] to 10% [CO_2] in N_2 and from 10% [CO_2] in N_2 to 100% [O_2] were investigated (Figure 3a). The reversibility of the system is good and, again, there is no cross interference from O_2. The response time is 0.7min and 2.0min for 90% signal change from N_2 to 10% [CO_2] in N_2 and from 10% [CO_2] in N_2 to N_2 respectively. The response of the system when subjected to different levels of carbon dioxide was investigated (Figure 3b) and this too was shown to be reversible. The increase in signal level with increasing carbon dioxide concentrations is non-linear (Figure 3c) with a 36% change of signal level from 0 to 10% [CO_2] in N_2.

A disadvantage in using a liquid medium as part of the detecting system for a fibre-optic gas sensor is that drift in output can arise from loss of solvent. We are currently investigating a liquid light-guide system which exploits the fact that the solvents have a high refractive index. They also have a lowvolatility so we are also investigating the use of high permeability membranes so that the sensor has a fast response time.

CONCLUSION

A sensing system for simultaneous determination of gaseous oxygen and carbon dioxide has been successfully developed. Possible applications include environmental and physiological monitoring of carbon dioxide over the range of 0-10% and oxygen, 0-100%.

References

1. Albery, W. J. and Barron, P., *J. Electroanal. Chem. Interfacial Electrochem.*, 1982, **138**, 79.
2. Arquint, Ph., van den Berg, A., van der Schoot, B. H., de Rooij, N. F., Bühler, H., Morf, W. E. and Dürselen, L. F. J., *Sens. Actuators, B*, 1993, **13-14**, 340.
3. Gehrich, J. L., Lübbers, D. W., Optiz, N., Hansmann, D. R., Miller, W. W., Tusa, J. K. and Yafuso, M., *IEEE Trans. Biomed. Eng.*, 1986, **BME-33**, 117.
4. Miller, W. W., Yafuso, M., Cheng, F. Y., Hui, H. K. and Arick, S., *Clin. Chem. (Winston-Salem, N.C.)*, 1987, **33**, 1538.
5. Wolfbeis, O. S., Weis, L. J., Leiner, M. J. P. and Ziegler, W. E., *Anal. Chem.*, 1988, **60**, 2028.
6. Yim, J. B., Khalil, G. E., Pihl, R. J., Huss, B. D. and Vurek, G. G., *U.S. Pat.* 5098659, 1992.
7. Tsubomura, H. and Mulliken, R. S., *J. Am. Chem. Soc.*, 1960, **82**, 5966.
8. Munck, A. U. and Scott, J. F., *Nature*, 1956, **177**, 587.

Use of GC-MS for the Detection of Metabolites in the Urine of Doped Horses: Application to Quinine Breakdown Products

Cevdet Demir and Richard G. Brereton

SCHOOL OF CHEMISTRY, UNIVERSITY OF BRISTOL, CANTOCK'S CLOSE, BRISTOL BS8 1TS, UK

1 INTRODUCTION

Methods for gas chromatographic - mass spectrometric (GC-MS) analysis of drugs in biological specimens have been developed over the last twenty years[1]. Many early efforts were directed towards derivatisation of functional groups for improved GC performance. Important developments include conversion of hydroxyl groups to trimethylsilyl (TMS) ethers, ketones to methyloximes and preparation of enol-TMS derivatives by silylation. Mass spectrometric methods include selective ion monitoring, electron impact (EI) and chemical ionization (CI)[2]. Complementary information such as molecular ions and fragmentation mechanisms can be obtained using a variety of approaches in addition to GC-MS. HPLC can also be employed to obtain further information[3-7], but in this paper we restrict results to GC-MS.

2 METABOLISM

Quinine is a member of the cinchona alkaloid family and has been used for the treatment of cardiac arrhythmias and malaria. Its structure is illustrated in Fig. 1. It is metabolized by oxidation of the quinoline and quinuclidine moieties to produce a series of phenolic and non-phenolic derivatives. Since the metabolites are detected from urine extracts, it is necessary to use chromatographic methods to achieve a specific determination.

There has been very little systematic work on metabolic pathways in herbivores. Traditional approaches, used for example in fungi, involving feeding with isotopically enriched substrates are not practicable in horses. Possible techniques include looking at changes in coupled chromatograms of urine or blood extracts after feeding with a potential drug.

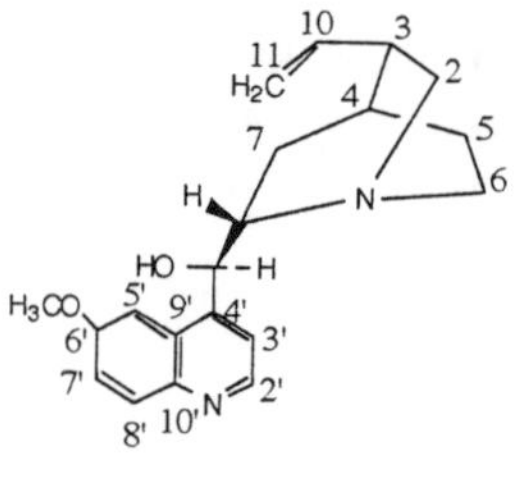

Figure 1 *The structure of quinine*

3 SAMPLE PREPARATION

Urine was collected from a horse 20 hours after oral administration 2 g of quinine. Blank and post-administration urines (5.0 ml) were enzymatically hydrolysed at pH 6.0 with 25 μl *Helix pomatia* juices and extracted by methods of Dumasia *et al.*[8] The dry extract was mixed with 50 μl bis(trimethylsilyl)trifluoroacetamide (BSTFA). The mixture was heated at 80°C for half an hour to obtain trimethylsilyl (TMS) derivatives; after this step the solution was evaporated to dryness in a stream of nitrogen and redissolved in a mixture of 1% MSTFA/toluene, about 50 μl, and 1μl sample injected into the GC-MS.

4 GC-MS CONDITIONS

Combined GC-MS was carried out on a Finnigan MAT TSQ-70 mass spectrometer in the electroionization (EI) and chemical ionization (CI) mode. A fused-silica capillary column (SE-54; 30 m × 220 μm I.D.; 0.25μm film thickness) was used with helium as carrier gas at 25 psi. The oven ternperature was programmed as follows: initial temperature, 100°C; initial hold, 4.0 min.; temperature programming rate, 8°C min^{-1}; final temperature, 300°C, final hold, 10 min. The transfer line temperature was 300°C, and the injector was programmed from 30°C to 300°C at 200°C min^{-1}. Under CI conditions, ammonia was used as a reagent gas with pressure of 0.6 torr. A Varian 3400 CC with SPI injector was used and injected with 1 μl. EI mass spectra were recorded at 70 eV.

5 BACKGROUND SAMPLES

A typical chromatogram of a blank urine sample is shown in Fig.2. Blank samples were extracted, derivatised and analysed as described above. The chromatograms obtained after the GC-MS analysis of blank urine samples were compared with post-administration samples. Table 1 lists the EI-GC-MS peaks from four background samples. Peaks were compared across the four samples; peaks with similar elution times and spectra were lined up. The data of Table 1 shows that there is quite a large statistical variability between samples; this might be due to time of day, feeding regime, metabolism of the horse and so on. However, there are some strong similarities. If a post-administration urine sample contains peaks that are not present in the blank samples or in other post-administration samples using different drugs, this provides preliminary evidence that a metabolite has been detected.

6 GC-MS OF THE POST-ADMINISTRATION SAMPLES

The GC-MS peaks for both EI and CI spectra are listed below. Several features are to be noted. First, the chromatograms in both modes are not identical, for two reasons. First, there is a shift in absolute time (but not relative times) because different offset acquisition times were used in this study. Second, the relative intensities of the peaks differ. This is because the chromatograms are presented as TIC's (total ion currents), i.e. the sum of intensities over all masses. CI and EI spectra differ considerably: the latter provide information on fragmentation, whereas the former is more diagnostic of molecular weights.

Table 1 *EI-GC-MS results of background samples (peak positions in datapoints)*

Blank 1	*Blank 2*	*Blank 3*	*Blank 4*
580			591
601		640	
		691	710
	812		834
857	866	882	
	916	899	
	964	964	962
974	978	977	976
988			
		1026	1029
1073	1061		
1083	1085	1079	
	1103	1104	1101
1151	1138	1174	1173
1209	1203	1204	1201
1217	1266		
1309	1297	1297	1295
1340	1327		1324
1399	1385		
1454	1441		
1510	1494		1470
1563	1547		
1614	1634	1637	1634
1653			
1728			
1767	1751	1751	

(Note: The rows correspond to peaks with similar mass spectra.)

Table 2 *EI-GC-MS results of post -administration samples (peak positions in datapoints)*

Sample A	*Sample B*
293	289
591	590
975	974
1294	1293
1398	1397
1446	1445
1488	1487
1633	1632

Table 3 *CI-GC-MS results of post - administration samples (peak positions in datapoints)*

Sample A	*Sample B*
238	238
468	468
502	501
943	944
1012	1013
1044	1044
1091	1093
1131	1132
1149	1151

7 INTERPRETATION

As an example we try to detect metabolites of quinine. Fig. 3a shows the EI chromatogram (TIC) of sample A. Fig. 3b is the corresponding CI chromatogram. Note the difference in the intensities of peaks. We can show that no intact quinine remains in the extract by co-injection with pure quinine.

Figs 4a and 4b show the selective ion chromatograms of sample A for m/z 136 in both EI and CI modes. The 136 ion is diagnostic of the quinuclidine moiety of the molecule. The EI and CI mass spectra of peaks at 1044 and 1091 (CI), and 1398 and 1446 (EI) are given in Figs 5a-d. For comparison, the EI mass spectrum of pure quinine is given in Fig. 6. Pseudomolecular ion peaks are present in the CI mass spectrum, but in the EI spectrum molecular ion peaks are not always observed, and there is little fragmentation. Therefore chemical ionization spectra usually contain abundant ions in the molecular ion region and a few fragment ions.

The spectra of Figs 5a-d are not present in the background samples or the other post-administration samples. The molecular ion in CI (m/z 485) is shifted by 89 units corresponding to OTMS. Since the m/z ion of 136 is unchanged, this suggests hydroxylation in the aromatic system. The EI fragmentation ion at 319 can be accounted for by the scheme in Fig. 7.

Hydroxylation in the quinuclidine (non-aromatic) moiety is also possible at the double bond between carbons 10 and 11. Fig. 8a is the CI mass spectrum of the peak at 1488, and Fig. 8b is the corresponding EI mass spectrum of the peak at 1131. The molecular weight remains at 485 as expected, but the 136 fragment ion is now shifted by 89 mass numbers, whereas the 261 peak remains unchanged.

Equivalent information can also be obtained from sample B. These spectra present strong evidence that there are metabolites hydroxylated in both halves of the quinine molecule.

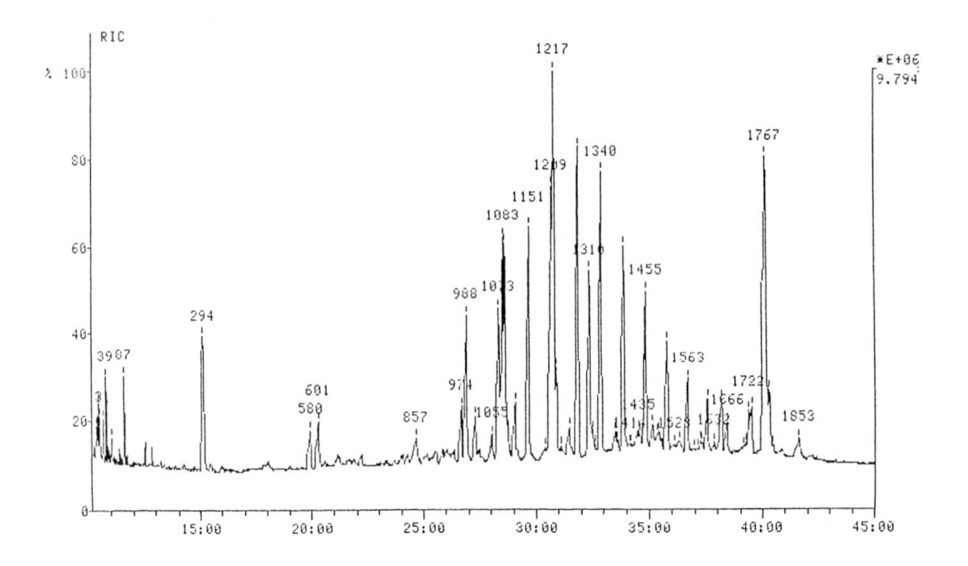

Figure 2 *A typical EI-GCMS chromatogram of a background sample*

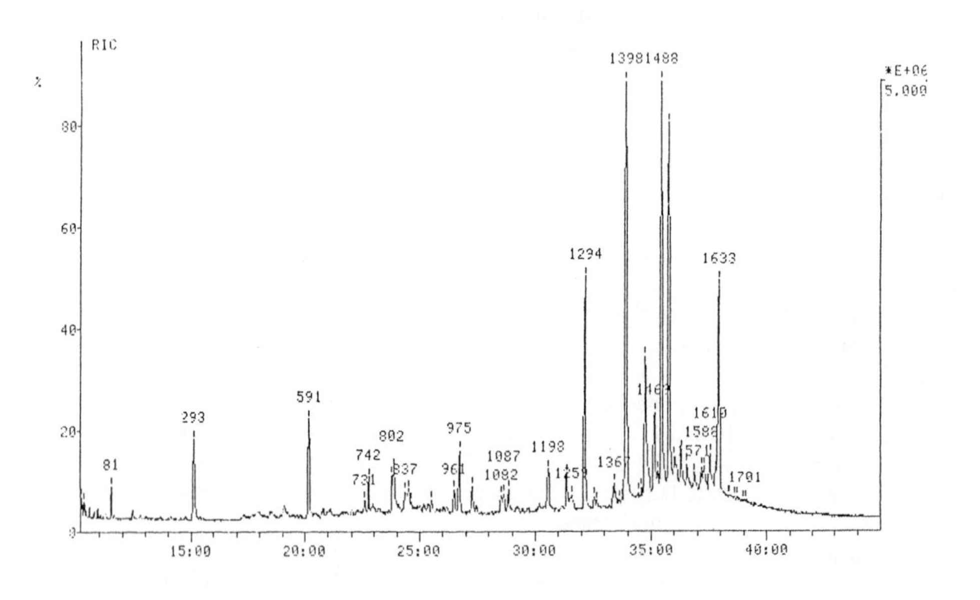

Figure 3a *EI-GCMS of sample A*

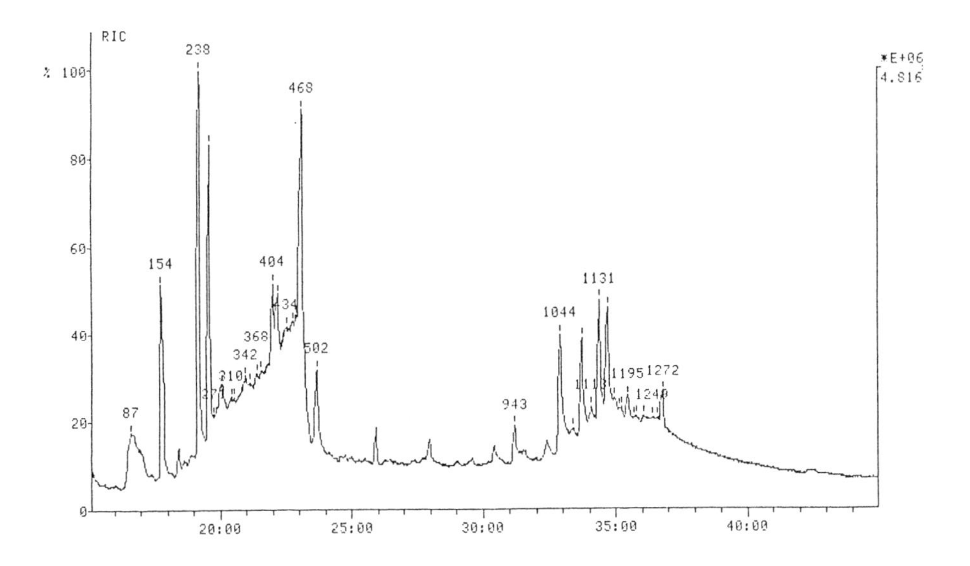

Figure 3b *CI-GCMS of sample A*

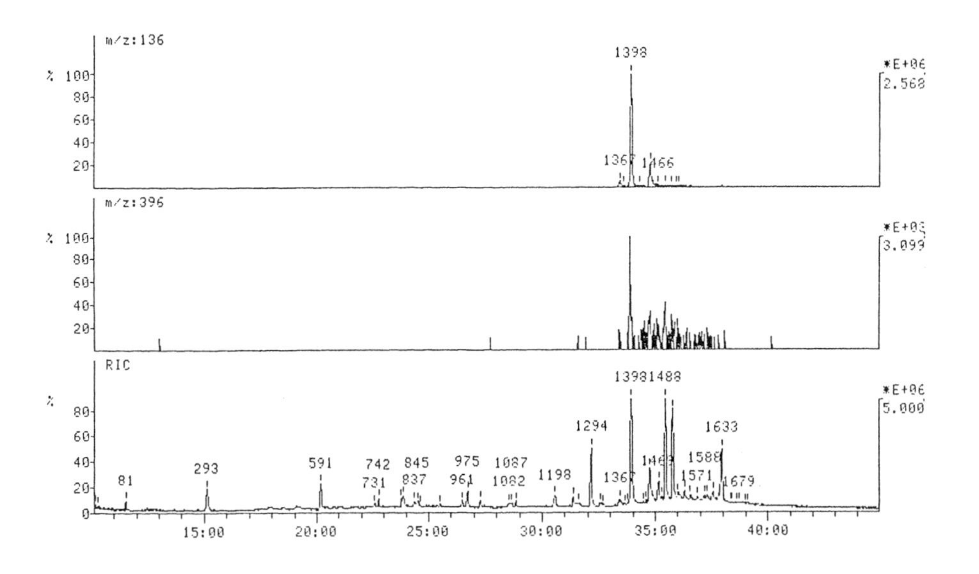

Figure 4a *EI selective ion chromatogram of sample A*

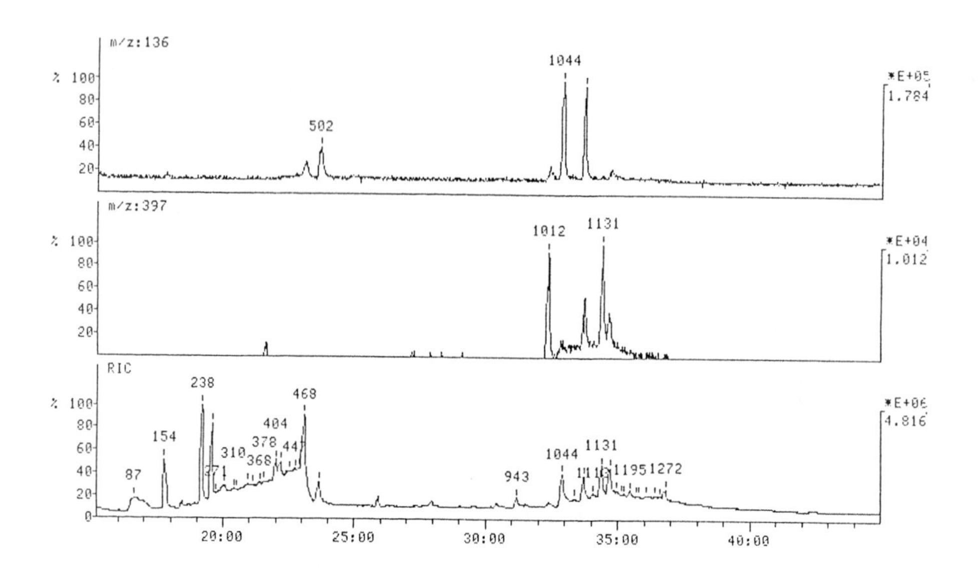

Figure 4b *CI selective ion chromatogram of sample A*

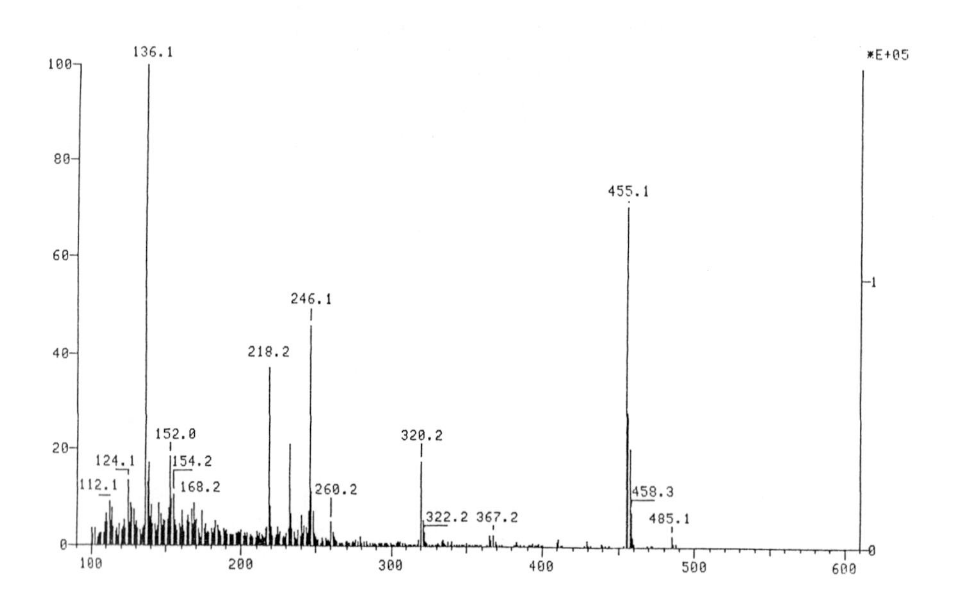

Figure 5a *CI mass spectrum of peak at 1044 of sample A*

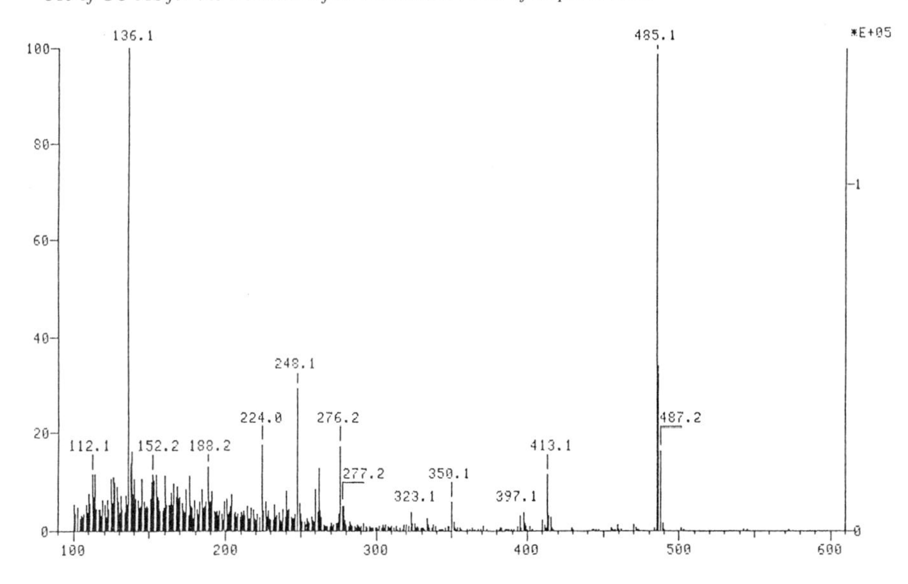

Figure 5b *CI mass spectrum of peak at 1091 of sample A*

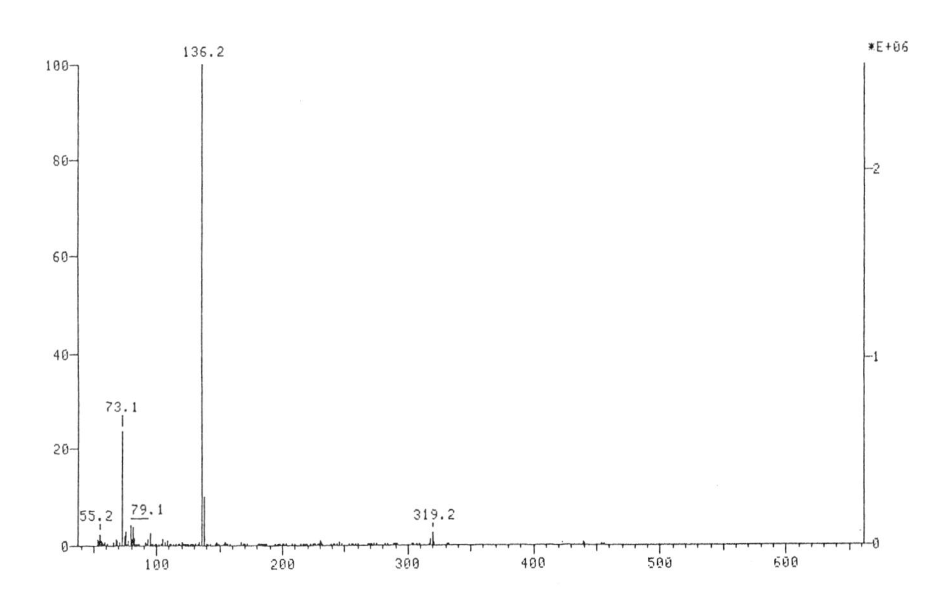

Figure 5c *EI mass spectrum of peak at 1398 of sample A*

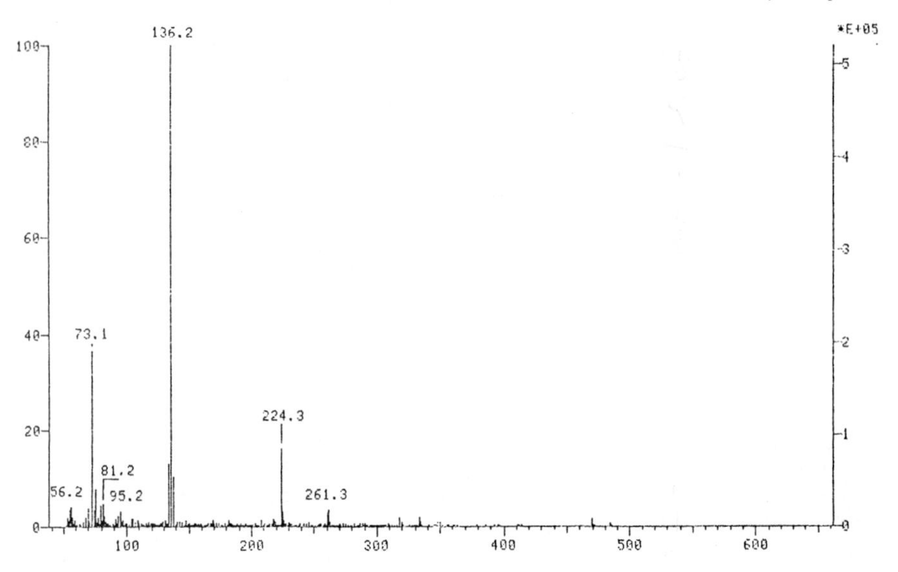

Figure 5d *EI mass spectrum of peak at 1446 of sample A*

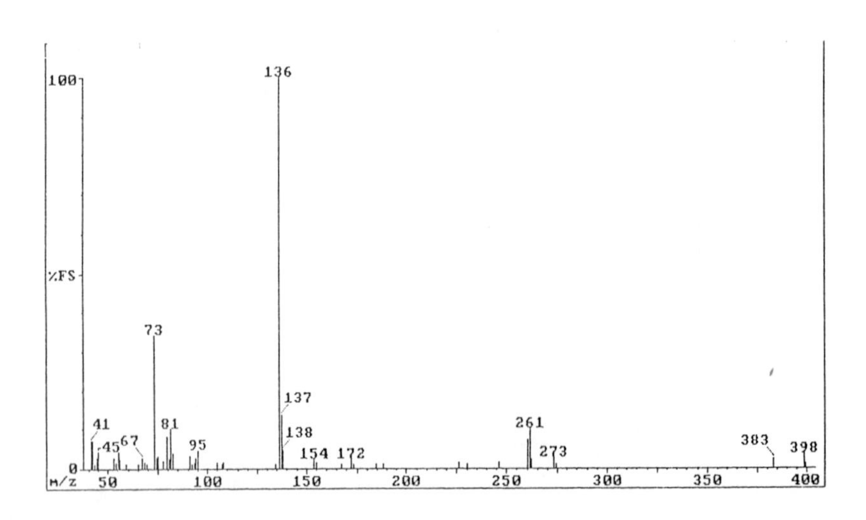

Figure 6 *EI mass spectrum of pure quinine*

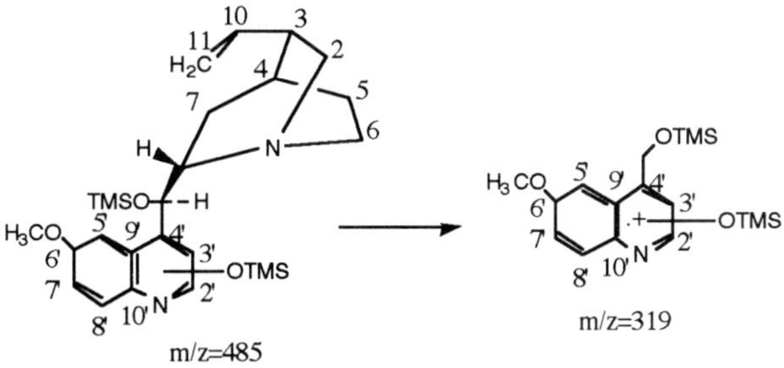

m/z=485 m/z=319

Figure 7 *EI mass spectrum fragmentation scheme*

8 CONCLUSION

GC-MS is of great importance in the investigation of the metabolism of drugs *in vivo*. Among the chromatographic methods using mass spectrometric detection, the selected ion monitoring mode is useful because of the high selectivity[9]. CI is used to yield information on the moleculer weights of unknowns. The EI spectra are more easily referenced to mass spectral databases, which assists in identifying unknown components in sample extracts. Additionally, the fragmentation produced in EI offers valuable information for structure elucidation. The determination of drugs and their metabolites in urine can also be carried out by liquid chromatography with thermospray mass spectra (TSP LC-MS) and atmospheric pressure ionization mass spectrometry (API LC-MS)[10], so a range of modern coupled mass spectrometric methods can be used to give insight into these breakdown pathways.

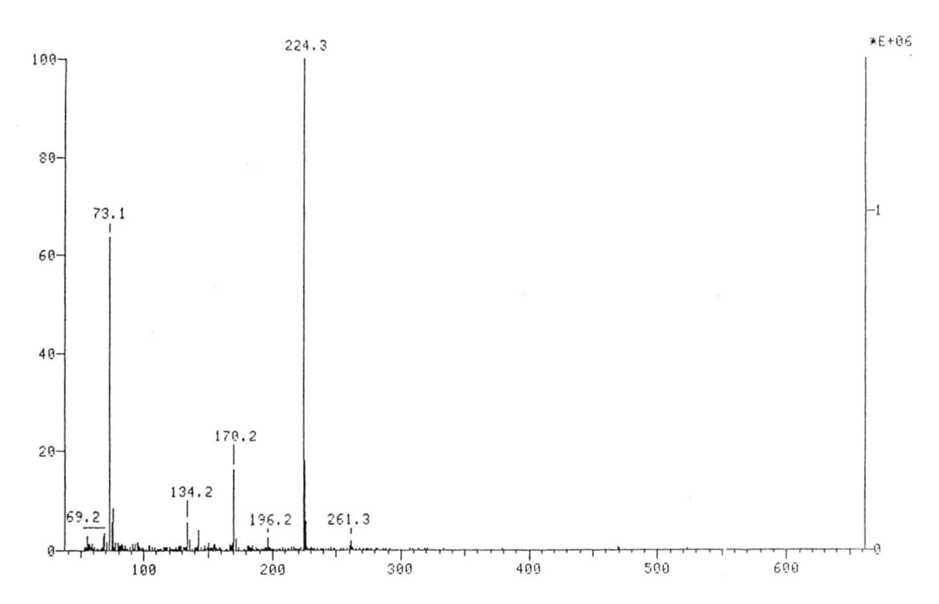

Figure 8a *EI mass spectrum of peak at 1488 of sample A*

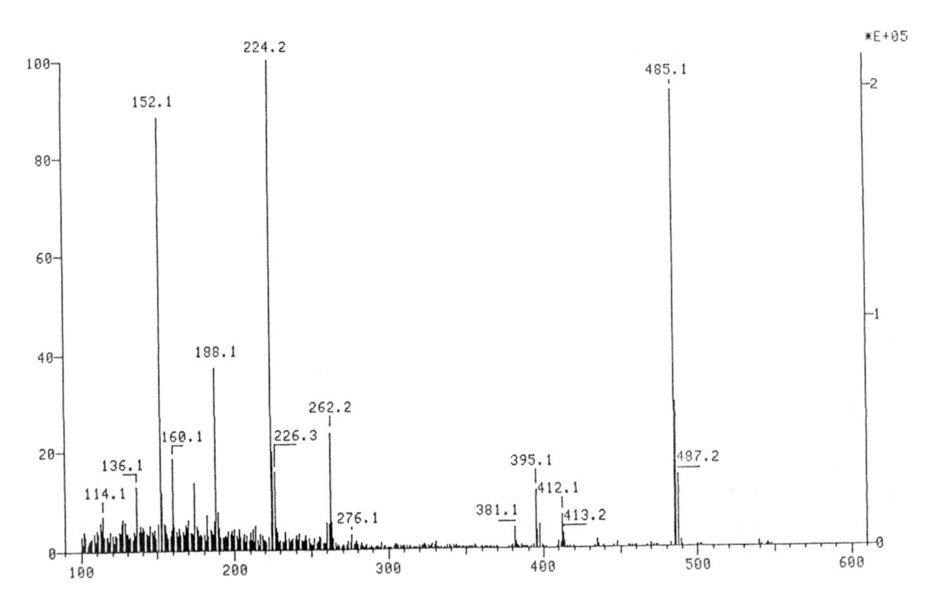

Figure 8b *CI mass spectrum of peak at 1131of sample A*

9 ACKNOWLEDGEMENTS

We thank Dr. S.Westwood of the Horseracing Forensic Laboratory, Newmarket for providing the samples discussed here and Mr. J. Carter for running the mass spectra.

References

1. S. Steffenrud, G. Maylin, *J. Chromatogr.*, 1992, **577,** 221.
2. J. A. van Rhijn, W. A. Tragg, H. H. Hekamp, *J. Chromatogr.*, 1993, **619**, 243.
3. C. Eddins, J. Hamann, K. Johnson, *J. Chromatogr. Sci.*, 1985, **23,** 308.
4. J. A. Hamann, K. Johnson, D. T. Jeter, *J. Chromatogr. Sci.*, 1985, **23,** 34.
5. S. E. Barrow, A. A. Taylor, E. C. Horning, M. G. Horning, *J. Chromatogr.*, 1980, **181**, 219.
6. G. W. Mihaly, K. M. Hyman, R. A. Smallwood, *J. Chromatogr.*, 1987, **415**, 177.
7. R. L. Furner, G. B. Brown, *J. Anal. Toxicol.*, 1981, **5**, 275.
8. M. C. Dumasa, E. Houghton, *J. Chromatogr.*, 1991, **564**, 503.
9. A. Polettini, M. C. Ricossa, A. Groppi and M. Montagna, *J. Chromatogr.*, 1991, **564**, 529.
10. R. D. Voyksner, *Environ. Sci. Technol.*, 1994, **28**, 118.

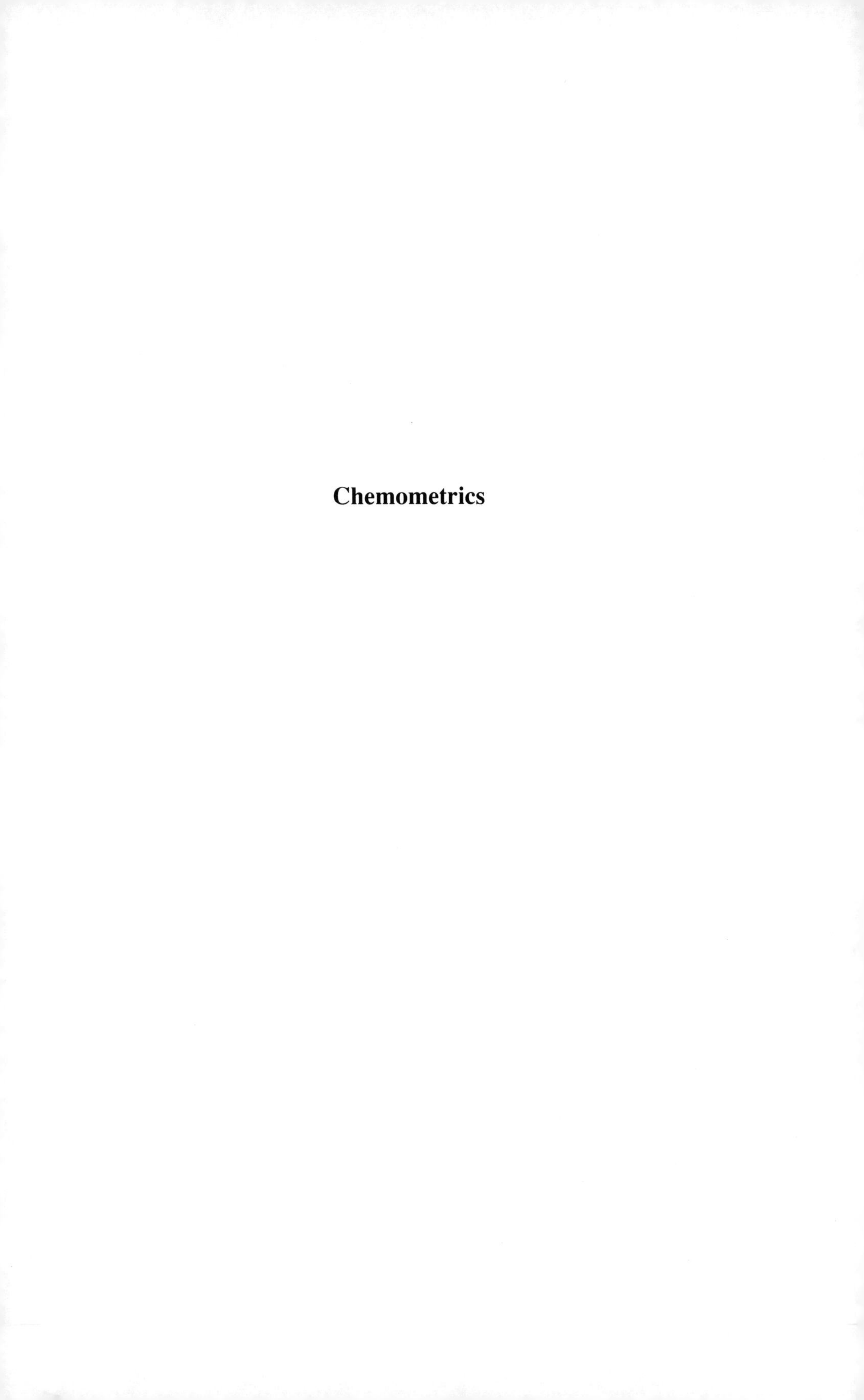

Chemometrics

Modelling Non-linear Data Using Neural Networks Regression in Connection with PLS or PCA

Claus Borggaard

DANISH MEAT RESEARCH INSTITUTE, MAGLEGAARDSVEJ 2, PO BOX 57, DK-4000 ROSKILDE, DENMARK

1 INTRODUCTION

When generating calibrations for our NIR-spectrometers the Danish Meat Research Institute (DMRI) uses a technique based on combining Artificial Neural Networks (NN) and PCA/PLS. In this method the spectra are first preprocessed with PCA or PLS in order to compress the data before they are passed on to the Neural Network which then performs the regression part of the calibration procedure. With this method we achieve results on calibration and classification problems, which often are much better than standard linear methods such as Principal Component Regression (PCR) or Partial Least Squares Regression (PLSR). Also,- the use of PCA or PLS to compress data makes it possible to use very small neural networks in the regression part, which in turn means that they can be trained in a very short time. For reasons not well understood by us, training neural networks on principal components (or PLS components) rather than on the spectra directly, is much less likely to result in the networks winding up in local minima.
In this paper we will first shortly describe the principles of the PCA/PLS data reduction. This is followed by an introduction to the principles of Neural Networks, a description of the DMRI implementation and finally the PCA-NN method is compared with PLS/PCR on a non-linear data set well known from the Chemometrics literature.

2 DATA REDUCTION WITH PCR/PLS

Spectroscopic data sets are often made up of measurements at many channels (wavelengths in the case of NIR and UV/VIS). In many applications the information content in one of these channels is highly correlated to the information content of another channel in the spectrum. In cases such as these, it is the goal of the chemometrician to throw away as much redundant information as possible from the spectra, before commencing with the extraction of useful information about the samples being measured.
For this purpose PCA and PLS have proven extremely useful. With these linear techniques the dimensionality of the calibration problem can be reduced say from 1000 data points per spectrum to about only 10 numbers (so called PCA or PLS scores), which in most cases will contain all the useful information from the entire spectrum.

With PCA this is done in the following way:

* We start off with a X-matrix containing the spectroscopic data with N rows (each row is a spectrum) and M columns (absorbance measurements).

* Next we calculate the directions in the M-dimensional space which accounts for most of the variation in the data. These directions which are orthogonal to one another are called PCA-loadings (eigenvectors to the correlation matrix $C = X^TX/(N-1)$). The K loadings (usually K={1-16}) with the largest eigenvalues are chosen and the rest are usually neglected as these are very susceptible to spectroscopic noise. Several procedures are available for finding these vectors[1-3];- the Unscrambler software package[4] uses the NIPALS algorithm which automatically calculates the loadings in descending order of variance being described.

* With these loadings it is now possible to compress the data in the X-matrix resulting in K numbers (called PCA- or PLS-scores) for each spectrum which are the projections of each spectrum in the X-matrix on to the loadings vectors:

Data reduction:

$$X \cdot p^T = T$$

Data expansion:

$$T \cdot p = X - E$$

X has been mean centred (the average spectrum is subtracted from each individual spectrum) and p is a (MxK) matrix containing the loading vectors and T is a (NxK) matrix containing the first K scores for each object.

E is an "Error" matrix (residual matrix); that is to say the part of the matrix X which is not described by the first K factors.

Thus, instead of operating with M data points per spectrum when performing a regression from the X-matrix to a matrix Y containing the laboratory reference values, we can work on the K PCA- or PLS-scores.

To see whether our PCA-model for reducing the dimensionality of X is actually working, we can take a new set of data (test set) acquired with the measuring device, and project them on to the loadings found during the modelling phase.

If the model has been made on a representative set of spectra, the residual E should be on the same level as the one seen during calibration.

In Figure 1 is shown a test set spectrum obtained on an Infratec 1225 (wavelength range 850nm - 1050nm). From a calibration set consisting of 100 samples, the first 3 principal components have been calculated. The test set spectrum has then been projected on to each of these directions giving the first four PCA scores (using the above expression for data compression). By using the above expression for data expansion, the original test spectrum can be reconstructed using from 1 to 3 principal components. These reconstructed spectra are shown on top of the original test spectrum. If the fourth principal component had been utilized in the reconstruction, we would not have been able to distinguish the original spectrum from the reconstructed spectrum. This clearly demonstrates the efficiency of the data compression technique.

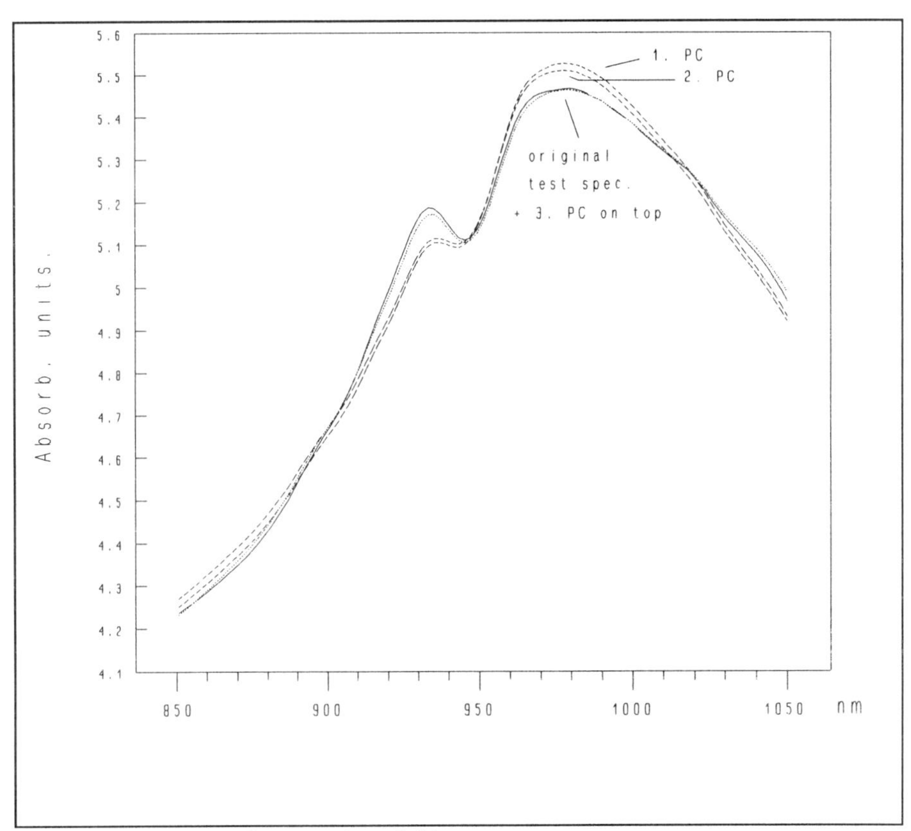

Figure 1 *Reconstructed spectra from PCA scores.*

PLS data reduction works the same way as PCA, the differences being:

1. The PLS-loadings are calculated in such a way that the information derived from the X-matrix is mostly taken from the parts of the spectra which relate linearly with the reference value in the Y-matrix[1-4].

2. PLS does not in general seek to model the entire X-matrix although in many cases this will be done quite well.
3. If data from all the channels in a measurement are completely independent of each other then PCA-regression will be worthless;- whereas PLS will do a very good job of finding the few relevant variables for the regression.

In our work with Neural Networks we use the PCA- or PLS-scores as input to the network hereby getting rid of all redundancies in the original data in the X-matrix.

3 ARTIFICIAL NEURAL NETWORKS

An artificial neural network consists of an input layer, one or more hidden layers, and an output layer. These layers are composed of individual neurons which in a computer are just locations in RAM-memory. In Figure 2 is shown a simple network which we will use later on in this discussion.

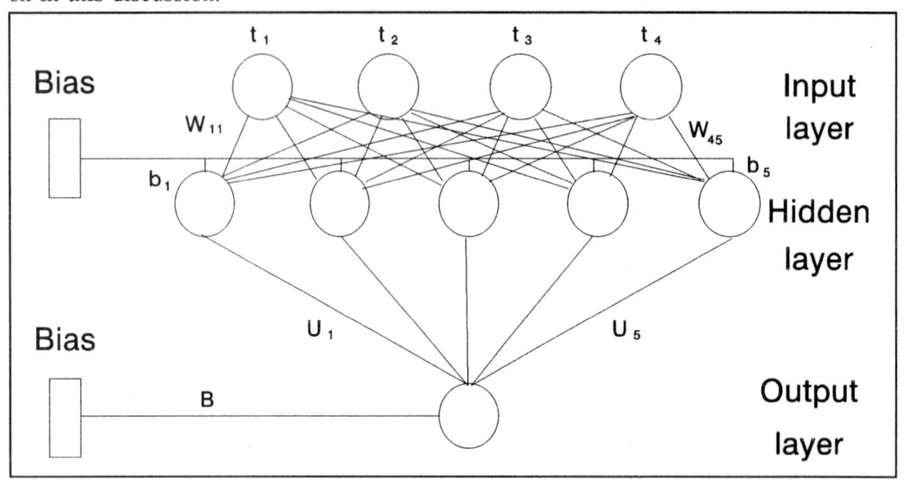

Figure 2 *A diagram of a simple artificial neural network.*

The above network is made up of 4 neurons in the input layer, 1 hidden layer with 5 neurons, and one output neuron. Each of the neurons in a layer is connected to all neurons in the previous layer (and in some of our applications also to the layers above the previous). Also this network contains two "bias" neurons. One of these is connected to the hidden layer and the next one is connected to the output neuron.

This means that data fed to the input layer is passed on to the next layer via the connections. During this transfer of data from "upstream" neuron i to "downstream" neuron j the numbers are multiplied by a weight w_{ij} (In Figure 2, u designates the weights of the connections from the hidden layer to output). The bias neurons can be looked upon as a neuron in the layer above giving out a constant=1 which will be multiplied by weights b_j and B. In the hidden layer all the information flowing to it from the previous layer is added up, and before it is passed on to the output, the data is passed through a sigmoid function:

$$\sigma(z) = \frac{1}{1 + e^{-z}}$$

Sometimes other transfer functions are chosen, e.g. hyperbolic tangents, but in our experience the sigmoid function is the most well suited. Some workers in the field of Neural Networks even use linear transfer functions in some of the hidden neurons. At the

DMRI we prefer to use direct connections from input layer to output neuron. This corresponds to superimposing the linear PCR/PLSR solution on top of the non-linear neural network solution.

The output from the network can then be calculated explicitly from the values that are given to the input neurons and the weights (here we use the sigmoid function at no direct connections):

$$output = B + \sum_{j=1}^{5} u_j \cdot \sigma \left(\sum_{i=1}^{4} (w_{ij} \cdot t_i) + b_j \right)$$

3.1 How to obtain the right weights for an application?

The neural network can only give a correct answer to the input stimuli if the weights have been adjusted to the correct values. This adjustment is done by presenting a training data set (calibration set) to the network. In the beginning the weights and biases are set at small random numbers. With these starting values, the network will most likely give a wrong answer to almost every example. The weights are then adjusted according to an algorithm known as back propagation of error [5]. In short this method utilizes the fact that the output value can, as previously shown, be written explicitly as a function of the weights. By differentiating the cost function:

$$cost = \sum_{i=1}^{N} (output_i - y_i^{ref})^2$$

with respect to the weights w, one will know, in which direction each weight has to be changed, in order to give a better result next time data is presented to the network.

3.2 Calibrations with Neural Networks

The DMRI has had a high degree of success with Neural Networks [6,7]. The reasons for this are the following:

1. We use PCA or PLS preprocessing of the data. This enables us to use a few scores as input to the network rather than the full spectrum. This in turn means that we can use very small networks resulting in a small risk of over-fitting and fast training times (from 10-60 min).

2. The software developed at the DMRI and now implemented in the Neural-Unscrambler [4] uses a test set validation during training. This is a very important feature when working with neural networks as the number of parameters to be estimated are quite large compared to standard PCR and PLSR. The test sets are true test sets, both with respect to the PCA/PLS preprocessing and the final regression performed by the Neural Network. The Neural Unscrambler software automatically saves the network with the best performance on the test set.

3. The Neural-Unscrambler software has the built in ability to handle a linear PCA- or PLS-model along side the non-linear processing system. This is made possible by the use of direct connections from the neurons in the input layer to the output neuron. This feature has proven to work so well that the results are always just as good as, or better than results obtained with PLSR.

Neural Networks can be made to perform the PCA data compression as well as the regression. It has been proven that if a feed-forward network, consisting of n-input neurons, one hidden neuron with a linear response function and n output neurons, is trained on e.g. a set of spectra and asked to predict the same spectra at the output neurons, the output from the hidden neuron will give the PCA score for the first principal component [8]. In this case the weights for each connection from the input layer to the hidden layer will correspond to the loadings vector defining the direction of the first principal component. If more hidden neurons are added to this architecture more principal components will be added. A network where sigmoid transfer functions are used at the neurons in the hidden layer, will produce a non-linear PCA solution. In our work with Neural Networks we have avoided this approach for the following reasons.

* Linear PCA/PLS is quick and will produce models with very small X-matrix residuals.

* If the spectra are to be feed directly into the Neural Network it will often be necessary to choose say every tenth wavelength, as the network size will otherwise become overwhelming. If this is done, it will usually be necessary to smooth the spectra prior to being presented to the input layer.

* It is our experience that training neural networks on data with a high degree of collinearity is extremely slow and often results in the network dropping into local minima.

Therefore we prefer to use standard PCA/PLS as a preprocessing step (data compression) before training the network for the regression part of the calibration problem.

4 NIR-DATA, ANALYSIS AND RESULTS

To demonstrate the ability of the PLS/PCR - NN approach, a set of NIR-spectra obtained on a Technicon 450 at the Norwegian Food Research Institute (MATFORSK) is used [9]. The data set is composed of 82 NIR-spectra of a mixture of water, fish meal and starch. Each spectrum contains absorbance values at 19 individual wavelengths in the region 1200 nm to 2400 nm. The wet chemistry references on the samples cover a range from 0% to 88% protein with the average being 34.18% and the standard deviation 21.66%.

The experimental design is given in ref. 8. The sum of the 3 ingredients add up to 100 %. It should therefore be possible to make a good PLS calibration with only two factors. However, this is not the case. A PLSR and PCR calibration using a jack-knife validation gives the following results:

- PLS uses 12 factors in the optimal model, result SEC = 1.32%, SEP = 2.08%, R = 0.995.

- PCR uses 15 factors in the optimal model, result SEC = 1.49%, SEP = 2.72%, R = 0.992.

It was quite surprising to us that we needed so many factors to make a reasonable calibration. We were even more astonished when we found that the residual variance of the X-matrix for both calibration objects and various randomly selected test sets went to almost zero after only 5 factors were included in the model.(In terms of Figure 1 the reconstructed spectra after 4-5 factors could not be distinguished from the original spectra.) This suggested that there was a non-linear relationship between spectroscopic data and sample composition which PCR and PLS-R was attempting to model using more factors.

After this preliminary study of the data, we decided to create 6 different partitionings of the data into calibration sets and test sets. Data was sorted according to increasing protein content, and the test sets were chosen in the following way:

- 6 test segments are made from the total data set:

  - 1. test segment is object 1, 7, 13,..., 79

  - 2. test segment is object 2, 8, 14,..., 80

  - 3. test segment is object 3, 9, 15,..., 81

    .

    .

  - 6. test segment is object 6, 12, 18,..., 78

With the Unscrambler package we now made PCR models for each of the 6 partitionings. The models were made on the calibration sets (68-69 objects) and validated on the test objects. The results of these calibrations are listed in Table 1, both for the optimal number of factors given by unscrambler and for 4 factors (which if the problem was linear should give an optimal model as the X-matrix residuals with 4 factors are negligible).

After this the Neural-Unscrambler package was used to make NN models on each of these separate data sets. As inputs we used the first 4 scores as calculated by Unscrambler in the PCA models. Training was performed on the same calibration objects as the PCA models above, and the networks were validated on the test objects.

PCR and Artificial Neural Networks are used as methods for modelling with results given in Table 1.

Table 1. *Comparison of results of modelling with PCR or Artificial Neural Network*

	Modelling method PCR				**Artificial Neural Network**	
Test segment	No. of Factors	SEP	SEP at 4 factors	SEC at 4 factors	SEC	SEP
1	16	1.73	4.15	2.76	0.58	0.82
2	15	2.60	4.02	3.45	0.67	0.75
3	10	1.55	3.89	4.19	0.60	0.76
4	9	3.74	3.23	6.69	0.54	0.85
5	15	1.51	4.06	3.29	0.53	0.88
6	16	1.02	4.06	3.23	0.63	0.73

The Artificial Neural Network has 4 PCA scores as input.

Finally we also performed PLS calibrations on the same data sets as were used with the PCR calibrations. Cross validated PLS models were made on the calibration objects, and the model with the optimal number of factors then tested on the test objects. The results of this analysis are shown in Table 2.

Table 2. *Comparison of results of modelling with PLS or Artificial Neural Network*

Test set no.	PLS (4 fac.) SEP	PLS (optimal no. of factors)	SEP
1	3.3	16	1.6
2	3.5	15	2.6
3	4.9	12	2.9
4	6.7	9	3.9
5	4.0	16	1.4
6	3.2	12	1.6

In Table 2 the models were based on a cross validation on the calibration objects for each partitioning. The optimal number of factors were then used to predict on the test sets. The predictions using 4 factors are put in for comparison with the results in Table 2.

5 CONCLUSION

On this data set it has been proven that Neural Networks are superior to both PCA and PLS in all the partitionings of the NIR-data set.

In this case the PCA-NN models made with Unscrambler and Neural-Unscrambler needed only the first 4 PCA factors as inputs, in order to give models which were much better than PCR or PLS, using from 9-16 factors. The SEP's were on the average improved by 66% compared to standard PCA. Therefore we find that NN offer the opportunity of modelling data that are non-linear without having to spend a lot of time finding the optimal scatter correction methods and so forth.

It can also be concluded that cross validated PLS models are very over-optimistic with respect to how they perform on a true test set, as it has been demonstrated that PLS on the whole data set gave a SEC=1.32% protein whereas PLS on independent test sets gave SEP's in the range 1.6% to 3.9%.

References

1. H. Martens and T. Næs, 'Multivariate Calibration', John Wiley & Sons,Chichester 1989.
2. C. Ridder, FIA og kemometri, Ph.D thesis, Kemisk Laboratorium A, Danmarks Tekniske Højskole, 1989.(Danish)
3. M. A. Sharaf, D. L. Illman and B.R. Kowalski, 'Chemometrics', John Wiley & Sons, New York, 1986.
4. UNSCRAMBLER, *User's Guide*, CAMO A/S, 1992.
5. P. R. Lippman, *IEEE ASSP mag.*, 1987, April,4-22.
6. C. Borggaard and H. H. Thoberg, *Anal. Chem.*, 1992, **64**, 545.
7. C. Borggaard and A. J. Rasmussen, Interpreting non-linear NIR-spectra using neural networks and PCA/PLS, in 'Near infra-red Spectroscopy Bridging the Gap between Data Analysis and NIR Applications', Editors: K. I. Hildrum, T. Isaksson, T. Næs and A. Tandberg, Ellis Horwood, Chichester, 1992, Chapter 11, p. 73.
8. P. Baldi, Hornik K., 'Neural Networks and Principal Component Analysis: Learning from Examples Without Local Minima. Neural Networks', Pergamon Press, New York, 1989, Vol. 2 , p. 53.
9. D. Oman, T. Næs and A. Zube, *J. Chemometrics*, 1993, **7**, 195.

Rugged Spectroscopic Calibration Using Neural Networks

Paul J. Gemperline

DEPARTMENT OF CHEMISTRY, EAST CAROLINA UNIVERSITY, GREENVILLE, NC 27858, USA

1 INTRODUCTION

Currently, one of the greatest obstacles to wide spread adoption of multivariate spectroscopic assays for routine analysis is the lack of ruggedness with respect to changes in instrument response. In this paper we report a strategy for developing rugged multivariate spectroscopic calibrations using artificial neural networks. For the purposes of this paper, we define a rugged multivariate spectroscopic calibration as one that is insensitive to slight variations in instrument response such as changes in baseline offset, changes in cell path length, cell positioning errors, wavelength calibration errors, and changes in instrument bandwidth. It is the hypothesis of this paper that methods with rugged characteristics should be easy to maintain and require infrequent re-calibration. Rugged calibrations should facilitate long term (weeks, months) use of stored calibration coefficients and eliminate the need for re-calibration after minor maintenance events like replacing a source lamp. Utilization of rugged calibrations should also make it easier to transfer methods developed on an instrument in a research laboratory to an instrument in the field without resorting to re-calibration.

Other researchers have recently reported methods to mathematically map the response function of a "slave" instrument to the response of a "master" instrument by using measurements of a set of standard reference materials from both instruments or by measuring a subset of calibration standards for the method to be transferred[1-3]. There have been some reports of successes and failures using these new methods.

1.1 Spectroscopic Calibration using Artificial Neural Networks

Many computational methods are useful for spectroscopic calibration including principal component regression (PCR), partial least-squares (PLS), and multiple linear regression (MLR). Recently, artificial neural networks have gained popularity for multivariate calibration[4-12]. Neural networks used for spectroscopic calibration typically consist of two or more layers of interconnected nodes that process the weighted inputs from the preceding layer and generate an output that is passed on to the next layer after transformation with some type of non-linear output function. In our work we use principal component scores of digitized ultraviolet, visible, near-infrared, or infrared absorbance or reflectance values as input values. These signals are passed to the hidden layer of two to twelve nodes through input nodes using a linear transfer function. Different numbers of linear and sigmoidal output functions are used in the hidden layer as determined by trial and error. One linear output node is used to estimate the concentration of one of the constituents in the mixture.

Information is encoded in the network by adjusting weighting coefficients during a "training" process where training patterns (e.g., digitized spectra, chromatograms or other measurements) are presented to the network along with their expected outputs. The input signals (e.g., absorbance values) are propagated through the network to compute one or more output values. The output values are compared with the expected values for the training pattern (e.g., class membership, functional group presence, concentration or biological activity) and the network weights are adjusted to minimize the error in the output.

In our work we have observed that neural network training does not converge to the same result when different random initializations are used. Under most circumstances this would be a disadvantage; however, it is possible to take advantage of this behavior by training many different neural networks to solve the same problem, then searching for one that is the most rugged.

1.1.1 Overfitting. In the course of applying neural networks to a wide range of calibration applications we have found that neural nets have a tendency to overfit calibration data sets[4]. Networks that overfit calibration data sets give poor predictive performance and exhibit greater sensitivity to small measurement errors on future samples. One way to characterize overfitting is to monitor network performance with the standard error of calibration (SEC, obtained from the calibration data set) and standard error of prediction (SEP, obtained from validation data sets). A value of SEC significantly lower than SEP is usually taken to be a strong indication of overfitting in our work.

1.1.2 Multiple Optima. We have recently described an advanced neural network training algorithm for spectroscopic calibration based on the quasi-Newton optimization technique[4]. This algorithm converge more quickly than other training strategies and gives consistently lower SEC and SEP values. Even though this optimization algorithm is well-known for its ability to converge to global optima we have observed that network training rarely converges to the same solution after initialization to new random values and re-training. We attribute this behavior to the very large number of parameters being optimized and random noise in input measurements. For example, a typical IR or NIR calibration problem might have 30 calibration spectra of 1024 highly correlated points, giving a total of 30 x 1024 = 30,700 points. A typical network used to solve the calibration problem could have 1024 input nodes, 5 nodes in the hidden layer and one output node, corresponding to a total of over 5,000 network weights to be adjusted. Many difficulties are to be expected in solving such a large optimization problem. Firstly, a prohibitive amount of computer time would be required to systematically explore such a large dimensional space for a global optimum. Secondly, since the noisy measurements are used as input variables, the network output error surfaces are likely to contain many local optima.

1.2 Rugged Calibrations

Our network training strategy attempts to develop rugged calibrations by taking advantage of the tendency of networks to converge to local optima. The strategy may be briefly summarized as follows: 1) Train many different neural networks to solve a calibration problem. 2) Monitor the network training progress at regular intervals by computing the standard error of calibration (SEC) using the calibration spectra and by computing the standard error of prediction (SEP) using a set of validation spectra. 3) Stop training the network when no further improvement in SEC is realized. 4) Perform a "sensitivity analysis" on 40 networks having the lowest SEP to identify the network that is least sensitive to small perturbations in response.

1.2.1 Training strategy. For each calibration problem, the following neural network configuration parameters were adjusted: the number of principal component scores used as input variables, the number of nodes in the hidden layer, and the number of sigmoid vs. linear transfer functions used for hidden layer nodes. Our neural network codes are unique when compared to standard codes because we allow the user to specify a mixtures of linear and sigmoid nodes in the hidden layer. We believe this permits the training algorithm to use a combination of linear and non-linear modeling more easily compared to networks with only sigmoid nodes in the hidden layer. We have observed that use of a combination of linear and sigmoid nodes in the hidden layer promotes rapid convergence in the early phases of network training, whereas networks with only sigmoid nodes in the hidden layer tend to converge more slowly in the early training phase. Each instance of a network configuration was trained at 5 different random initializations to reduce problems caused by convergence to local minima. Depending on the number of PC scores used more than 7000 networks were trained per calibration problem.

```
for num_pc_scores = 1 to 11
    for num_hidden_nodes = 1 to num_pc_scores+1
        for num_sigmoid_nodes = 0 to num_hidden_nodes
            for trial = 1 to 5
                {Initialize and train one neural network}
            end
        end
    end
end
```

Neural network training programs were in the MATLAB programming language for an IBM RS/6000 workstation. Using the quasi-Newton training algorithm, typical training times for a single neural network ranged from a few seconds to 5 minutes, depending on the complexity of the network being trained. The average training time for a network was about one to two minutes.

1.2.2 Sensitivity analysis. Sensitivity analysis was performed on the "best" 40 networks which gave the lowest standard error of prediction on a validation data set. Sensitivity analysis was performed by perturbing the validation data set. Three kinds of instrument response errors were simulated by the perturbations. Baseline shifts or offsets were simulated by adding a constant to each spectrum (±1%, ±2% of the mean absorbance). Path length changes were simulated by multiplying each spectrum by a constant (±1%, ±2% of the mean absorbance). Wavelength calibration errors were simulated by FFT interpolation at ±1/4, ±1/2 of the digitization interval. Table 1 shows the experimental design used for simulating these effects and the sample output of the sensitivity analysis for one calibration.

2 EXPERIMENTAL

NIR process data was obtained from Lois Weyer, Hercules, Inc. The goal of the project was to replace a GC assay for acetone, t-butyl alcohol, and isopropyl alcohol in an aqueous process stream with an NIR assay. Residual amounts of four different products may be present in the solvent mixtures. Four data sets were acquired representing four different process streams. Samples were taken on February 22, 23, 24, March 4, 5, 30 and June 10 and 11, 1993. NIR spectral data were acquired using a single beam Guided Wave 200 instrument equipped with fiber probe. The February and March data was split into two parts. One part was kept for calibration. The other part was combined with the March 4 and 5 data to create a validation set for ruggedness testing. The June data was used as a test set to monitor the training progress. First derivative spectra of the process mixtures are shown in Figure 1. The samples of process stream A had poor signal to noise ratios and did not exhibit sufficient variability in composition to provide useful calibration results and is not discussed further in this paper.

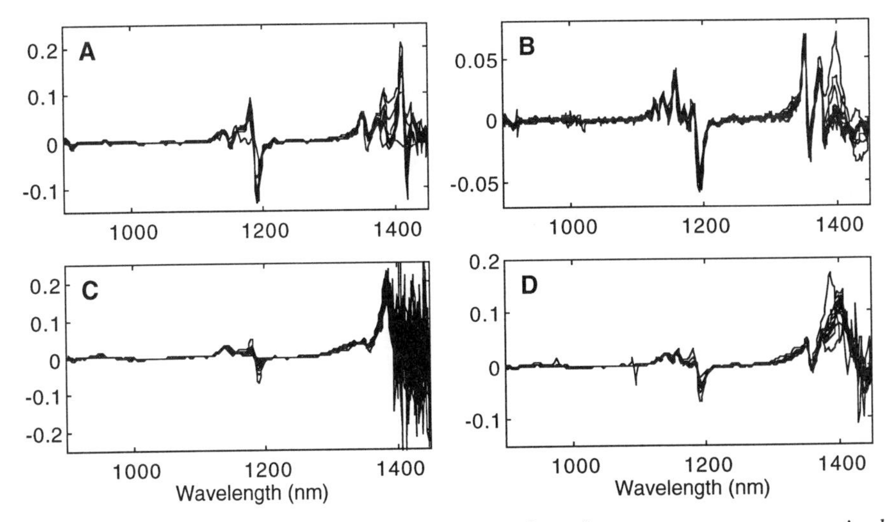

Figure 1 *First derivative NIR spectra of mixtures from four process streams acquired using a Guided Wave model 200 spectrophotometer equiped with a wand fiber probe.*

3 RESULTS

Three (B, C and D) of the above four sets of NIR process data were processed by our rugged neural network calibration software. For comparison, the same data sets were processed by partial least-squares. An illustration of the experimental design used for the sensitivity analysis and sample results for the NIR calibration of t-butyl alcohol in process stream B are shown in Table 1.

Table 1 *Illustration of sensitivity analysis results for NIR calibration of t-butyl alcohol in process stream B. The best PCA-neural network calibration is less sensitive to measurement errors compared to the best PLS calibration.*

Wavelength calibr. error (spl. interval)	*Pathlength error (% of mean)*	*Baseline error (% of mean)*	*SEP*	
			PLS	*PCA-NN*
0	0	0	1.358	0.132
-0.25	-1	-1	1.555	0.490
-0.25	-1	1	1.555	0.490
-0.25	1	-1	1.532	0.477
-0.25	1	1	1.532	0.477
0.25	-1	-1	1.173	0.434
0.25	-1	1	1.173	0.434
0.25	1	-1	1.142	0.290
0.25	1	1	1.142	0.290
0	0	-2	1.358	0.132
0	0	2	1.358	0.132
-0.5	0	0	1.699	0.937
0.5	0	0	0.960	0.587
0	-2	0	1.385	0.183
0	2	0	1.331	0.228
		RMS:	1.162	0.617

Small changes in spectroscopic response were simulated by perturbing the test set according to the values shown in the table. The effects simulated include wavelength calibration errors, path length errors, and baseline shifts. The column labeled "SEP" shows the standard error of prediction for the perturbed test data. As expected, baseline offset errors were completely compensated by use of first derivative spectra. Path length errors seemed to introduce less error compared to wavelength calibration errors. In fact, the PCA-NN calibration seems remarkably insensitive to simulated path length errors. Overall, the best PCA-neural network (5 PC input scores, 5 sigmoid hidden nodes) calibration is much less sensitive to measurement errors compared to the best PLS (6 factors) calibration.

A summary of the rugged the neural network calibrations with comparisons to PLS is given in Table 2. The rows labeled "no model" report the standard deviation of the labeled constituent. Substantially lower values of SEC or SEP than the "no model" row indicate useful predictive capabilities. The column labeled "Parms" is used to report the configuration of the best PLS or neural network calibration. For example, "4pc 3s 21", indicates a neural network with four PC scores as input variables, three sigmoid nodes in the hidden layer and two linear nodes in the hidden layer. For stream C, poor signal-to-noise ratios in the spectra were compensated by using FFT coefficients 5 through 15 as input variables. These terms encode most of the low frequency information and facilitate the exclusion of high frequency noise from the calibration models.

Table 2 *Comparison of rugged neural network calibrations with best PLS calibrations. See text for explanation.*

Stream B: water				
Method	*Parms.*	*SEC*	*SEP*	*June*
no model		0.34	0.31	0.07
PLS	5 fac	0.08	0.09	0.47
PC-NN	4pc 4s	0.04	0.05	0.06

Stream B: acetone				
Method	*Parms.*	*SEC*	*SEP*	*June*
no model		1.91	1.75	0.16
PLS	4 fac	0.12	0.21	1.91
PC-NN	9pc 9s	0.07	0.19	0.14

Stream B: t-butyl alcohol				
Method	*Parms.*	*SEC*	*SEP*	*June*
no model		1.68	1.55	0.15
PLS	6 fac	0.06	0.18	1.36
PC-NN	5pc 5s	0.08	0.11	0.13

Stream C: water				
Method	*Parms.*	*SEC*	*SEP*	*June*
no model		6.24	6.63	10.62
PLS	1 fac	4.74	7.52	11.89
FFT-NN	5:15 8s	0.08	1.52	0.83

Stream D: water				
Method	*Parms.*	*SEC*	*SEP*	*June*
no model		6.63	3.08	1.82
PLS	1 fac	6.02	2.87	1.57
PC-NN	4pc 3s21	0.25	1.15	0.63

Figure 2 shows the results for the calibration of water in stream B in more detail. The best PLS calibration used 5 factors. The best neural network used 4 PC scores as input variables and 4 sigmoid nodes in the hidden layer. SEC and SEP for the PLS and PCA-NN models are nearly identical in both of these models; however, the PLS model does a very poor job of predicting the June samples. Although the spread of the June samples is low, it seems that the predictive performance of this PCA-NN model on future samples may be more rugged than PLS.

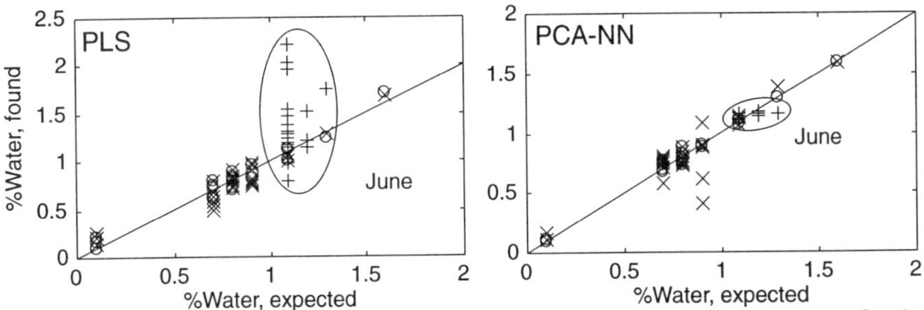

Figure 2 *Rugged calibration curves for water in stream B. Calibration samples (o). Validation samples (x). June data (+). The PCA-NN calibration model is less sensitive to the differences between the samples measured in the winter and summer months (February and March vs. June)*

Figure 3 shows the results for water in stream D. The best PLS calibration used 1 factor. PLS models with more than one factor gave worse SEP values. The best neural network used 4 PC scores as input variables, 3 sigmoid variables in the hidden layer and 1 linear node in the hidden layer. PLS was unable to compensate for two "outliers" in the calibration data set. Inclusion of these two spectra caused addition of more factors to the PLS model to ruin its predictive ability. The rugged PCA-neural network does not show sensitivity to the two "outliers" in the calibration data set. Overall its performance on the validation data set and the June data set was significantly better than PLS.

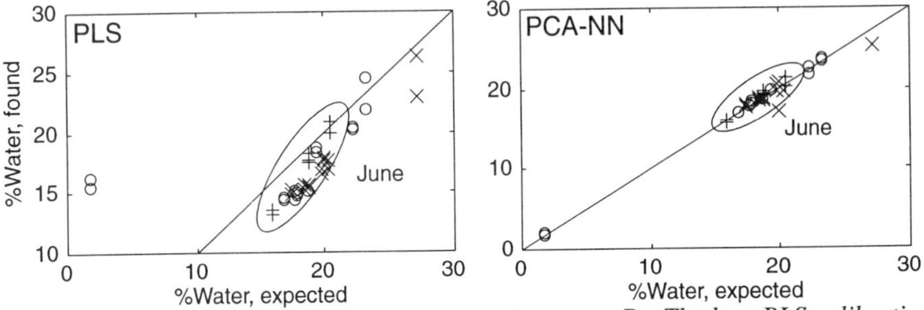

Figure 3 *Rugged calibration curves for water in stream D. The best PLS calibration used 1 factor. Calibration samples (o). Validation samples (x). June data (+). Two "outliers" in the calibration data set ruin the performance of PLS whereas the PCA-NN calibration gives acceptable predictions.*

4 CONCLUSIONS

These preliminary results demonstrate that artificial neural networks can be trained to produce rugged, multivariate spectroscopic calibrations. Special training strategies must be devised to take advantage of the peculiar aspects of neural networks to realize the goal of developing rugged calibrations. One strategy has been described in this paper. The rugged calibrations reported in this paper seem insensitive to slight variations in instrument response due to changes in the instrument, environment and sample stream. All of the rugged calibration results were obtained without the need for so called "bias" and "skew" adjustments to existing calibrations. Hopefully, these kinds of rugged calibrations will facilitate the long-term maintenance and use of multivariate spectroscopic calibrations.

5 ACKNOWLEDGEMENTS

This research was supported in part by the Measurement and Control Engineering Center at the University of Tennessee and the National Science Foundation.

REFERENCES

1. Y. Wang, B.R. Kowalski, *Anal. Chem.*, 1993, **65**, 1174-1180.
2. Y. Wang, B.R. Kowalski, Appl. Spectrosc., 1992, 46, 764-771.
3. P. Dardenne, R. Biston, G. Sinnaeve, In 'Near Infra-Red Spectroscopy', K.I. Hildrum, T. Isacksson, T. Naes, A. Tandberg, Eds., Horwood: Chichester, UK, 1992, pp 453-8.
4. P.J. Gemperline, *submitted to J. Chemom.*
5. J.R. Long, V.G. Gregoriou, P.J. Gemperline, *Anal. Chem.*, 1990, **62**, 1791-1797.
6. T.B. Blank, S.D. Brown, *Anal. Chem.*, 1993, **65**, 3081-3089.
7. Gemperline, P.J., Long, J.R., Gregoriou, V.G., *Anal. Chem.*, 1991, **63**, 2313-2323.
8. C. Borggaard, H.H. Thodberg, *Anal. Chem.*, 1992, **64**, 545-551.
9. Gemperline, P., *Chemom. Intell. Lab. Syst.*, 1992, **15**, 115-126.
10. Kvall, H., Naes, T., Isaksson, T., Ellekjaer, M.R., *J. Near Infrared Spectrosc.*, 1993, **1**, 97-101.
11. Naes, T., Knut, K., Isaksson, T., Charles, M., *J. Near Infrared Spectrosc.*, 1993, **1**, 1-11.
12. Sekuic, S., Seasholtz, M.B., Wang, Z., Kowalski, B.R., *Anal. Chem.*, 1993, **65**, 835a-845a.

The Regression Model Comparison Plot (REMOCOP)

Paul Geladi

DEPARTMENT OF ORGANIC CHEMISTRY, INSTITUTE OF CHEMISTRY, UNIVERSITY OF UMEÅ, S 901 87 UMEÅ, SWEDEN

1 INTRODUCTION

Regression methods are used for modeling linear and nonlinear relations between a block of predictor variables (X) and a block of response variables (Y). With the calibration data for **X** and **Y**, a model is built and then this model is used for the prediction of **Y** in cases where only **X** can be measured. Over the years, a number of methods have been developed for carrying out this linear regression. All these methods have their virtues and drawbacks. Many methods build a linear regression model of the type:

$$\mathbf{y} = \mathbf{X}\mathbf{b} + \mathbf{f} \qquad \text{(eq1)}$$

y: a vector (Nx1) of response variables.
X: a matrix (NxK) of predictor variables.
b: a vector (Kx1) of regression coefficients.
f: a vector (Nx1) of residuals.
k: index for X variables k=1,...,K.
n: index for objects n=1,...,N.

The method of Ordinary Least Squares (OLS) tries to find a **b** that minimizes the Euclidian norm of **f**. It does not work very well in many situations and as an alternative, so-called "biased" or "regularized" regression methods have been introduced (1-3). Some of the most frequently used ones are: Partial Least Squares Regression, Principal Component Regression, Ridge Regression. These methods are explained in the literature[1-3].

A digitized image is a matrix **X** or **Y** and a multivariate image is a 3-way array $\underline{\mathbf{X}}$ or $\underline{\mathbf{Y}}$. In some cases it is easier to handle images in this way. In other cases, it is better to reorganize the arrays. For setting up the regression equation, a reorganization of the matrix **Y** [IxJ] in a vector **y** [IJx1] and of $\underline{\mathbf{X}}$ [IxJxK] in a matrix **X** [IJxK] is more convenient.

Interpreting regression coefficient vectors **b** and comparing vectors **b** from different models has always been something desirable. It is well known that interpreting large and small values in **b** is very dangerous. The importance of the values in **b** can not easily be deducted from the sizes of the coefficients unless an orthogonal design in coded factors is used. It is also known that comparing vectors **b** by their model error or prediction error is not very satisfactory. Many methods for comparing regression models are univariate in nature because they only consider a number representing prediction quality. A more visual technique for comparing regression models is needed.

A technique for comparing regression vectors for multivariate image regression has been introduced[4]. It consists of collecting the regression vectors for all the used models in a matrix and plotting the principal component scores of that matrix in scatter plots. In these plots, each model is a point and the relationship in space of the models can be studied.The technique was used on a multivariate microscopic image of a piece of chinaware.The example was a 6x512x50 calibration X-image with a 512x50 Y-image. This example is used again. A second example[5] is a 5x32x32 portion of a satellite image with a 32x32 Y-image of integer values. The test image is 5x100x100. In this case, the measured Y-values are available as a 100x100 image.

It is shown in detail how the REMOCOP method is applied and some comments and a future outlook are given. Besides the classical PCR, PLS and Ridge Regression models, some new models are created as designs in the REMOCOP space.The regression methods used are described in ref. 9.

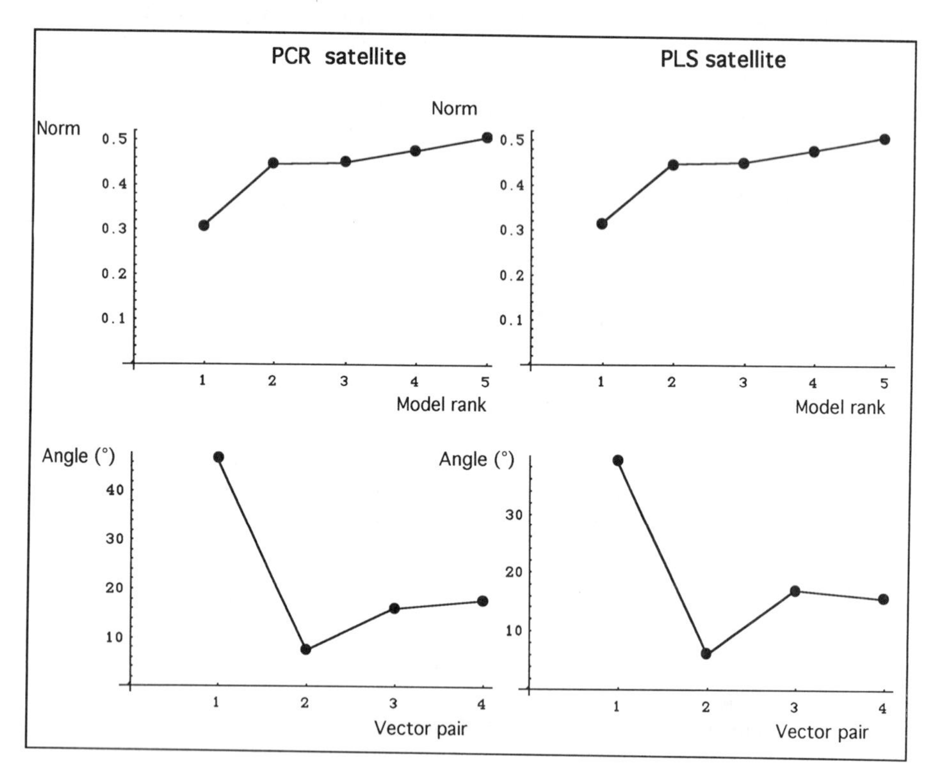

Figure 1. *Norm of the regression vector and angle between 2 subsequent pairs of vectors as a function of rank, both for PCR and PLS. The example is from the satellite image*[5]. *Vector pairs are: 1* $(b_1\text{-}b_2)$, *2* $(\mathbf{b}_2\text{-}\mathbf{b}_3)$, *3* $(\mathbf{b}_3\text{-}\mathbf{b}_4)$, *4* $(\mathbf{b}_4\text{-}\mathbf{b}_5)$.

2 EXPERIMENTAL

The multiwavelength microscopic image has been described[4]. The object studied is a piece of painted and glazed chinaware. The total image is 6x512x512 and the used wavelengths are: 460-500-540-580-630-680 nm. The calibration part is 6x512x50 with as Y-image the binary result from a multivariate segmentation. It was shown[4] by cross validation that PLS with 3-4 components forms an adequate regression model.

A 7-wavelength satellite image[5] of size 512x512 was used for testing principal component regression on multivariate images. A local calibration model was built for a 32x32 pixels region with 5 wavelength bands (bands 1-5) as X-variables and band 7 as Y-variable. The total image was used for comparing predictions of models of different rank. In (3,5) it was shown by cross-validation that 2-3 components form an adequate regression model and that increasing the rank does not help very much. For the present study, PLS and PCR models of ranks 1 to 5 were calculated. Rank 5 PLS and PCR models correspond to the OLS model. This is also shown in ref. 9. For ridge coefficients, the values of 100, 200, 300, 400, 1K, 10K, 30K, 60K, 100K, 300K and 1000K were used. Also some ridge calculations are given in the appendix. A test image of 5x100x100 was chosen outside the calibration region for testing the models. The Y-values for this region are known, so some sort of SEP (Standard Error of Prediction) can be calculated.

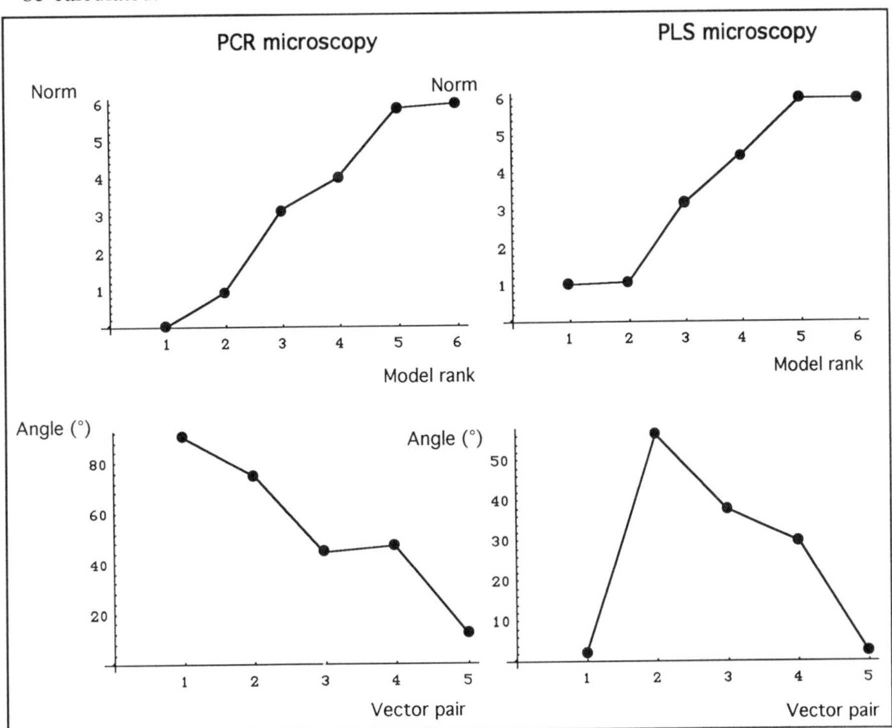

Figure 2. *Norm of the regression vector and angle between 2 subsequent pairs of vectors as a function of rank, both for PCR and PLS. The example is from the microscope image[4]. Vector pairs are: 1 (b_1-b_2) , 2 (b_2-b_3) , 3 (b_3-b_4) , 4 (b_4-b_5) and 5 (b_5-b_6).*

All image operations were carried out in ERDAS 7.4 on a 486 PC and in Imagine 8.0 on a Unix workstation. Also in-house programs written in the Erdas Toolkit were used. Non-image calculations were done in Mathematica on an Apple Macintosh computer. The regression calculations are explained in ref. 9.

3 REMOCOP

The principles of REMOCOP (Regression Model Comparison Plot) are very simple. The

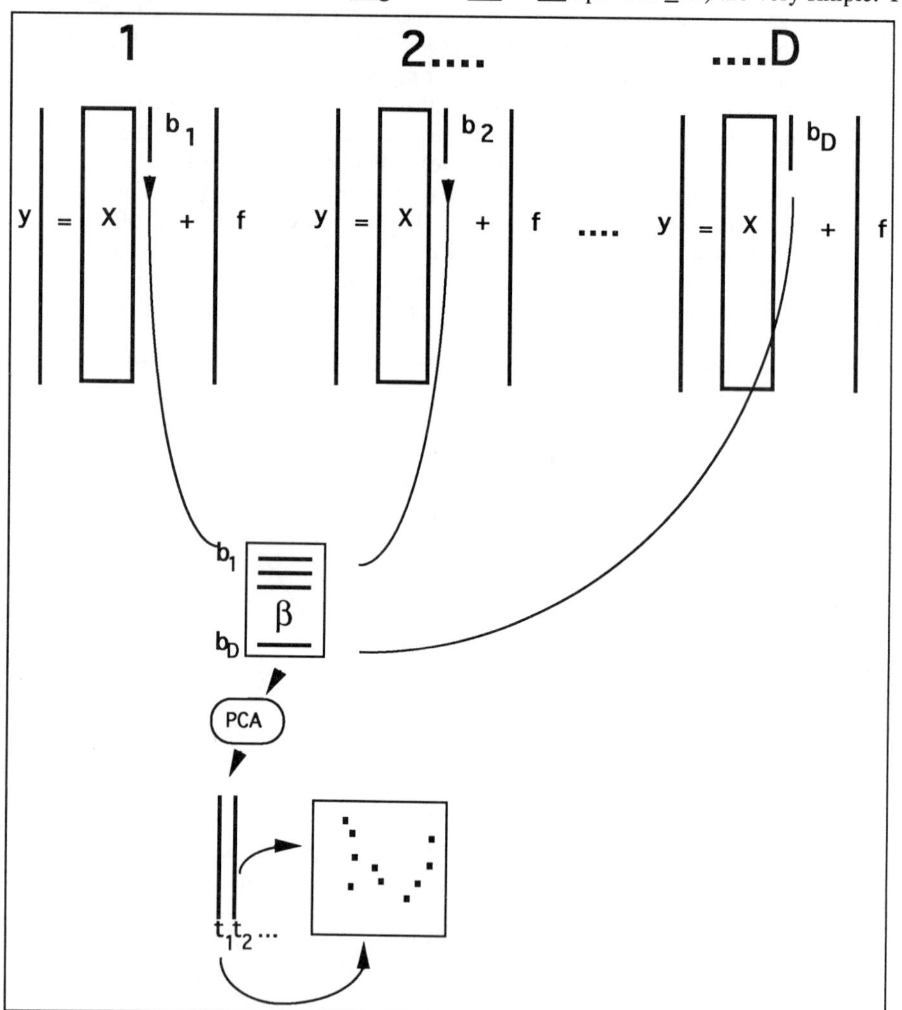

Figure 3. *The principle of REMOCOP. Regression vectors from a number of models are collected to give a data matrix* **b**. *In* **b**, *the objects (numbered d=1,..,D) represent the models. After a PCA or SVD on this matrix, the score plots give an overview over similarity or dissimilarity of the regression models.*

basic idea is that the vector of regression coefficients **b** is the only governing factor of a linear prediction and that all vectors **b** that work well should have something in common. This was shown in (5) by comparing norms of and angles between of vectors **b** for PCR models of different rank. It was shown that the **b** that work well have about the same norm and that the angles between them are small. This leads to the conclusion that there is a simple underlying structure (a latent structure) describing the essential properties of a group of regression vectors **b**. The plots of the norm of the regression vectors and the angles between subsequent vectors are given in figures 1 and 2. The results for the satellite image (figure 1) show that there is a jump in the norm of **b** between rank 1 and rank 2 models. After that, the norm of **b** stabilizes and the angle between $\mathbf{b}_2$ and $\mathbf{b}_3$ becomes rather small. The results are very similar for PCR and PLS. For the microscopy image (figure 2) the picture is less clear. PCR and PLS do not show as much similarity in this case. This may be due to the artificial situation of using a binary Y-image and trying to do discriminant regression with it.

To get a better idea of the space spanned by the regression vectors, one can collect all regression vectors (from PLS, PCR, RR) in a matrix **b**.In this matrix **b** the rows or objects are regression models. Studying the spatial structure of **b** can be done by carrying out a principal component analysis on it. The principle of this is shown in figure 3. The sizes of the singular values show which dimensions are really important and a plot of the scores shows the relationships (similarity/dissimilarity) of the regression vectors:

$$\mathbf{b} = \mathbf{UDV'} \qquad \text{(eq2)}$$

D is diagonal and has the singular values $l^{1/2}$ on the diagonal
U has the normalized scores as columns $\mathbf{u}_1$, $\mathbf{u}_2$, $\mathbf{u}_3$...
V has the loadings as columns

For the microscopy image[4], 3 components explained 98% of the total sum of squares of the **b** for PLS and PCR models and 99.4% of the **b** for RR models. The modified screeplot for the satellite image is shown in figure 4. The plot is for PLS and PCR regression vectors of ranks 1 to 5 and for 11 RR regression vectors (see experimental section). It is easily seen that a large part of the data structure of **b** is in the first 3 components.

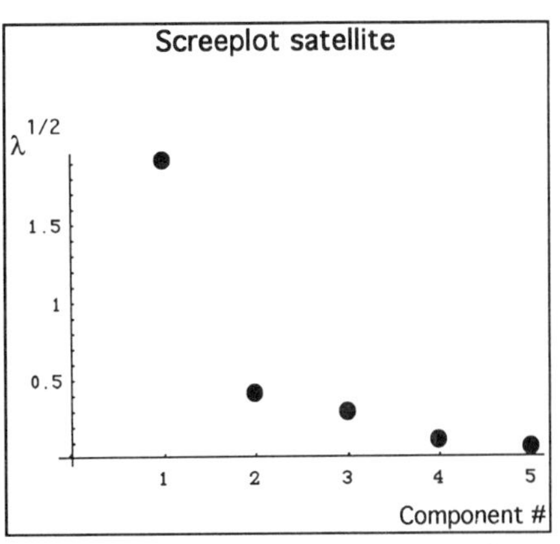

Figure 4. *The screeplot for the models of the satellite image shows that 3 components give the main contributions for spanning the space of the regression vectors for all the regression models.*

The REMOCOP plots are shown in figure 5 for the SVD score vector pairs $\mathbf{u}_1$-$\mathbf{u}_2$, $\mathbf{u}_1$-$\mathbf{u}_3$ and $\mathbf{u}_2$-$\mathbf{u}_3$. It can be seen that the useful PLS and PCR models of rank 2 and 3 cluster together in space and that also RR models for k=10K and k=30K are very close. As shown previously[4] there are regions of overfitting, underfitting and

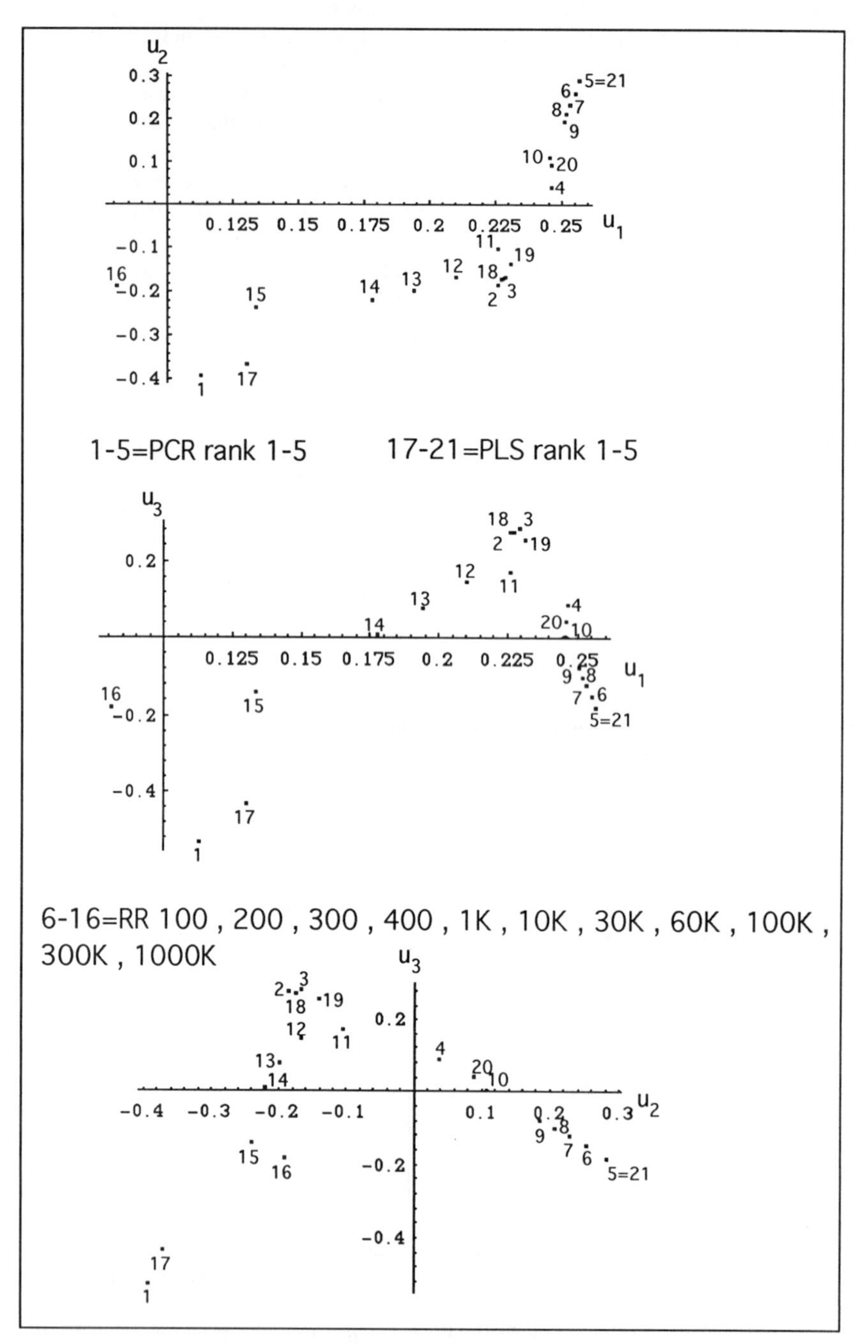

Figure 5. *The remocop plot for the satellite image example.* $\mathbf{u}_1$-$\mathbf{u}_2$, $\mathbf{u}_1$-$\mathbf{u}_3$ *and* $\mathbf{u}_2$-$\mathbf{u}_3$ *are plotted. The regression models are identified by numbers.*

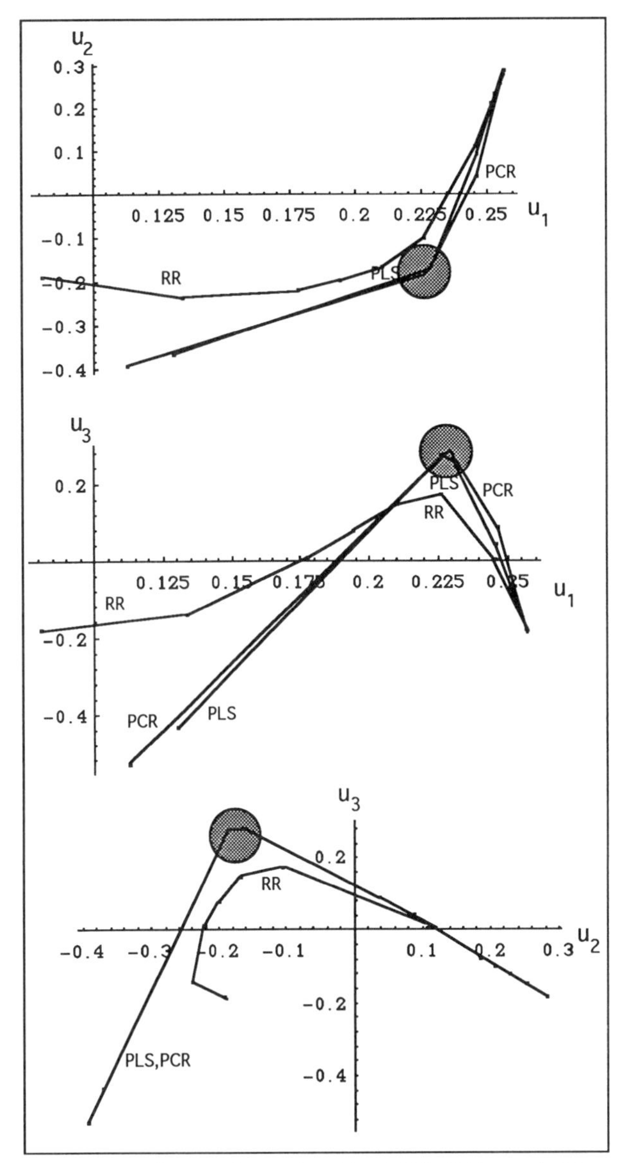

Figure 6. *The same plot as figure 5. The PLS, PCR and RR models are connected by straight lines, forming traces of increasing rank or decreasing ridge parameter. The traces all end up in the OLS model. It can be seen that the traces for PCR and PLS fall very close together. The trace for RR should be a curved one. The region for good (as found by cross validation) PLS and PCR models is indicated by a circle. The trace for the RR models can come very close to this circle, indicating that there are values of the ridge parameter k that can give good models.*

"good" regression models. The REMOCOP plot can also be shown with traces by connecting the locations of the PLS, PCR and RR models. PLS and PCR models have the discrete parameter "rank". RR models form a continuum, but it is very difficult to calculate all values forming this continuum. The traces are shown in figure 6. For the microscope image, the traces are shown in figure 7. PLS and PCR models do not agree that well here, but are distinct from the RR models. This confirms the findings in figure 2.

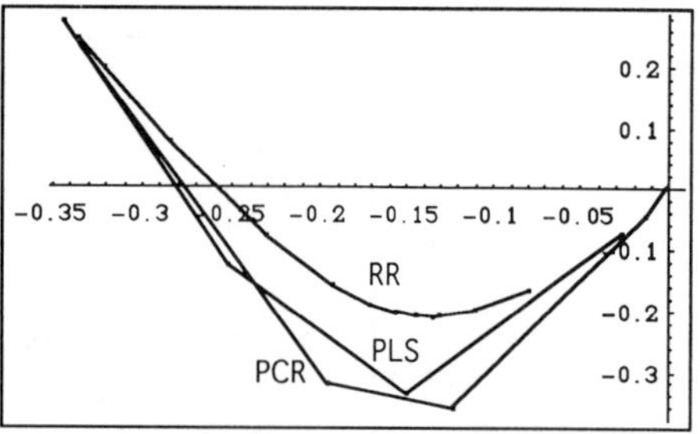

Figure 7. *The traces in the REMOCOP plot for the microscopy data set. PLS and PCR rank 1 starts on the right, rank 5 is on the left. RR models left to right for k=3K, 9K, 30K, 90K, 180K, 270K, 360K, 450K, 540K, 600K, 900K and 1800K.*

For the satellite image, a region of 100x100 was chosen as a test image. The regression models were all used to predict the Y-values for this test region. Figure 8 shows the results of this calculation in the form of image statistics. Mean and standard deviation are given as a percentage of the true measured values. It can be seen that rank 1 PLS and PCR models and RR models with a too high ridge coefficient k give extremely bad results.

The quality of a model is predictive quality. When a test set is left out of the calibration process, it can be used for assessing the predictive quality of the model. The real test of a model is the calculation of SEP (Standard Error of Prediction). For images, it is easier to calculate:

$$|\mathbf{Y}_{meas} - \mathbf{Y}_{calc}| \quad \text{(eq3)}$$

These calculated quantities are images. A way of approaching SEP is to calculate the standard deviation of the image $|\mathbf{Y}_{meas} - \mathbf{Y}_{calc}|$. A smaller value for the

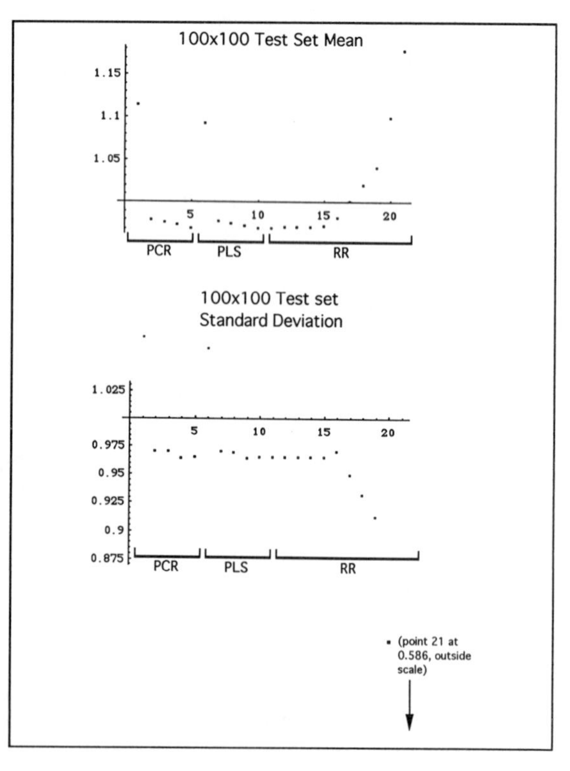

Figure 8. *Mean and standard deviation for the 100x100 test set of the satellite image. The horizontal line is the value for the measured data. The points are for different PCR,PLS and RR models. The PLS and PCR models are for increasing rank. The RR models are given for increasing ridge parameter.*

standard deviation of | $\mathbf{Y}_{meas}$ - $\mathbf{Y}_{calc}$ | indicates a better model for prediction in the test set. The statistic defined here will be called SEP*.

Figure 9 gives the SEP* for the calculated models. Underfitting (low rank, high ridge coefficient) is very bad. Overfitting is not disastrous, but models should be kept as simple as possible, so the lowest rank that gives no underfitting is the best one. The predicted values do not perfectly match the measured ones. This means that the selected test set is not perfect for the calibration set. This is not unexpected.

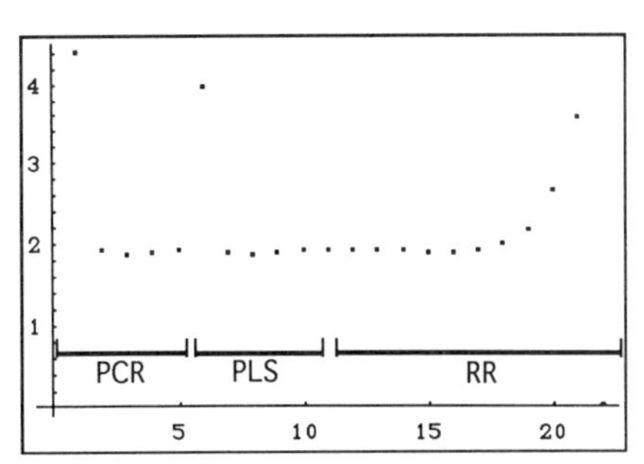

Figure 9. *The value of SEP* for the different regression models.The PLS and PCR models are for increasing rank. The RR models are given for increasing ridge parameter.*

In Figure 10, a detail of the REMOCOP plot is shown in the region of the "good" models. The model names are indicated. The values for SEP* are also shown in the plot. The mimimum SEP* is 1.87 for PCR and PLS of rank 3 and 1.88 for RR with coefficient 10K. SEP* goes up quickly in the underfitting region, but remains rather low in the overfitting region.

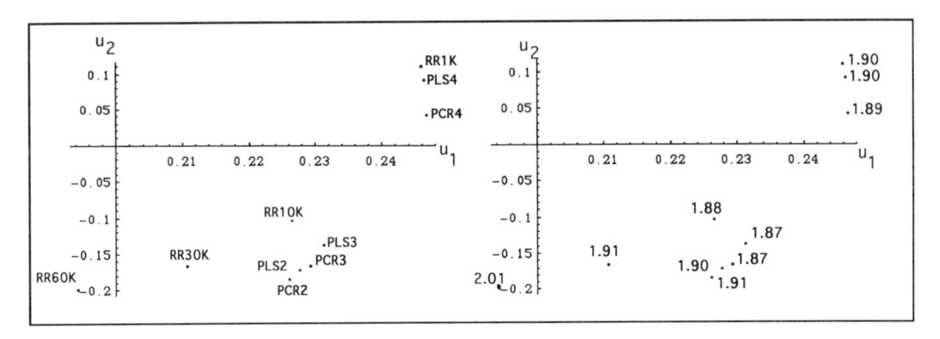

Figure 10. *A detail of the REMOCOP plot of* $\mathbf{u}_1$ *and* $\mathbf{u}_2$ *with the model names (left) and the SEP* values (right) filled in.*

3.1 Constructing new regression vectors.

It is possible to select values of interest in the REMOCOP plots and to construct regression vectors with these values. This is shown in ref. 9. The case shown here is that of a design. It is assumed that a 3-component representation of **b** is sufficient. An interesting point in the REMOCOP plot with values $\mathbf{u}_1$= 0.225, $\mathbf{u}_2$= -0.20 and $\mathbf{u}_3$= 0.20 was chosen as a center point of a 2^3 design with values $\mathbf{u}_1$: 0.2 , 0.25 $\mathbf{u}_2$: -0.17 , -0.23 and $\mathbf{u}_3$: 0.15 and 0.25. The design is shown in Table 1, together with the obtained responses, which are regression vectors. The calculations are explained in ref. 9.

Table 1. *A design in $\mathbf{u}_1$,$\mathbf{u}_2$ and $\mathbf{u}_3$ and the responses as regression vectors (rounded to 4 significant decimals).*

Name	Design ($\mathbf{u}_1$, $\mathbf{u}_2$, $\mathbf{u}_3$)	Regression Coefficients
b1	{0.2 , -0.17 , 0.15}	{0.08541, 0.05294, 0.1366, -0.07129, 0.3441}
b2	{0.2 , -0.17 , 0.25}	{0.09681, 0.04139, 0.1351, -0.09354, 0.3424}
b3	{0.2 , -0.23 , 0.15}	{0.09899, 0.04125, 0.1297, -0.05848, 0.3530}
b4	{0.2 , -0.23 , 0.25}	{0.11030, 0.02970, 0.1282, -0.08073, 0.3513}
b5	{0.25 , -0.17 , 0.15}	{0.09288, 0.07878, 0.1763, -0.0898, 0.4245}
b6	{0.25 , -0.17 , 0.25}	{0.1043, 0.06723, 0.1748, -0.1120, 0.4228}
b7	{0.25 , -0.23 , 0.15}	{0.1065, 0.06709, 0.1694, -0.07704, 0.4334}
b8	{0.25 , -0.23 , 0.25}	{0.1179, 0.05555, 0.1678, -0.09928, 0.4317}
bc	{ 0.225, -0.20, 0.20}	{0.1016, 0.05424, 0.1522, -0.08529, 0.3879}

The regression vectors obtained in the design can be used to calculate SEP* for the test data. The design and its SEP* values are plotted in the REMOCOP plot in figure 11. It is easily noticed in figure 11 that the design was chosen with too large ranges for the u-values. A new, narrower 2^3 design was tried with as center point $\mathbf{u}_1$= 0.23, $\mathbf{u}_2$= -0.15 and $\mathbf{u}_3$= 0.24. The design is shown in Table 2. The design and its SEP* values are also shown in figure 11. It can be noticed that there is a region where low SEP* values are found everywhere. The rank 3 PLS and PCR models still give a lower SEP*. This can be attributed to the data reduction from rank 5 to rank 3. Maybe the small contribution of components 4 and 5 is important to get the absolutely best regression vector. It should also be mentioned that small differences are not meaningful and that all SEP* values between 1.87 and 1.90 are reasonable. Plots like this make it easy to see where good models can be found and how leaving the region of the design may affect the results. Leaving the design area by the lower right corner quickly leads to large SEP* values. Leaving the design area at the upper left corner is rather risk-free.

Table 2. *A narrower design in $\mathbf{u}_1$,$\mathbf{u}_2$ and $\mathbf{u}_3$.*

Name	Design		
	$\mathbf{u}_1$	$\mathbf{u}_2$	$\mathbf{u}_3$
b1	0.227	-0.13	0.23
b2	0.227	-0.13	0.25
b3	0.227	-0.17	0.23
b4	0.227	-0.17	0.25
b5	0.233	-0.13	0.23
b6	0.233	-0.13	0.25
b7	0.233	-0.17	0.23
b8	0.233	-0.17	0.25
bc	0.230	-0.15	0.24

Conclusions

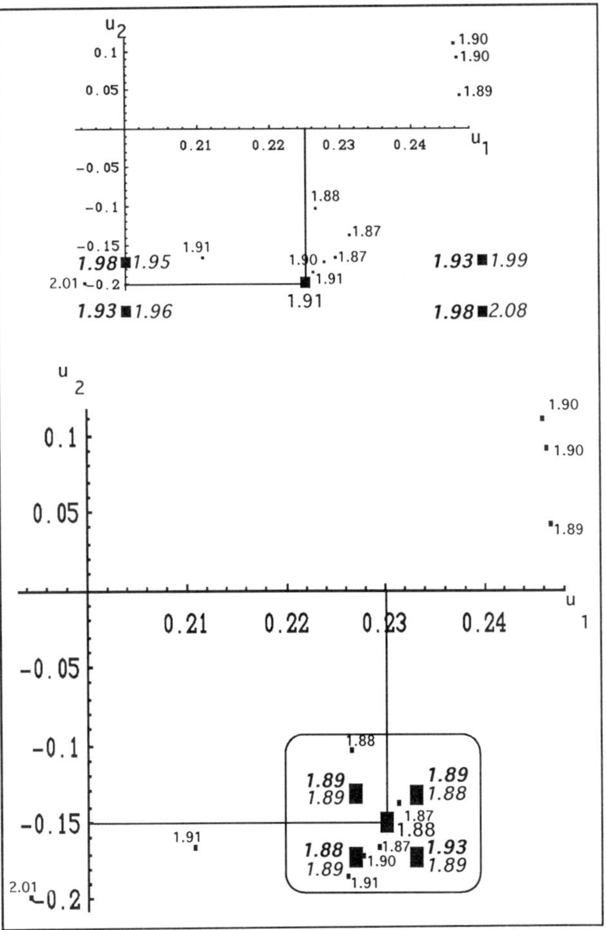

Figure 11. *The designs filled in in the REMOCOP plot of figure 10 with SEP*. Boldface italics = high level for* $\mathbf{u}_3$, *italics = low level for* $\mathbf{u}_3$. *The values of figure 10 are also shown for comparison. Upper part : the wide design. Lower part: the narrow design.*

The conclusions given here are only valid for data sets with many objects and few variables. Other data sets and extension to many variables will have to be studied in the future. It is expected that some of the properties found will generalize to other data sets.

1) It was shown that it is possible to visualize the space of the regression vectors in the REMOCOP plot. The REMOCOP plots show the relationships between regression models in a way that is unavailable with univariate statistics such as SEP.
2) In the space of the regression vectors, a small and irregularly shaped subspace of the useful or "good" models can be found.
3) For the examples studied, PLS and PCR seem to follow each other closely, while the RR models form their own trace. There is usually an RR model that comes pretty close to a "good" PLS or PCR model.
4) A data reduction in the subspace of useful models is possible. Using only 3 of 5 possible components from the SVD of **b** does not detoriorate the regression vectors in the example studied.
5) New regression vectors can be constructed in the subspace of the useful models. By using experimental designs around "interesting" points, new regression vectors can be studied systematically.
6) The point made above and the traces in REMOCOP space allow the construction of PCR and PLS models of "fractional" rank e.g. 2.5, 3.25 etc.

7) In future applications, a more detailed study of the designs in the REMOCOP space may prove useful. It may also be instructive to study the vectors in **V** (eq 2), especially when many variables are involved.

Acknowledgements

Fredrik Lindgren of the Joint Research Center in Ispra, Italy and Svante Wold, Stefan Rännar and Anders Berglund of Umeå University are acknowledged for advice and discussion.

References

1. H. Martens & T. Næs, 'Multivariate Calibration', Wiley, Chichester, 1989.
2. P. Brown, 'Measurement, Regression and Calibration', Oxford Science Publications, Oxford, 1993.
3. F. Lindgren, 'Third Generation PLS. Some Elements and Applications', PhD thesis, Umeå University, Sweden, 1994.
4. P. Geladi, J. Swerts & F. Lindgren, *Chemometrics and Intelligent Laboratory Systems*,1994, **24**, 145.
5. P. Geladi & K. Esbensen, *Journal of Chemometrics*, 1991, **5**, 97.
6. S. Wolfram, 'Mathematica. A System for Doing Mathematics by Computer 2nd ed.', Addison-Wesley, Reading MA, 1991.
7. F. Lindgren, P. Geladi & S. Wold, *Journal of Chemometrics*, 1993, **7**, 45.
8. S. de Jong & C. Ter Braak, *Journal of Chemometrics*, 1994, **8**, 169.
9. Supplementary material is available from the Books Department of The Royal Society of Chemistry or via E-mail address CUBITTA@RSC.ORG.

Studies in Near Infrared Spectroscopic Qualitative Analysis

Michel Coene, Roger Grinter, and Anthony M. C. Davies[1]

SCHOOL OF CHEMICAL SCIENCES, UNIVERSITY OF EAST ANGLIA, NORWICH NR4 7TJ, UK

[1] NORWICH NEAR INFRARED CONSULTANCY, 75 INTWOOD ROAD, CRINGLEFORD, NORWICH NR4 6AA, UK

1 INTRODUCTION

The use of near infrared (NIR) spectroscopy for qualitative analysis has become an important application in the pharmaceutical and chemical industries; however there are few published studies on the how best to apply NIR data to this important task. The first study was carried out by Rose[1] who used NIR data from a Technicon 400 (19 filter) instrument to discriminate between different penicillins using Canonical Variate Analysis (CVA). In 1985 a number of groups reported work on discriminant analysis using spectra from scanning instruments. Davies and McClure[2] used Fourier transformation (FT) to compress NIR data into 25 pairs of Fourier coefficients which they used to separate normal and decaffeinated instant coffees using Principal Component Analysis (PCA) for additional data compression, followed by CVA. Bertrand, Robert and Loisel[3] used PCA to separate different varieties of wheat while Mark and Tunnell[4,5] introduced the use of Mahalanobis distance (MD). The MD technique has gained many applications and is the basis of commercial software such as PRISM developed by researchers at Glaxo Laboratories[6]. As a result of these developments it has become accepted that complete spectra are required for qualitative analysis. However work in this department[7] demonstrated that a simple NIR instrument using 14 interference filters could successfully discriminate between isomers of bromo- and chloro-benzoic acids. We found this to be a surprising result and have continued the investigation of the minimum number of data points required for discrimination between similar compounds. This work has been facilitated by the availability of a scanning spectrometer which obtained scans of the 1100-2498 nm region containing 700 data points. Spectra were obtained on the same benzoic acid derivatives as in the earlier study and also on a group of sugars. Data points were systematically combed from the original spectra and the reduced spectra were tested in a discrimination process using PCA followed by CVA.

2 METHODS AND MATERIALS

2.1 Instrumentation

Samples were scanned using a NIRSystems 6500 spectrometer (NIRSystems Inc., Silver Spring, MD, USA) in the reflectance mode. The volume of sample required to fill the standard sample cup was reduced to about 1g by placing an aluminium disk, with a central hole of 3 cm diameter, into the sample cup. Spectra were acquired over the range 1100-2498 nm at 2 nm intervals, 32 co-added spectra were referenced to 32 co-added spectra of a ceramic reference material and stored on the hard disk of the controlling computer.

2.2 Software

Spectral data were transferred to a spreadsheet program (EXCEL) and the number of data points in each spectrum were systematically reduced from 700 to 4. These data were transferred to the SYSTAT (Evanston, IL, USA) environment for data transformations, graph plotting, PCA[8] and CVA[9]. Spectral data were also transferred to the operating environment of the NCSTATE CSAS[10] software for analysing NIR spectra which was used to plot the spectra shown in the figures. The number of principal components (PCs) retained for CVA was reduced from 15 PCs (or that maximum number possible when using data with a smaller number of input variables). With the sugar data the number of PCs was reduced for each sized data set until the CVA produced several errors in identification. The results of the CVA are shown graphically and in tables known as "confusion" tables in which the total number of times a sample type was identified as being a particular type of sample is recorded against the actual sample. If all samples are correctly identified then all off-diagonal results will be zero and the numbers on the diagonal will be the actual number of samples of that type.

2.3 Samples

Reagent grade samples of:- 2-bromo-benzoic acid, 3-bromo-benzoic acid, 4-bromo-benzoic acid, 2-chloro-benzoic acid, 3-chloro-benzoic acid, 4-chloro-benzoic acid, fructose, galactose, glucose, sorbose, and sucrose were recrystallised under conditions to produce large crystals. The samples were scanned and additional spectra were recorded while grinding the sample between each recording, thus producing a set of spectra for each sample for particle sizes ranging from about pea size to fine powders. This produced 11 spectra of each bromo-benzoic acids, 13 spectra of each chloro-benzoic acids and 8 spectra of each sugar.

3 RESULTS

3.1 NIR Spectra

A typical set of spectra for one sample type is shown in Figure 1, which shows the effect of particle size on NIR spectra. The set of six halogenated benzoic acid isomers is shown in Figure 2 for varying particle sizes of each sample. Figure 3 shows representative spectra of each sugar.

3.2 Halogenated Benzoic acid Isomers

The complete data set contained 72 spectra with 700 data points. Four versions of this data were produced containing 50, 25, 15, 7 data points at equal intervals. The results of the CVA are shown graphically, for the first three canonical variates (CVs), in Figure 4. All of these tests gave perfect results using five CVs as is demonstrated by the results for the seven data point set shown in Table 1.

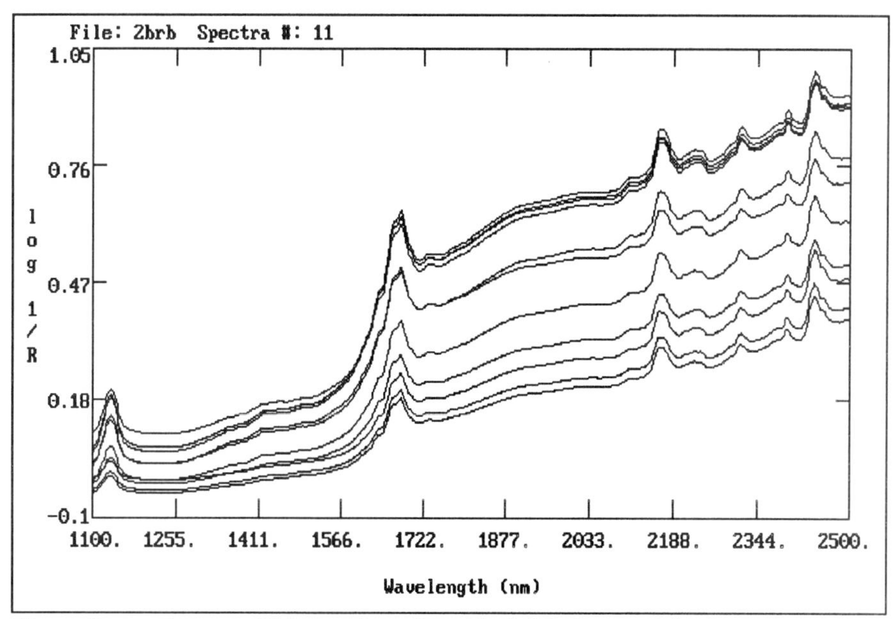

Figure 1. *Spectra of 2-bromo-benzoic acid of differing particle size.*

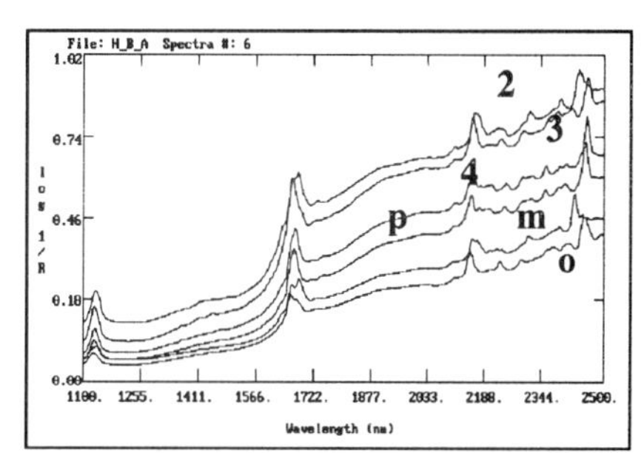

Figure 2. *Spectra of halogenated benzoic acid isomers. For key see Figure 4.*

Figure 3. *Spectra of the five sugars. For key see Figure 5.*

3.3 Sugars

The original data set contained 40 spectra of 700 data points. Versions of this data containing 4, 5, 6, 7, and 8 data points at equal intervals were tested using varying numbers of PCs for the CVA. Typical results for successful and unsuccessful separations are shown graphically in Figure 5, results for an individual data set, (7 data points) are given in Tables 2 and 3. A summary of all results is given in Table 4.

Table 1. *Results of CVA identification of halogenated benzoic acid isomers from 7 data points using 6 PCs.*

Benzoic Acid Derivative	Tested as:-					
	2-Br	3-Br	4-Br	2-Cl	3-Cl	4-Cl
2-bromo-	**11**	0	0	0	0	0
3-bromo-	0	**11**	0	0	0	0
4-bromo-	0	0	**11**	0	0	0
2-chloro-	0	0	0	**13**	0	0
3-chloro-	0	0	0	0	**13**	0
4-chloro-	0	0	0	0	0	**13**

Table 2. *Results of CVA identification of sugars from 7 data points using 6 or 5 PCs.*

Sugar Identity	Using 6 PCs and tested as:-					Using 5 PCs and tested as:-				
	Fruc	**Gal**	**Glu**	**Sor**	**Suc**	**Fruc**	**Gal**	**Glu**	**Sor**	**Suc**
Fructose	**8**	0	0	0	0	**8**	0	0	0	0
Galactose	0	**8**	0	0	0	0	**8**	0	0	0
Glucose	0	0	**8**	0	0	0	0	**8**	0	0
Sorbose	0	0	0	**8**	0	0	0	0	**8**	0
Sucrose	0	0	0	0	**8**	0	0	0	0	**8**

Table 3. *Results of CVA identification of sugars from 7 data points using 4 or 3 PCs.*

Sugar Identity	Using 4 PCs and tested as:-					Using 3 PCs and tested as:-				
	Fruc	**Gal**	**Glu**	**Sor**	**Suc**	**Fruc**	**Gal**	**Glu**	**Sor**	**Suc**
Fructose	**6**	0	0	0	**2**	**5**	0	0	0	**3**
Galactose	0	**8**	0	0	0	0	**6**	0	0	**2**
Glucose	0	0	**8**	0	0	0	0	**8**	0	0
Sorbose	0	0	0	**8**	0	0	0	**4**	**4**	0
Sucrose	**1**	0	0	0	**7**	**2**	0	0	0	**6**

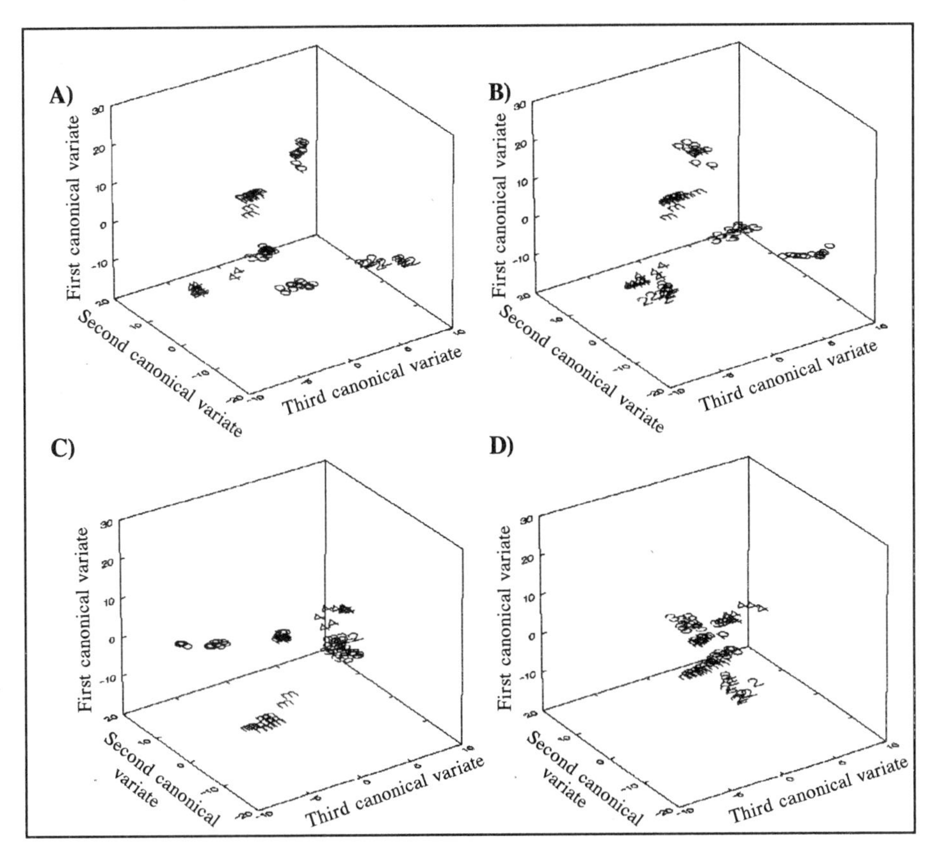

Figure 4. *Separation of halogenated benzoic acid isomers A) 50, B) 25, C) 15 and D) 7 data points and 15 PCs (A-C) , 6 PCs (D).* **2**, *2-bromo-*; **3**, *3-bromo-*; **4**, *4-bromo-*; **o**, *2-chloro-*; **m**, *3-chloro-*; **p**, *4-chloro-*

4 DISCUSSION

The spectra of the halogenated benzoic acids demonstrate two facets of NIR spectroscopy. The fact that particle size has such a marked effect on the absorption and that they often appear to be mainly composed of featureless, sloping baselines. The sugar spectra may appear to be visually more interesting but there are few Lorentzian or Gaussian shaped absorption peaks. However, the results of the PCA/CVA processing have again demonstrated how few data points are required to characterise these compounds. In the previous study[7], 14 log 1/R values were used, but these were measured at selected wavelengths which had been chosen on the basis of years of research into quantitative NIR analysis. The results from the present study are even more surprising because it has been demonstrated that only seven or eight data points were required to characterise two very different groups of compounds and the data points were unselected, being taken at equal intervals across the 700 available points. The fact that this is surprising may be

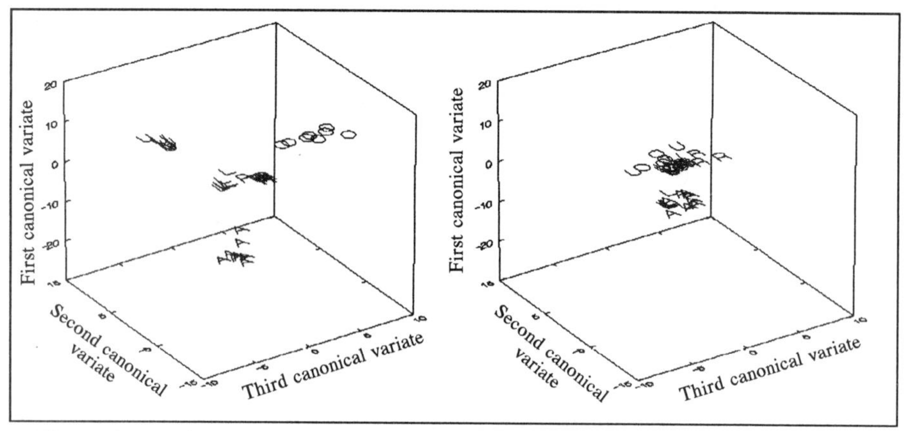

Figure 5. *Separation of sugars by PCA and CVA processing, using eight (left) or five (right) equally spaced data points from NIR spectra containing 700 data points. A; GAlactose; L, GLucose; O, SOrbose; R, FRuctose;*

connected with the way we view spectra, a suggestion made by Hirschfeld in one of his last Coblentz Society Newsletters[11]. Hirschfeld remarked that before we "see" anything the data from the retina undergoes a considerable amount of processing, over which we have no control. One of the consequences of this is that we find baselines uninteresting. If all NIR spectra had the same baseline then we would be correct to ignore them, but we know that this is not true. If it is assumed that baselines in NIR spectra have characterising information equal to that of absorption peaks, it then becomes less surprising that compounds can be characterised by eight data points. Computers have neither the benefits nor the disadvantages of human visual systems and so what becomes important is the information content of the data. NIR spectrometers have good wavelength accuracy and also very good signal to noise (S/N) characteristics; typical noise levels are less than 0.000100 log 1/R units. This means that the data is precise on both x (wavelength) and y (absorption) scales.

Table 4. *Number of errors in CVA of sugar spectra.*

Number of Data Points	Number of Factors 7	6	5	4	3
8	0	0	0	0	4
7		0	0	3	11
6			0	0	1
5				7	6
4					13

Estimation of the number of different compounds that could be characterised by eight variables requires estimates not only of the noise in the measurement and the effects of particle size but also of the distribution of values along each variable. We have data for S/N levels and the results from this study indicate that more than five data points (variables) are required before any characterisation is achieved. We do not have any information about the distribution of the data but if we suggest that one additional variable might enable distinction to be made between 20 different compounds, then seven variables would resolve 400 and eight would resolve 8000. A more pessimistic forecast is that eight variables would be sufficient to characterise 1000 compounds. In industrial use NIR spectroscopy is used to check the identity of samples in closed populations of a few hundred compounds, a figure well inside the estimate for eight variables. The requirement to have only a few variables for the task of identity testing could be of considerable interest to designers of new instrumentation. It should be possible to build very fast, solid state systems giving data at eight NIR wavelengths.

5 CONCLUSIONS

Classification of compounds of very similar structures can be obtained from eight equally spaced data points taken from NIR spectra over the range 1100-2498 nm.
Further research is required to determine if this is generally applicable.

Acknowledgements

The School of Chemical Sciences, University of East Anglia is grateful to NIRSystems, Inc. for the loan of the 6500 NIR spectrometer.

References

1. J. Rose, Paper 182 Pittsburg conference 1982.
2. A.M.C. Davies, and W.F. McClure, 1985, *Anal.Proc.*, **22**, 321.
3. D. Bertrand, P. Robert & W. Loisel, *J. Sci. Food Agric.*, 1985, **36**, 1120.
4. H.Mark & D.A. Tunnell, *Anal. Chem.*, 1985, **57**, 1449.
5. D.A. Tunnell, in (Eds. C.S. Creaser and A.M.C. Davies) 'Analytical Applications of Spectroscopy', Royal Society of Chemistry, London, 1988, p457.
6. B. Davies & C.M. Harland, in (Eds. C.S. Creaser and A.M.C. Davies) 'Analytical Applications of Spectroscopy', Royal Society of Chemistry, London, 1988, p462.
7. H.K. Pradhan, A.M.C. Davies, and R. Grinter, in 'Making Light Work' (Eds. I. Murray and I.A. Cowe), VCH, Weinheim , 1992, 55.
8. W.W. Chatfield, and A.J. Collins, 'Introduction to Multivariate Analysis', Chapman & Hall, London, 1980, p. 57.
9. W.W. Chatfield, and A.J. Collins, 'Introduction to Multivariate Analysis', Chapman & Hall, London, 1980, p. 153.
10. A. Hamid, and W.F. McClure, 'Software for an On-line Computerized Spectrometer', NCARS Bulletin No. 252, North Carolina State University, Raleigh, USA, 1978.
11. T.B. Hirschfeld, *Appl. Spectrosc.*, 1986, **40**, 121.

Author Index

Albert, K., 86
Andrews, D.L., 123
Araujo, P.W., 136

Baker, C., 61
Belt, S.T., 94
Borggaard, C., 209
Braumann, U., 86
Brereton, R.G., 136, 196
Bricker, T.M., 109
Budevska, B.O., 31

Chai, C., 61
Choi, M.F., 189
Coene, M., 237
Cooke, D.A., 94
Cowe, I.A., 175

David, A.R.J., 129
Davies, A.M.C., 51, 175, 237
Demidov, A.A., 123
Demir, C., 196
Dhanoa, M.S., 117
Donkin, P., 94
Dyer, C.D., 42

Eddison, C.G., 51, 175
Evans, J.D., 66

Fish, D.J., 66
Freshwater, R.A., 66

Gachanja, A.N., 164
Galkin, A.A., 157
Geladi, P., 225
Gemperline, P.J., 218
George, M.W., 13
Glavin, G.G., 157
Griffiths, P.R., 31
Grinter, R., 237

Harris, R.K., 77
Hawkins, P., 189
Hay, J.N., 184
Hewitt, N., 175
Hird, S.J., 94
Houk, R.S., 109

Ibbett, R., 100

James, M., 100
Jones, R.L., 66
Jones, W.J., 20

Kavianpour, K., 136
Kihle, J., 56

Limero, T.F., 147
Lister, S.J., 117

Mantoura, R.F.C., 171
Maso, G.N., 157
McCormack, T., 129
Mueller-Harvey, I., 117

Parker, S.F., 61, 184
Pharr, C.M., 31
Poliakoff, M., 13
Price, D., 171

Reed, J.D., 117
Roscoe, H.K., 66
Rowland, S.J., 94

Sheppard, N., 3
Strong, E.K., 66

Taylor, W.H., 66

Turner, J.J., 13

Whitley. A., 42
Williams, K.P.J., 42
Worsfold, P.J., 129, 164, 171
Wraige, E.J., 94

Subject Index

AAS. *See* Atomic absorption spectroscopy
Adsorbates. *See* Surface spectroscopy
AFS. *See* Atomic fluorescence spectroscopy
Aldehyde detection, 164-170
Array detectors, 66-71, 136-143, 154. *See also* Charge-coupled detectors
Atmospheric spectroscopy, 66-74, 147-150, 152-154
Atomic absorption spectroscopy (AAS), 136-143
Atomic fluorescence spectroscopy (AFS), 157-163
Attenuated total internal reflection (ATR), 8

Beer-Lambert relationship, 20, 25, 52, 191
Biosensors, 189

Canonical variate analysis (CVA), 237-243. *See also* Multivariate analysis
Carbon dioxide, detection, 189-195
Carbon fibres, 46. *See also* Composite materials
Carbon monoxide, adsorbed, 4, 6
Carboxylic acid detection, 164-170
Catalysis, 3-9
Charge-coupled detectors (CCD), 7, 43, 66-71, 165
Chemiluminescence, 164-170, 171-174
Chemisorption, 3-9
Chemometrics, 51-55, 117-122, 175-183, 209-243
Chlorine, oxides, 66, 68
Chlorophyll, 123-128, 136-143
Chromatography. *See entries for specific types*
Co-master spectrometry, 175-183, 218
Composite materials, 184-188
Computer simulations, 123, 127-128
Concentration-modulated absorption spectroscopy (COMAS), 20-30

CRAMPS. *See* Nuclear magnetic resonance, combined rotation and multiple-pulse spectroscopy
CVA. *See* Canonical variate analysis
Cyclic voltammetry, 40

Diamond films, 46-48
Diode array detectors (DAD). *See* Array detectors
Difference frequency generation, 15, 16
Drug detection, 196-206
Dynamic mechanical analysis, 100-105
Dysprosium analysis, 157-163

Electrochemically modulated infrared spectroscopy (EMIRS), 31-41
Electron energy loss spectroscopy (EELS), 6
Electron tunnelling spectroscopy, inelastic (IETS), 5, 8
Energy migration, 123-128
Erbium analysis, 157-163
Ethylidyne, 4
Europium analysis, 157-163

Femtosecond spectroscopy, 15
Fibre optic detection, 189-195
Flow injection (FI) procedures, 129-135, 165-166, 172
Fluorescence:
- excitation-emission micro-spectrometry (FLEEMS), 56-60
- kinetics, 123-128

Förster energy transfer, 123-128
Fourier transform:
- infrared (FTIR) spectroscopy: 4, 6, 13, 16-17
  - step-scan, 17, 31-41
  - stroboscopic, 17

Fourier transform:
infrared (FTIR) spectroscopy *(cont.)*
subtractively normalized interfacial (SNIFTIRS), 32-33
Raman spectroscopy, 42-43

GAMS. *See* Mass spectrometry, gas analyser
Gas chromatography-mass spectrometry (GC-MS), 147-154, 196-206
Grain analysis, 51, 53, 176, 178-182, 237
crop residue, 117-122

Hard disk contaminants, 48-49
Hexacyanoferrate redox couple, 33-41
High performance liquid chromatography (HPLC), 86, 136-143
Hydrocarbons, 56-60, 94-99, 164-170
Hydrogen peroxide detection, 171-174

ICP. *See* Inductively coupled plasma
IETS. *See* Electron tunnelling spectroscopy, inelastic
IMS. *See* Ion mobility spectrometry
Inductively coupled plasma (ICP), 24
atomic fluorescence spectroscopy, 157-163
mass spectrometry (ICP-MS), 109-116
Infrared microscopy. *See* Micro-spectroscopy
INS. *See* Neutron scattering, inelastic
Iodine, molecular, 27-28
Ion mobility spectrometry (IMS), 153

Lasers, 14, 20-25, 27-29, 43, 59, 60, 61
scanning optical microscopy (LSOM). *See* Micro-spectroscopy
Latex, polymer, 100-105
Liquid chromatography (LC), 165. *See also* High performance liquid chromatography
Lithium analysis, 24-27
Luminescence spectroscopy, 56-60. *See also* Fluorescence

Magnesium analysis, 136-143
Marine spectroscopy, 94, 95-98, 129-135, 136, 171-174
Mass spectrometry, 94, 95, 98, 147-156
chemical ionization (CI), 197-206
gas analyser (GAMS), 152
gas chromatography (GC-MS), 147-154, 196-206
inductively coupled plasma, 109-116
magnetic sector, 149, 151-152
quadrupole, 149
Metal carbonyls, 13-16
Metalloproteins, 17, 110-115, 123-128, 136-143
Metal surfaces: 4-8, 31-41
selection rule, 6
Microdensitometers, 67
Micro-spectroscopy, 42-50, 56-60, 61-65
MLR. *See* Regression, multiple linear
Multivariate analysis, 117-121, 175, 218, 225-236, 237-243

Near infrared (NIR) spectroscopy, 3, 51-55, 117-122, 175-183, 209, 214, 220-221, 237-243
Neural networks, 209-217, 218-224
Neutron scattering, inelastic, 5, 184-188
Nitrate analysis, 129-135
Nitrogen, oxides, 66, 68. *See also* Nitrate analysis
Nuclear magnetic resonance (NMR), 77-105
combined rotation and multiple-pulse spectroscopy (CRAMPS), 77
correlation spectroscopy (COSY), 91-92, 97
cross-polarisation, 77, 78-84
exchange spectroscopy (EXSY), 83
heteronuclear correlation (HETCOR), 83-84
magic-angle spinning (MAS), 77, 78-85
multinuclear, 78-85
proton decoupling, 77, 78-84, 96, 97
solids, 77-85, 100-105
spin relaxation, 88, 101-105
two-dimensional, 83-84, 91-92

Oil analysis. *See* Hydrocarbons
On-line analysis, 86-93
Optogalvanic spectroscopy, 24

Organometallic intermediates, 14-16
Oxygen detection, 189-195
Ozone, atmospheric, 66-74

Partial least squares (PLS) analysis, 209-217, 218, 225-236
PCA. *See* Principal component analysis
PET. *See* Polyethylene terephthalate
Photomultiplier tubes (PMT), 168, 172
Photosynthetic systems, 17, 123-128, 171
Plasticisers, 89-90
PLS. *See* Partial least squares analysis
PMT. *See* Photomultiplier tubes
Polyethylene, 61-65. *See also* Polymer layers
Polyethylene terephthalate (PET), 45
Polyimides, 185-188
Polymer layers, 45. *See also* Latex, polymer
Principal component analysis (PCA), 118-121, 209-217, 218, 225-236, 237, 238
Process analysis, 220-223. *See also* On-line analysis

RAIRS. *See* Reflection-absorption infrared spectroscopy
Raman scattering: 4, 7
- imaging, 42-50
- microscopy, 42-50
- resonance, 7
- stimulated electronic, 16
- surface enhanced, 5, 31

Rare earth elements, 157-163
Reflection-absorption infrared spectroscopy (RAIRS), 6
Regression, 225-236
- multiple linear (MLR), 218

Riverine spectroscopy, 129-135
Rovibronic spectroscopy, 28

Seawater analysis. *See* Marine spectroscopy
Selenium analysis, 109-116
SEC. *See* Size exclusion chromatography
SERS. *See* Raman scattering, stimulated electronic *and* Raman scattering, surface enhanced
Serum analysis, 109-116
SFC. *See* Supercritical fluid chromatography
Silicon, 44
Single potential alteration infrared spectroscopy (SPAIRS), 32
Size exclusion chromatography (SEC), 109-116
SNIFTIRS. *See* Fourier transform infrared spectroscopy, subtractively normalized interfacial
Space, the final frontier, 147-156
SPAIRS. *See* Single potential alteration infrared spectroscopy
Spectral hole-burning, 25
Spectroelectrochemistry, 31-41
Spectrophotometric detection, 129-135, 189-195
Star-pointing spectrometers, 66-74
Step-scan FTIR. *See* Fourier transform infrared spectroscopy, step-scan, 17
Stroboscopic FTIR. *See* Fourier transform infrared spectroscopy, stroboscopic, 17
Sugars, 239-242
Sum frequency generation, 7, 15, 16
Supercritical fluid chromatography (SFC), 86-93
Surface spectroscopy, 3-9, 31-41. *See also* Raman spectroscopy, surface enhanced

TGA. *See* Trace gas analysis
Thermal analysis, 100-105
Time-resolved measurements, 13-19
Trace gas analysis (TGA), 152
Transferability, 175-183

Upconversion. *See* Sum frequency generation

Vibrational electron energy loss spectroscopy (VEELS). *See* Electron energy loss spectroscopy